J.W. Einax, H.W. Zwanziger, S. Geiß

Chemometrics
in Environmental Analysis

A Wiley company

Jürgen W. Einax, Heinz W. Zwanziger
Sabine Geiß

Chemometrics in Environmental Analysis

Prof. Dr. Jürgen W. Einax
Friedrich-Schiller-Universität Jena
Institut für Anorganische und
Analytische Chemie
Lessingstraße 8
D-07743 Jena

Prof. Dr. Heinz W. Zwanziger
Fachhochschule Merseburg
Fachbereich Chemie- und
Umweltingenieurwesen
Geusaer Straße
D-06217 Merseburg

Dr. Sabine Geiß
Thüringer Landesanstalt
für Umwelt Jena
Zentrallabor
Prüssingstraße 25
D-07745 Jena

This book was carefully produced. Nevertheless, authors and publisher do not warrant the information contained therein to be free of errors. Readers are advised to keep in mind that statements, data, illustrations, procedural details or other items may inadvertently be inaccurate.

Editorial Director: Dr. Steffen Pauly, Cornelia Clauß
Production Manager: Claudia Grössl

Library of Congress Card No. applied for

British Library Cataloguing-in-Publication Data:
A catalogue record for this book is available from the British Library

Die Deutsche Bibliothek – CIP-Einheitsaufnahme
Einax, Jürgen:
Chemometrics in environmental analysis / J. W. Einax ; H. W. Zwanziger ; S. Geiss. –
Weinheim : VCH, 1997
ISBN 3-527-28772-8
NE: Zwanziger, Heinz:; Geiss, Sabine:

© VCH Verlagsgesellschaft mbH, D-69451Weinheim (Federal Republic of Germany), 1997

Printed on acid-free and low-chlorine paper

All rights reserved (including those of translation into other languages). No part of this book may be reproduced in any form – by photoprinting, microfilm, or any other means – nor transmitted or translated into a machine language without written permission from the publishers. Registered names, trademarks, etc. used in this book, even when not specifically marked as such, are not to be considered unprotected by law.
Cover illustration: Susanne Baum
Composition: Filmsatz Unger & Sommer GmbH, D-69469 Weinheim
Printing: betz-druck GmbH, D-64291 Darmstadt
Bookbinding: Großbuchbinderei J. Schäffer, D-67269 Grünstadt
Printed in the Federal Republic of Germany

Dedicated to

Professor Dr. KLAUS DOERFFEL
1925 – 1995

who encouraged not only the authors
to use statistical and chemometric methods
in Analytical Chemistry

Preface

The young scientific discipline "Chemometrics" has rapidly developed in the past two decades. This enormous increase was initiated by advances in intelligent instruments and laboratory automation as well as by the possibility of using powerful computers and user-friendly software. So, chemometrics became a tool in all parts of quantitative chemistry, but particularly in the field of analytical chemistry. Nowadays, the analyst is increasingly faced with the need to use mathematical and statistical methods in his daily work.

Another fact is that in a comparable time period mankind has also become increasingly aware of problems resulting from environmental pollution. The pathways and the effects of pollutants in environmental compartments and on human beings are very complex and often largely unknown. The need to obtain more and deeper knowledge on environmental pollutant loading led to the active development of environmental analytical methods. On the one hand, the capability of newer, mostly multicomponent or multielement analytical methods produces an enormous flood of data. On the other hand, environmental data are characterized by their extremely varying character. These two facts and the complexity of the objects under investigation urgently require the application of chemometric methods particularly in the field of environmental research.

The book consists of two main parts. The first part has a more methodological character, it is technique-oriented. In the sections of this part mathematical fundamentals of important newer chemometric methods are comprehensively represented and discussed, and illustrated by typical and environmental analytical examples which are easy to understand. The second part, which has been written in a more problem-orientated format, focuses on case studies from the field of environmental analysis. The discussed examples of the investigation of the most important environmental compartments, such as the atmosphere, hydrosphere, and pedosphere, demonstrate both the power and the limitations of chemometric methods applied to real-world studies.

The main aim of the book is to introduce the reader to the foundations of relevant chemometric methods and likewise to show for selected and typical case studies what can be achieved by using these methods in environmental analysis. It cannot be the scope of such a book, written for nonchemometricians, to give more than an introduction to chemometrics. So, highly sophisticated chemometric methods are not treated.

The authors of the book come from different fields of analytical chemistry ranging from environmental monitoring to university research. Learning chemistry as their profession they have started to use and apply chemometrics in their own working field as self-made men (and woman). This may explain the enthusiasm with which the book was written and also its limitations. So, we hope that the mathematicians and the statisticians will excuse some of our simplifications. The environmental specialist may possibly criticize the choice of case studies discussed. Again we beg the pardon of the reader. We have mainly selected case studies from our own working field. It seemed to be more useful to represent examples where we are familiar with the environmental and experimental background of the data. Last, but not least, the authors have to excuse the German style of English. We undertook the hard task of writing in English to disseminate the subject matter as widely as possible beyond the boundaries of our own language.

We hope that our general aim may be reached: to arouse the interest of the environmental scientist who has not yet much concerned himself with chemometric methods and the use of the tools of chemometrics in his own field of activity.

The investigation and evaluation of the case studies discussed in the book was the result of highly constructive and interdisciplinary cooperation with innumerable collegues and collaborators. It is a pleasure for us to thank them all very heartily. Different publishers, cited at the appropriate positions in the text, are gratefully thanked for their permission to use ideas and parts of some of our own papers published previously. The authors gratefully acknowledge the staff at VCH for their kind patience during the production of this book. Particularly, we thank Mr. Ian Davies, Cambridge, for constructive cooperation and help during the process of language editing.

Finally, we hope you will enjoy the book and we will be grateful to any reader who either draws our attention to errors in the book or suggests any improvements.

Jena and Merseburg, January 1997

Jürgen W. Einax
Heinz W. Zwanziger
Sabine Geiß

Contents

Symbols and Abbreviations XVII

1	**Introduction**	1
1.1	Development of Chemometrics	1
1.2	Definition of Chemometrics	2
1.3	Classification of Chemometric Methods	3
1.4	What is Environmental Analysis?	6
1.5	The Need to Apply Chemometric Methods	8
1.6	What do we Expect of Chemometrics in Environmental Analysis? ..	13
1.7	Recommended Reading and Useful Software	14
1.8	Overview of Chemometric Literature	15
	References	20

Part A   Essential Chemometric Methods

2	**Measurements and Basic Statistics**	25
2.1	Basic Statistics	25
2.1.1	Introductory Remarks	25
2.1.2	Frequency Distribution of Observations	25
2.1.3	The Normal Distribution Model	27
2.1.4	Parameters of Distributions	28
2.1.5	Computation of some Basic Parameters	29
2.1.6	Confidence Intervals	32
2.2	Hypotheses and Tests	35
2.2.1	Hypotheses, Tests, Decisions, and their Risks	35
2.2.2	How to Estimate Correct Sample Sizes	40
2.2.3	Parameter Tests or Outlier Tests?	41
2.2.4	Repeatability, Reproducibility, and Cooperative Tests	43
2.3	The Principle of Analysis of Variance	46
2.4	Causal Modeling, Regression, and Calibration	47
2.4.1	Correlation or Regression?	47
2.4.2	Other Notions, their Symbols and Meanings	47
2.4.3	The Model and the Prediction	50
2.4.4	Conditions to be met for Linear Regression Analysis	51

2.4.5	Alternative Linear Regression Models	52
2.4.6	Regression Errors and Tests of the Coefficients	54
2.4.7	Weighted and Robust Regression	56
2.4.8	Nonlinear Models	59
2.4.9	Measures of Goodness of Fit	60
2.4.10	Analysis of Variance for Regression Models	62
2.4.11	Regression Functions and Confidence Regions	64
2.4.12	Limits of Decision, of Detection, and of Determination	66
	References	69
3	**Remarks on Experimental Design and Optimization**	71
3.1	Basic Notions and Ideas	71
3.2	Factorial Designs	73
3.3	Example: Set-up and Interpretation Steps of Factorial Designs	76
3.3.1	Fixing the Levels of Factors	76
3.3.2	Performing the Measurements	77
3.3.3	Generalizing the Design	77
3.3.4	Constructing the General Computation Scheme	80
3.3.5	Extending for More Factors	80
3.3.6	Computing the Main Effects	81
3.3.7	Testing the Model Adequacy and the Coefficients	83
3.3.8	Interpreting the Results	85
3.3.9	Analysis of Variance	86
3.3.10	Analysis of Covariance	88
3.4	From Factorial to Sequential Designs	90
3.5	Sequential Methods for Optimum Search	91
3.6	Factorial or Sequential Methods?	92
	References	93
4	**Sampling and Sampling Design**	95
4.1	Introduction	95
4.2	Basic Considerations	97
4.3	Theoretical Aspects of Sampling	101
4.3.1	Number of Individual Samples Required	101
4.3.2	Mass of Individual Samples Required	103
4.3.3	Minimization of the Variance of Sampling and Analysis	109
4.3.4	Investigation of the Origin of Variance	111
4.3.5	Sampling Location and Frequency	112
4.4	Geostatistical Methods	113
4.4.1	Intrinsic Hypothesis	114
4.4.2	Semivariogram Analysis	114
4.4.3	Estimation of New Points in the Sampling Area – Kriging	117
4.4.4	Cross-Validation	119

4.5	Sampling Plans and Programs	120
4.5.1	Basic Considerations	120
4.5.2	Purpose of Sampling and the Chemometric Methods Applicable	121
4.5.3	Sampling Plans	122
4.5.3.1	Basic Types of Sampling	122
4.5.3.2	Grid Plans	124
4.5.3.3	Primary Sampling	127
4.5.3.4	Secondary Sampling	128
4.5.4	Sampling Programs	129
	References	133
5	**Multivariate Data Analysis**	139
5.1	General Remarks	139
5.2	Graphical Methods of Data Presentation	140
5.2.1	Introduction	140
5.2.2	Transformation	140
5.2.3	Visualization of Similar Features – Correlations	144
5.2.4	Similar Objects or Groups of Objects	145
5.2.4.1	Nesting Techniques	145
5.2.4.2	Star Plots	147
5.2.4.3	Pictoral Representation	148
5.2.4.4	Functional Representation	148
5.2.5	Representation of Groups	150
5.2.5.1	Box-Whisker Plots	150
5.2.5.2	Multiple Box-Whisker Plots	151
5.2.6	Limitations	152
5.3	Cluster Analysis	153
5.3.1	Objectives of Cluster Analysis	153
5.3.2	Similarity Measures and Data Preprocessing	153
5.3.3	Clustering Algorithms	156
5.3.4	CA Calculations Demonstrated with a Simple Example	159
5.3.5	Typical CA Results Illustrated with an Extended Example	161
5.4	Principal Components Analysis and Factor Analysis	164
5.4.1	Description of Principal Components Analysis	165
5.4.2	PCA Calculations Demonstrated with a Simple Example	168
5.4.3	Description of Factor Analysis	171
5.4.4	Typical FA Results Illustrated with an Extended Example	175
5.5	Canonical Correlation Analysis	179
5.5.1	Description of Canonical Correlation Analysis	179
5.5.2	Typical CCA Results Illustrated with an Extended Example	180
5.6	Multivariate Analysis of Variance and Discriminant Analysis	182
5.6.1	General Description	182
5.6.2	DA Calculations Demonstrated with a Simple Example	189

5.6.3	Typical DA Results Illustrated with an Extended Example	192
5.7	Multivariate Modeling of Causal Dependencies	195
5.7.1	Multiple Regression	196
5.7.2	Partial Least Squares Method	199
5.7.3	Simultaneous Equations and Path Analysis	201
	References	202
6	**Basic Methods of Time Series Analysis**	**205**
6.1	Introduction	205
6.2	Example: Nitrate Loadings in a Drinking Water Reservoir – Description of the Problem	206
6.3	Plotting Methods	208
6.3.1	Time Series Plot	208
6.3.2	Seasonal Sub-Series Plot	208
6.4	Smoothing and Filtering	209
6.4.1	Simple Moving Average	209
6.4.2	Exponential Smoothing	211
6.4.3	Simple and Seasonal Differencing and the CUSUM Technique	214
6.4.4	Seasonal Decomposition	216
6.5	Regression Techniques	217
6.5.1	Trend Evaluation with Ordinary Least Squares Regression	217
6.5.2	Least Squares Regression with an Explanatory Variable	219
6.5.3	Least Squares Regression with Dummy Variables (Multiple Least Squares Regression)	220
6.6	Correlation Techniques	222
6.6.1	Autocorrelation, Autoregression, Partial Autocorrelation, and Cross-correlation Function	222
6.6.2	Autoregression Analysis – Regression with an Explanatory Variable	225
6.6.3	Multivariate Auto- and Cross-correlation Analysis	228
6.7	ARIMA Modeling	234
6.7.1	Mathematical Fundamentals	234
6.7.2	Application of ARIMA Models	237
6.7.2.1	Specification of ARIMA Models	237
6.7.2.2	Application of the ARIMA Modeling to the Example Time Series	240
6.7.3	Forecasting with ARIMA	246
	References	246

Part B   **Case Studies**

7	**Atmosphere**	251
7.1	Sampling of Emitted Particulates	251
7.2	Evaluation and Interpretation of Atmospheric Pollution Data	252
7.2.1	Chemometric Characterization of the Impact of Particulate Emissions in Monitoring Raster Screens	252
7.2.1.1	Problem and Experimental	252
7.2.1.2	Analytical Results and Chemometric Interpretation	253
7.2.1.2.1	Univariate Aspects	254
7.2.1.2.2	Cluster Analysis	256
7.2.1.2.3	Multivariate Analysis of Variance and Discriminant Analysis, and PLS Modeling	258
7.2.1.2.4	Factor Analysis	264
7.2.1.3	Conclusions	268
7.2.2	Chemometric Characterization of the Temporal Course of the Impact of Particulate Emissions at One Sampling Location	269
7.2.2.1	First Example	269
7.2.2.1.1	Experimental	269
7.2.2.1.2	Data Preparation and Univariate Aspects	270
7.2.2.1.3	Combination of Cluster Analysis, and Multivariate Analysis of Variance and Discriminant Analysis	271
7.2.2.1.4	Factor Analysis	273
7.2.2.2	Second Example	275
7.2.2.2.1	Experimental	275
7.2.2.2.2	Multivariate Autocorrelation Analysis	276
7.2.2.2.3	Factor Analysis	278
	References	282
8	**Hydrosphere**	285
8.1	Sampling in Rivers and Streams	285
8.1.1	Sampling Strategies in Rivers	285
8.1.1.1	Experimental	285
8.1.1.2	Interpretation of the Results	286
8.1.1.2.1	Factor Analysis	286
8.1.1.2.2	Multivariate Analysis of Variance and Discriminant Analysis	286
8.1.1.3	Conclusions	289
8.1.2	Representative Sampling Distance along a River	291
8.2	Analytical and Chemometric Investigations of Polluted River Waters and Sediments	292
8.2.1	Problem and Experimental	292
8.2.2	Results from River Water Investigations, and their Interpretation	294
8.2.2.1	Scaling in Factor Analysis for River Assessment	294

8.2.2.2	Application of Factor Analysis to Samples Taken from the River Saale in the Summer	295
8.2.3	Chemometric Description of River Sediments	297
8.3	Speciation of Heavy Metals in River Water Investigated by Chemometric Methods	298
8.3.1	Comparing Investigations on Sediment Loadings by Means of Chemical Differentiation and Multivariate Statistical Data Analysis	299
8.3.1.1	Experimental	299
8.3.1.2	Results of Sequential Leaching	300
8.3.1.3	Factor Analysis of the Total Concentrations of Heavy Metals in River Sediments	302
8.3.1.4	Comparison of Chemical and Multivariate Statistical Investigations	302
8.3.2	Chemometric Methods in Combination with Electrochemical Analysis Applied to the Investigation of Heavy Metal Binding Forms in Waters	303
8.3.2.1	Experimental	303
8.3.2.2	Results and Discussion	306
8.3.3	Investigations of the Interaction Between River Water and River Sediment	311
8.3.3.1	Problem and Experimental	311
8.3.3.2	Variation of pH Value	313
8.3.3.3	Variation of Organic Loading	314
8.3.3.4	Variation of the Concentration of Condensed, Organic, and Ortho Phosphates	314
8.3.3.5	Variation of the Concentration of Suspended Material	315
	References	316
9	**Pedosphere**	**319**
9.1	Representative Soil Sampling	319
9.1.1	Problem	319
9.1.2	Soil Sampling and Heavy Metal Determination	319
9.1.3	Chemometric Investigation of Measurement Results	320
9.1.3.1	Univariate Analysis of Variance	320
9.1.3.2	Multivariate Data Analysis	321
9.1.3.2.1	Cluster Analysis	321
9.1.3.2.2	Multivariate Analysis of Variance and Discriminant Analysis	323
9.1.3.2.3	Principal Components Analysis	323
9.1.3.3	Autocorrelation Analysis	324
9.1.3.3.1	Univariate Autocorrelation Analysis	324
9.1.3.3.2	Multivariate Autocorrelation Analysis	327
9.1.4	Conclusions	328
9.2	Multivariate Statistical Evaluation and Interpretation of Soil Pollution Data	329

9.2.1	Studies on Heavy Metal Pollution of Soils at Different Locations	329
9.2.1.1	Experimental	329
9.2.1.2	Analytical Results and Chemometric Interpretation	330
9.2.1.2.1	Univariate Aspects	330
9.2.1.2.2	Cluster Analysis	331
9.2.1.2.3	Multivariate Analysis of Variance and Discriminant Analysis	332
9.2.1.2.4	Factor Analysis	335
9.2.2	Soil Depth Profiles	336
9.2.2.1	Problem and Experimental	336
9.2.2.2	Results and Discussion	337
9.2.2.3	Conclusions	341
9.3	Use of Robust Statistics to Describe Plant Lead Levels Arising from Traffic	341
9.3.1	Problem	341
9.3.2	Mathematical Fundamentals	342
9.3.3	Experimental	343
9.3.4	Analytical Results and Chemometric Evaluation	343
9.4	Geostatistical Methods	347
9.4.1	Introduction	347
9.4.2	Experimental	347
9.4.3	Results and Discussion	349
9.4.3.1	Semivariogram Estimation and Analysis	349
9.4.3.2	Kriging	350
9.4.3.3	Assessment of the Polluted Area	354
9.4.3.4	Determination of the Minimum Number of Samples for Representative Assessment of the State of Pollutant Loading	354
9.4.4	Conclusions	355
	References	356
10	**Related Topics**	359
10.1	Foods	359
10.1.1	Problem	359
10.1.2	Experimental	359
10.1.3	Results and Discussion	360
10.1.3.1	Univariate Aspects	360
10.1.3.2	Multivariate Statistical Data Analysis	361
10.1.3.2.1	Principal Components Analysis	361
10.1.3.2.2	Multivariate Analysis of Variance and Discriminant Analysis	361
10.1.4	Conclusions	362
10.2	Empirical Modeling of the Effects of Interference on the Flame Photometric Determination of Potassium and Sodium in Water	363
10.2.1	Problem	363
10.2.2	Theoretical Fundamentals	364

10.2.3	Experimental	365
10.2.4	Results, Mathematical Modeling, and Discussion	366
10.2.4.1	Test for Variance Homogeneity	366
10.2.4.2	Calculation and Testing of the Regression Coefficients	367
10.2.5	Conclusions	370
	References	371

Appendix

Appendix A	Selected Quantiles $k(P)$ of the GAUSSian (Normal) Distribution	373
Appendix B	Selected Quantiles $k(f; P)$ of the STUDENT t-Distribution	374
Appendix C	Selected Values for Two-Sided Confidence Intervals of Standard Deviations	375
Appendix D	Selected Quantiles $k(f_1; f_2; P)$ of the FISHER F-Distribution	376
Appendix E	Selected Critical Values for DIXON Outlier Tests	378

Index 379

Symbols and Abbreviations

List of Important Symbols

Some important symbols and abbreviations are listed below. Symbols and abbreviations the usage of which is restricted to special sections are defined in those sections.

a	regression coefficient
b	regression coefficient
c	concentration
C	constant
d	diameter, distance
df	discriminant function
$disp$	dispersion
e	error, difference between measured and estimated values, residual
e	eigenvector
f	degree of freedom
F	value of the F-distribution
I	unity matrix
l	distance in geostatistical analysis
L	range
m	number of features
med	median
n	number of objects, experiments or observations
P	probability
q	quantile
r	correlation coefficient
R	multivariate correlation coefficient
s	standard deviation
s^2	variance
SS	sum of squares
t	value of the t-distribution
u	value of the standardized normal distribution
w	weight, weighting factor
x	variable
$\bar{x}$	mean value
X	matrix of variables

y	variable, response
$\mathbf{Y}$	matrix of variables
z	space-dependent or standardized variable
Z	space-dependent random function
α	probability of an error of the first kind, risk
β	probability of an error of the second kind
γ	semivariance
Δ	difference
λ	lag
λ	eigenvalue
μ	true mean value
ρ	density
σ	true standard deviation
τ	lag
Ψ	direction

Symbols in bold letters: Small letters – vectors
Capital letters – matrices

Superscript indices:

$\wedge$	– estimated value
$-$	– mean value
$\sim$	– median

Subscript indices:

A	– analysis
crit	– critical
exp	– experimental
i	– running index
j	– running index
max	– maximum
min	– minimum
rel	– relative
S	– sampling

List of Abbreviations

AAS	atomic absorption spectrometry
ACF	autocorrelation function
ANOVA	analysis of variance
AR	autoregression
ARIMA	autoregressive integrated moving average
ARMA	autoregressive moving average
CA	cluster analysis
CCA	canonical correlation analysis
DA	discriminant analysis
fa	fulvic acid
FA	factor analysis
GFAAS	graphite furnace atomic absorption spectrometry
ICP-OES	optical emission spectroscopy with inductively coupled plasma
LMS	least median squares
LS	least squares
MA	moving average
MACF	multivariate autocorrelation function
MANOVA	multivariate analysis of variance
MDA	multivariate discriminant analysis
MVDA	multivariate analysis of variance and discriminant analysis
PACF	partial autocorrelation function
PC	principal component
PCA	principal components analysis
PLS	partial least squares
RLS	reweighted least squares
TSP	total sedimented airborne particulates

1 Introduction

1.1 Development of Chemometrics

For a long time the very broad range of mathematical and statistical methods has provided an excellent opportunity for the quantitative description of experimental results and effects in natural sciences. It is, therefore, not surprising that statistical methods are applied, in particular, in those scientific disciplines which investigate the effects resulting from many vague influences.

The aim of applying mathematical and statistical methods is first of all to detect and secondly to describe the interrelations between influencing forces on the one hand, and the interrelations between these influences and the resulting effects on the other hand. Subsequently, new scientific subdisciplines such as biometrics, psychometrics, medical statistics, econometrics, etc. have been developed. These developments have been reflected by the publication of corresponding journals, for example "Biometrika" (1901), "Psychometrika" (1936), "Technometrics" (1959) [GELADI, 1995]. Chemometrics was, therefore, not the first discipline of its kind.

In the field of chemistry, the processes requiring investigation have become increasingly complex. Consequently, the relevance of analytical chemistry has increased rapidly. Initially the acquisition of data was a severely limiting step in the analytical process. This situation has changed considerably since the 1950s because many new instrumental methods have been introduced in analytical chemistry. This has meant a dramatic change in the "production of analytical information" [VANDEGINSTE, 1987] in recent decades.

The resulting flood of data required more and more reduction, clear representation (in the sense of visualization), and extraction of the relevant information. This abundance of data also provided the possibility of more detailed and quantitative description of reaction mechanisms and structure-activity relationships.

Parallel to the rapid development of instrumental analytical chemistry came the explosive development of computer science and technology, a very powerful tool for the solution of the above problems. It became easier for chemists, and especially analytical chemists, to apply computers and advanced statistical and mathematical methods in their own working fields. It is, therefore, not surprising that these two revolutionary developments led to the formation of a new chemical subdiscipline, called chemometrics.

The name chemometrics was first coined by the young Swedish scientist SVANTE WOLD in the early 1970s. His cooperation with the American analytical chemist BRUCE R. KOWALSKI, who at the time was working on methods for pattern recognition in chemistry, resulted in the foundation of the International Chemometrics Society in 1974. The inaugural symposium of the "Chemometrics Working Group" in the Chemical Society of the G.D.R. was held in Leipzig on January 25, 1984; since German reunification this group has been a part of the "Working Group Chemometrics and Data Processing" within the German Society of Chemists (1992). In the meantime, the Royal Society of Chemistry also formed a Chemometrics Group within its Analytical Division.

An important part of chemometrics is, incidentally, statistics which in some countries has a long history. For example the American Statistical Society in Washington, D.C. celebrated its 150th anniversary in August 1989.

In the past decade, the development of chemometrics has found its expression in the publication of two journals specializing in chemometrics: "Chemometrics and Intelligent Laboratory Systems" (1986) and "Journal of Chemometrics" (1987); the increasing importance of chemometric methods in the field of environmental research is apparent from the publication of the journal "Environmetrics" (1990). As a consequence of the rapidly increasing number of papers dealing with chemometrics, a reference journal "Windows on Chemometrics" has been published since 1993.

1.2 Definition of Chemometrics

The current definition, based on the classical definition of FRANK and KOWALSKI [1982], is [MASSART et al., 1988]:

Chemometrics "can be defined as the chemical discipline that uses mathematical, statistical, and other methods employing formal logic

- to design or select optimal measurement procedures and experiments, and
- to provide maximum relevant chemical information by analyzing chemical data".

FRANK and KOWALSKI [1982] also state: "Chemometric tools are vehicles that can aid chemists to move more efficiently on the path from measurements to information to knowledge". Another formulation is given by KATEMAN [1988]: Chemometrics is the "... nonmaterial part of analytical chemistry". Expressed in other words [BRERETON, 1990]: "Chemometrics is a collection of methods for the design and analysis of laboratory experiments, most, but not all, chemically based. Chemometrics is about using available resources as efficiently as practicable, and arriving at as useful a conclusion as possible taking into account limitations of cost, manpower, time, equipment etc.".

The present state of chemometrics and the differing points of view are illustrated by the definition of DANZER [1990]: "Chemometrics is the linking element between chemistry (not only analytical chemistry), mathematics, and hard- and software".

It is obvious that every scientist whether in the field of chemistry, mathematics, statistics, or computer science has a different view of the importance of chemometrics. The advantages of applying chemometric methods are more valuable than discussion of the origin and the size of the roots of chemometrics. It should, therefore, be stressed that chemometrics has an interdisciplinary nature. Some scientists also use the expression "environmetrics" for the application of chemometric methods in the field of environmental science.

Using a definition similar to that of the International Chemometrics Society, chemometrics can therefore be summarized as follows:

Chemometrics in its present state is a chemical subdiscipline which deals with the application of mathematical, statistical, and other methods employing formal logic

- to evaluate and to interpret (chemical, analytical) data
- to optimize and to model (chemical, analytical) processes and experiments
- to extract a maximum of chemical (and hence analytical) information from experimental data

1.3 Classification of Chemometric Methods

Chemometric methods range from simple statistics to highly sophisticated data analysis. The large variety of methods calls for some classification to enable user orientation.

1. One possibility is to classify the methods used in the various steps of the analytical process (see also Section 1.5). Note that the analytical process usually starts with the definition or selection of the matter to be investigated. Here it is very important to realize that every sample or object of investigation has a **history**. Because this history may cause severe systematic errors, THIERS [1957] explains: "Unless the complete history of any sample is known with certainty, the analyst is well advised not to spend his time in analyzing it." Clearly, in such circumstances one should be extremely cautious about drawing conclusions from chemometric interpretations even where data are available.

We will not go into further detail, but rather we will discuss the basic steps and the generally accepted distinction today of the notions "principle", "method", and "procedure". The main steps of the analytical process are **sampling, sample pretreatment, measurement,** and **interpretation** of the results (the collected data) (Fig. 1-1). **Procedure** means all activities from sample definition to the extraction of information by interpreting the data. **Methods** may be defined as the processes carried out between sample pretreatment and interpretation of the results. And finally **principle** describes the process in which analyte matter produces a signal that is further treated.

By executing the steps of the analytical process, we can take advantage of most of the **basic methods of chemometrics**, e.g., statistics including analysis of variance, experimental design and optimization, regression modeling, and methods of time series analysis.

2. If we are primarily interested in problems concerning data analysis, we have to focus on **advanced methods of chemometrics**. In order to classify these methods, we may

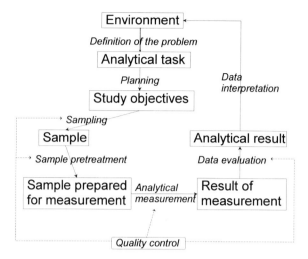

Fig. 1-1. Steps of the analytical process

start with the following consideration: measurements are assumed to be connected with **reality**, to reflect reality without distortion. We then derive **data** from **measurements**. This data has to be subjected to **data analysis** in order to extract **information**.

Important topics in chemometrics are: principles of sampling, experimental design, choosing and optimizing analytical conditions, univariate and multivariate signal processing (including methods of calibration), and data analysis (pattern cognition and recognition).

In general, the following topics also belong to the working field of chemometrics: process control and optimization, rational analysis and laboratory organization (including laboratory information and management systems — LIMS), library search, and principles of artificial intelligence (including expert systems, neural networks, and genetic algorithms).

The future development of chemometrics may be characterized by the balancing of opposites [GELADI, 1995]: combination of hard and soft information, using *a priori* information when available and combining with *a posteriori* information. So, a holistic strategy (Fig. 1-2) with different chemometric methods in each of its steps may help to solve problems of the "real world of chemistry and other disciplines".

Because data analysis is of central interest, particularly in the application of chemometric methods in the field of environmental research, a rough list of important multivariate statistical methods is given below (Tab. 1-1).

Characteristic of all the methods listed in Tab. 1-1 is that comprehensive consideration of the data set serves as the method of data representation. In accordance with the manifold capabilities of the methods, it is obvious that some of the methods presented in Tab. 1-1 can also be used for solving other problems.

1.3 Classification of Chemometric Methods

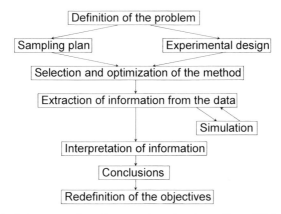

Fig. 1-2. A generic holistic strategy for a chemometric project (according to [GELADI, 1995])

Tab. 1-1. Important methods of multivariate statistics

Method	Solving the problem

Unsupervised learning methods
– cluster analysis
– display methods
  – nonlinear mapping (NLM)
  – minimal spanning tree (MST)
  – principal components analysis (PCA)
	Finding structures/similarities (groups, classes) in the data

Supervised learning methods – multivariate analysis of variance and discriminant analysis (MVDA) – k nearest neighbors (kNN) – linear learning machine (LLM) – BAYES classification – soft independent modeling of class analogy (SIMCA) – UNEQ classification	Quantitative demarcation of *a priori* classes, relationships between class properties and variables
Factorial methods – factor analysis (FA) – principal components analysis (PCA) – partial least squares modeling (PLS) – canonical correlation analysis	Finding factors (causal complexes)
Correlation and regression analysis – with direct variables – with latent variables	Quantitative description of the relationships between variables

1.4 What is Environmental Analysis?

The study of our environment includes a very broad range of subjects. There are several very different environmental compartments, beginning with nonliving media e.g. the atmosphere, water, and soil, ranging to living compartments such as plants, animals, and human beings. To understand our environment we must realize that it is never static. Both the state (seen as an instantaneous exposure) and the dynamic changes in and between these parts of the environment are determined by physical, chemical, and biological processes. Because environmental processes and reactions are often nonstationary, irreversible, and take place in (open) systems which are difficult to define, it is impossible in practice to describe them by using deterministic models. The very stochastic character of environmental processes predominates.

The increasing interest in the environment arises from concern that natural cycles and processes are being disrupted by mankind to such an extent that the quality of life, and sometimes even life itself, is being threatened. Current pollution and potential or real damage require the urgent need for monitoring and control. This implies not only simple monitoring of the state of pollution in selected environmental compartments, but also more intensive study of the physical, chemical, and biological processes in and between different media.

The different steps of environmental analysis may also be expressed by the following topics [REEVE, 1994]:

– Recognition of the problem
– Monitoring to determine the extent of the problem
– Determination of control procedures
– Legislation to ensure the control procedures are implemented
– Monitoring to ensure the problem has been controlled

Analytical chemistry is an essential component in almost all aspects of scientific investigations of the environment, the problems caused by mankind and their possible solutions.

The question is now: What are the characteristics of **environmental analysis**?

1. Trace analytes
In practice in all cases under investigation, environmental problems are connected with problems of trace or ultratrace analysis. Only very seldom are major components of interest.
Trace analysis requires a high degree of accuracy and precision, since contaminants measured at ppb or even ppt levels, and sometimes lower, may pose human and/or ecological health hazards [WENNING and ERICKSON, 1994].

2. Occurrence of many pollutants

Often, many simultaneously occurring pollutants or contaminants determine an environmental problem. In industry, agriculture, and households, products are often mixtures of many compounds. The process of production and consumption is accompanied by emissions and consequently by contamination. One example is the use of toxaphene in the past, a very complex mixture of polychlorinated camphenes, as a pesticide. Technical toxaphene consists of more than 175 individual compounds. A second example is industrial and domestic emissions resulting from the combustion of fossil fuels. The emissions contain both a mixture of gases (SO_2, NO_x, CO_2, etc.) and airborne particulate matter which itself contains a broad range of heavy metals and also polycyclic aromatic hydrocarbons (PAH).

3. Multiple pathways in the environment

The mechanism of transport of the majority of pollutants in the environment is, in general, unknown. This transport is often associated with changes in the chemical character and the concentration of individual chemical species in the environment (washing-out, rain-out, metabolism, "aging", etc.). The mechanisms of the resulting effects (e.g. damage to the environment) are also unknown. Fig. 1-3 demonstrates part of the heavy metal cycle in the environment where different sources and sinks exist, and these are connected by several different, simultaneous, and sometimes interfering pathways.

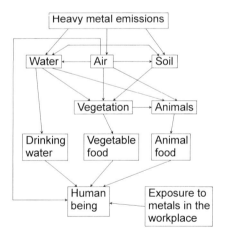

Fig. 1-3. The human being as a part of the anthropogenically influenced heavy metal cycle

1.5 The Need to Apply Chemometric Methods

The characteristic peculiarities of environmental analysis discussed at the end of Section 1.4 lead to the following problems which urgently require the application of chemometric methods.

1. Environmental data is strongly characterized by inherent variability. Only limited understanding of the environmental distribution of contaminants can, therefore, be gained from a single analysis [WENNING and ERICKSON, 1994]. If we consider such data, we must realize that variability normally has several different origins:

- Variability caused by the environment

– Natural variability
  As demonstrated in Fig. 1-4 for the moderately homogeneous surface of an agricultural area without any strong contaminating influences, variability over the area under investigation is mainly determined stochastically. The natural variability in other environmental media, e.g. the atmosphere and hydrosphere, is usually higher than in the pedosphere.

– Anthropogenic variability
  A further source of variability which interferes with natural variability is the influence of anthropogenic activities. Tab. 1-2 shows the highest and the lowest values of some selected heavy metals measured in the surface soil of an area strongly influenced by particulate emissions from a large metallurgical factory. The variability amounts to an order of magnitude of between 1 and 2.

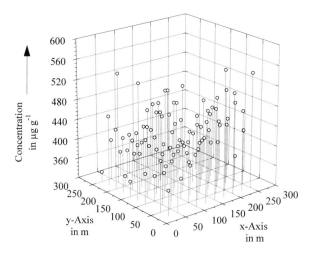

Fig. 1-4. Zinc distribution in the surface soil of an agricultural area

Tab. 1-2. Heavy metal content of surface soil in the area surrounding a large metallurgical factory (values in mg kg^{-1})

Element	Lowest value	Highest value
Cd	0.19	8.26
Co	0.12	22.5
Cr	2.5	62.7
Cu	2.6	48.2
Fe	1083	12477
Mn	43.1	3149
Ni	0.7	35.6
Pb	7.6	240
Zn	11.2	1186

- Spatial and/or temporal variability
  The variability of environmental data must also be regarded as being dependent on space and/or time. As an example, the temporal variability is demonstrated for the occurrence of volatile chlorinated hydrocarbons in river water (Fig. 1-5). The very different pattern for the time functions of the selected volatile chlorinated hydrocarbons at two sampling locations 40 km apart shows that the concentration fluctuations are quite random.

- Variability caused by experimental error
  The steps of the analytical process are illustrated in Fig. 1-1. It is well known that in each step of this process experimental errors are possible. Furthermore, each analytical result contains at least some experimental error; the relative size of this error increases considerably as the analyte concentration in the sample decreases. This general relationship is demonstrated in Fig. 1-6 for some selected environmentally relevant compounds. Because the majority of studies concerning the environment deals with trace or ultratrace analysis, this fact is very evident and important.

To conclude, the stochastic and error-influenced character of environmental data requires the use of mathematical and statistical methods for further analysis.

2. The nature of study objectives in environmental research is often multivariate. Several pollutant patterns from different, sometimes unknown, sources may occur. The state of pollution of a sampling point, line, or area in any environmental compartment, whether atmosphere, water, soil, or biota, depends mostly on the nature of the different sources of pollution. Stack emissions are characterized by a multi-element pattern. Waste water effluents contain different contaminants, ranging from heavy metals to cocktails of organic compounds.

If we investigate the transport of pollutants in an environmental compartment (e.g. the transport of airborne particulate matter through the atmosphere or soluble effluents in a river) or the pathway from one compartment to the other (e.g. the transmission of con-

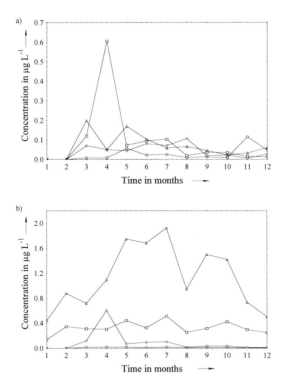

Fig. 1-5. Seasonal fluctuations of volatile chlorinated hydrocarbons in the river Weiße Elster (Germany): a) sampling location above Greiz, b) sampling location below Gera. ○ tetrachloromethane, □ bromdichloromethane, ◇ trichloroethene, △ tetrachloroethene

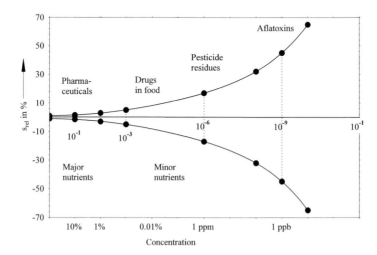

Fig. 1-6. Dependence of relative standard deviation on analyte concentration in food (according to [HORWITZ et al., 1980])

taminants from soil to plants or into groundwater or the sedimentation-remobilization process between the river water body and the river sediment), we also have to take into account the multivariate character of these processes. It is a characteristic of environmental processes that the pathways of pollutants in the environment are often manifold (see also Fig. 1-3 for demonstration) and sometimes unknown.

Pollutant patterns change radically during the process of duration and transport in the environment, particularly as a result of dilution and reactions with each another and with naturally occurring components. In particular, organic contaminants are often changed by abiotic and/or biotic activities ("aging", weathering, metabolism, degradation, etc.).

To consider all contaminants, and their potential reactions and interactions, simultaneously, the application of multivariate methods is required.

3. A third and often neglected reason for the need for careful application of chemometric methods is the problem of the type of distribution of environmental data. Most basic and advanced statistical methods are based on the assumption of normally distributed data. But in the case of environmental data, this assumption is often not valid. Figs. 1-7 and 1-8 demonstrate two different types of experimentally found empirical data distribution. Particularly for trace amounts in the environment, a log-normal distribution, as demonstrated for the frequency distribution of NO_2 in ambient air (Fig. 1-7), is typical.

Before using parametric methods, the data have to be transformed and tested to see if normal distribution occurs. Sometimes multimodal distributions exist, as illustrated in Fig. 1-8 for the occurrence of particulate matter in the atmosphere. A further difficulty in this case is that the type of distribution may change during the transmission process in the air. If normal

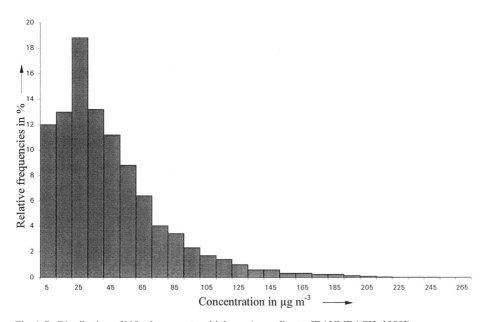

Fig. 1-7. Distribution of NO_2 data near to a highway (according to [BAUMBACH, 1990])

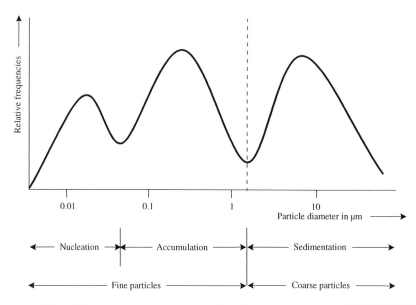

Fig. 1-8. Specific particle surface as a function of particle diameter (according to [WHITBY, 1987])

data distribution cannot be provided by data transformation algorithms, methods of nonparametric statistics, e.g. robust statistics, have to be applied (see also Section 9.3).

The specific problems discussed above emphasize that "environmental chemistry poses considerably harder problems to the chemometrician than straight analytical chemistry" [BRERETON, 1995]. The current state of environmental analysis often involves empirical planning of experiments and monitoring, as well as expensive and time-consuming analysis, with the result that only simple statistics are applied to the data obtained. In practice, simple comparison of data averaged in time or space with legally fixed thresholds or limits is often performed at the end of the environmental analysis process. Because environmental data contains so much information, chemometric methods should be used to extract the latent information from these data.

In summary, the application of chemometric methods may be very useful for the solution of the following environmental problems:

- Planning and optimization of the whole environmental analytical process, starting with the planning and experimental design of environmental sampling or experiments, optimization of the analytical procedure in the laboratory, including modern chemometric methods for signal detection and treatment
- Compression of large data sets, elimination of redundancy and noise
- Visualization of complex, often high-dimensional quantitative relationships
- Detection and identification of emitters and dischargers (in general: origins)
- Detection and quantification of differences in loadings
- Investigation of spatial and temporal relationships between environmental data and their changes

- Investigation of different species in the environment with respect to their electronic state, binding forms and their reactions (speciation)
- Investigation of interactions between pollutants and components of environmental compartments
- Environmental impact assessment

1.6 What do we Expect of Chemometrics in Environmental Analysis?

When applying chemometric methods the general question is: Why do we want information? First of all, it should be pointed out that data are only series of numbers. These numbers are worthless because they have no meaning. This data are only meaningful if used in a human context.

One very important (in most cases the ultimate) goal is the support of **decisions**. So, with the aid of decisions, we come to conclusions concerning reality – and this is what we really need.

We should remember that an ordered set of data can be written in the form of a matrix, which in most instances contains the values of measured or known properties (variables, features) listed in columns. The rows of the matrix are then associated with the objects (e.g. samples at different locations or times) under study.

Basically there are two situations:
(1) We do not have, or do not want, to use additional or *a priori* information about the data measured.
(2) We have, and use, some *a priori* knowledge about the data.

Typical questions are, for example:
(1.1) Are there **relationships between objects**?
(1.2) Are there **relationships between variables**?
(2.1) Are there **"dependencies" of** (grouped) **objects**?
(2.2) Are there **dependencies of** one or several **variables on** one or several **others**, and can they be modeled?

Under (1) assume that you have collected and analyzed a number of water samples of different origin for acidity (pH), hardness (H), and lead and cadmium content (Pb, Cd).

For the moment we do not use the information about the origin. Then special questions may be:

(1.1.1) Can one see a grouping of the objects, e.g. ten water samples, in respect of their origin using only pH, H, Pb, and Cd?
(1.2.1) Do pH and Cd correlate significantly?
(1.2.2) Can we see sets (clusters) of features, e.g. cluster {pH, H} or cluster {Pb, Cd}?

The answers to these questions will usually be given by so-called **unsupervised learning** or **unsupervised pattern recognition methods**. These methods may also be called grouping methods or automatic classification methods because they search for classes of similar objects (see cluster analysis) or classes of similar features (see correlation analysis, principal components analysis, factor analysis).

Under (2) assume that you now include the information about sample origin: tap water (T), ground water (G), and sewage water (S). Then special questions may be:

(2.1.1) Are the results from T, G, and S samples statistically different?
(2.1.2) In a clustering step answering (1.1.1) you may have found that most of the T and G samples form one group which appears different from the S sample group. Then question (2.1.1) refers to groups one and two.
(2.1.3) Does an unknown sample belong to the first or to the second group?
(2.2.1) Can we predict the results of Pb from the results of pH? What is the functional relationship for that dependency?

Questions of type (2.1) may be answered by analysis of variance or by discriminant analysis. All these methods may be found under the name **supervised learning** or **supervised pattern recognition methods**. In the sense of question (2.1.3) one may speak of supervised classification or even better of re-classification methods. In situations of type (2.2) methods from the large family of regression methods are appropriate.

The scheme given is for rough orientation only, because several methods are useful for quite different purposes. As an example, principal components analysis may be applied in answering both (1.1) and (1.2) type questions and is sometimes necessary in data processing prior to (2.2) type questions.

1.7 Recommended Reading and Useful Software

Section 1.8 gives an overview of monographs which deal with basic and advanced statistics, and some hints are given about recommended books and journals. Important books on general aspects of chemometrics include: SHARAF et al. [1986], MASSART et al. [1988], and BRERETON [1990]. Books which deal with the application of chemometric methods in environmental research are: BREEN and ROBINSON [1985], DEVILLERS and KARCHER [1991], and EINAX [1995a; 1995b].

Scientific articles on applied chemometrics can also be found in the periodicals: "Journal of Chemometrics and Intelligent Laboratory Systems" and "Journal of Chemometrics", and with emphasis on the environment: "Environmetrics". Many analytical journals also publish papers on chemometrics; these include "Analytical Chemistry", "Analytica Chimica Acta", "Analytical Proceedings", "Analyst", "Fresenius' Journal of Analytical Chemistry", "Journal of Environmental Analytical Chemistry", etc.

A brief overview of new literature in the field of chemometrics is available in the journals: "Windows on Chemometrics" and "Analytical Abstracts".

Because of the interdisciplinary character of chemometrics, it is obvious that other periodicals which concentrate on biological, geoscientific, and ecological subjects also publish articles on the application of chemometrics.

Because of the explosive development of computer sciences, the number of software programs and packages has increased dramatically. Tab. 1-3 presents a list of some programs and software packages for table calculation, and methods of basic and advanced statistics. It is clear that in such a booming market, only a relatively subjective overview can be given.

Tab. 1-3. Software programs and packages for table calculation and statistics

ABSTAT	ARTHUR	BMDP
CLEOPATRA	CLUSTAN	CSS: STATISTICA
DADiSP	EXCEL	FANTASIA
GENSTAT	GLIM	ISP
LOTUS	MAPLE	MINITAB
MULTIVAR	NCSS	NIPALS
PARVUS	POLET	PIROUETTE
P-STAT	QPRO	SAS
SCAN	SIMCA	SPECTRAMAP
SPIDA	S-PLUS	SPSS
SQUAD	STASY	STATA
STATGRAPH	STATISTICA	STATMOST
STAVE	SYSTAT/SYGRAPH	TARGET
TESTIMATE	UNISTAT	UNSCRAMBLER

Users of chemometric methods should select their own software tools, on the one hand according to their desires and the tasks which are to be served, and on the other hand, according to personal experience, costs, and the hardware available.

The authors of this book have found LOTUS and EXCEL satisfactory for data preprocessing and STATGRAPH, SPSS, and STATISTICA suitable for the application of statistical methods. UNSCRAMBLER is recommended if applying PLS modeling.

1.8 Overview of Chemometric Literature

General Overview on Statistics and Chemometrics

Adams, M.J.: Chemometrics in Analytical Spectroscopy, Royal Society of Chemistry, Cambridge, 1995
Adler, B.: Computerchemie – eine Einführung, Deutscher Verlag für Grundstoffindustrie, Leipzig, 1986
Aitchison, J.: The Statistical Analysis of Compositional Data, Chapman and Hall, London, 1986
Armanino, C. (Ed.): Chemometrics and Species Identification, Topics in Current Chemistry, Vol. 141, Springer, Berlin, Heidelberg, New York, London, Paris, Tokyo, 1987
Brereton, R.G.: Chemometrics. Applications of Mathematics and Statistics to Laboratory Systems, Ellis Horwood, Chichester, 1990

Buydens, L.M.C., Melssen, W.J. (Eds.): Chemometrics: Exploring and Exploiting Chemical Information, University of Nijmegen, 1994
Caulcutt, R., Boddy, R.: Statistics for Analytical Chemists, Chapman and Hall, London, 1983
Chatfield, C., Collins, A.J.: Introduction to Multivariate Analysis, Chapman and Hall, London, 1989
Davis, J. C.: Statistics and Data Analysis in Geology, 2nd Ed., Wiley, New York, 1986
Dillon, W.R., Goldstein, M.: Multivariate Analysis: Methods and Applications, Wiley, New York, Chichester, Brisbane, Toronto, Singapore, 1984
Doerffel, K.: Statistik in der analytischen Chemie, 5. Aufl., Deutscher Verlag für Grundstoffindstrie, Leipzig, 1990
Doerffel, K., Eckschlager, K., Henrion, G.: Chemometrische Strategien in der Analytik, Deutscher Verlag für Grundstoffindustrie, Leipzig, 1990
Eason, G., Coles, C.W., Gettinby, G.: Mathematics and Statistics for the Bio-Sciences, Ellis Horwood, Chichester, 1980
Ebert, K., Ederer, H., Isenhour, T.L.: Computer Applications in Chemistry, VCH, Weinheim, 1989
Enke, H., Gölles, J., Haux, R., Wernecke, K.-D. (Eds.): Methoden und Werkzeuge für die exploratorische Datenanalyse in den Biowissenschaften, Gustav Fischer, Stuttgart, Jena, NewYork, 1992
Fahrmeir, L., Hamerle, A.: Multivariate statistische Verfahren, Walter de Gruyter, Berlin, New York, 1984
Fax, L.: Angewandte Statistik, Springer, Berlin, Heidelberg, New York, 1978
Flury, B., Riedwyl, H.: Multivariate Statistics: A Practical Approach, Chapman and Hall, 1988
Goldstein, M., Dillon, W.R.: Multivariate Analysis: Methods and Applications, Wiley, New York, 1984
Graham, R.C.: Data Analysis for the Chemical Sciences: A Guide to Statistical Techniques, VCH, New York, Weinheim, Cambridge, 1993
Hartung, J., Elpelt, B.: Multivariate Statistik, 4. Aufl., R. Oldenbourg, München, Wien, 1992
Hartung, J., Elpelt, B., Klösener, K.-H.: Statistik, 8. Aufl., R. Oldenbourg, München, Wien, 1991
Henrion, R., Henrion, G.: Multivariate Datenanalyse. Methodik und Anwendung in der Chemie und verwandten Gebieten, Springer, Berlin, Heidelberg, New York, London, Paris, Tokyo, Hong Kong, Barcelona, Budapest, 1994
Henrion, G., Henrion, A., Henrion, R.: Beispiele zur Datenanalyse mit BASIC-Programmen, Deutscher Verlag der Wissenschaften, Berlin, 1988
Johnson, R.A., Wichern, D.W.: Applied Multivariate Statistical Analysis, 2nd Ed., Prentice-Hall, Englewood Cliffs, New Jersey, 1988
Kowalski, B.R. (Ed.): Chemometrics: Mathematics and Statistics in Chemistry, NATO ASI Series C, Mathematical and Physical Sciences, Vol. 138, Reidel, Dordrecht, 1984
Kowalski, B.R. (Ed.): Chemometrics: Theory and Application, ACS Symposium Series, Vol. 52, American Chemical Society, Washington, D.C., 1977
Kramer, R.: Basic chemometrics. A Practical Introduction to Quantitative Analysis, Wiley, New York, 1995
Krzanowski, W.J.: Principles of Multivariate Analysis, Oxford Science Publications, The Universities Press Ltd., Belfast, 1990
Lebart, L., Morineau, A., Fenelon, J.-P.: Statistische Datenanalyse, Methoden und Programme, Akademie-Verlag, Berlin, 1984
Lewi, P.J.: Multivariate Analysis in Industrial Practice, Research Studies Press, Letchworth, 1982
Lohse, H., Ludwig, R., Röhr, M.: Statistische Verfahren für Psychologen, Pädagogen und Soziologen, 2. Aufl., Volk und Wissen, Berlin, 1986
Manly, B.F.J.: Multivariate Statistical Methods: A Primer, Chapman and Hall, London, 1986
Massart, D.L., Vandeginste, B.G.M., Deming, S.N., Michotte, Y., Kaufman, L.: Chemometrics: A Textbook, Elsevier, Amsterdam, 1988
Meier, P.C., Zünd, R.E.: Statistical Methods in Analytical Chemistry, Wiley, New York, Chichester, Brisbane, Toronto, Singapore, 1993
Meloun, M., Militzky, J., Forina, M.: Chemometrics for Analytical Chemistry, Vol. I, PC-Aided Statistical Data Analysis, Ellis Horwood, Chichester, 1992
Miller, J.C., Miller, J.N.: Statistics for Analytical Chemistry, 3rd Ed., Ellis Horwood, PTR Prentice Hall, New York, 1993

Sachs, L.: Angewandte Statistik, 7. Aufl., Springer, Berlin, Heidelberg, New York, 1992
Sharaf, M.A., Illman, D.L., Kowalski, B.R.: Chemometrics, Wiley, New York, 1986
Srivastava, M.S., Carter, E.M.: An Introduction to Applied Multivariate Statistics, Elsevier, North-Holland, New York, Amsterdam, Oxford, 1983
Varmuza, K.: Pattern Recognition in Chemistry, Springer, Berlin, Heidelberg, New York, 1980
Weber, E.: Grundriß der biologischen Statistik, 9. Aufl., Gustav Fischer, Jena, 1986

Cluster Analysis

Anderberg, M.R.: Cluster Analysis for Applications, Academic Press, New York, 1973
Everitt, B.S.: Cluster Analysis, Heinemann, London, 1974
Massart, D.L., Kaufman, L.: The Interpretation of Analytical Chemical Data by the use of Cluster Analysis, Wiley, New York, 1983
Mucha, H.-J.: Clusteranalyse mit Mikrocomputern, Akademie-Verlag, Berlin, 1992
Steinhausen, D., Langer, K.: Clusteranalyse – Einführung in Methoden und Verfahren der automatischen Klassifikation, Walter de Gruyter, Berlin, 1977
Willett, P.: Similarity and Clustering in Chemical Information Systems, Research Studies Press, Letchworth, 1987

Correlation and Regression Analysis

Doerffel, K., Wundrack, A.: Korrelationsfunktionen in der Analytik in: Fresenius, W., Günzler, H., Huber, W., Lüderwald, I., Tölg, G., Wisser, H. (Eds.): Analytiker-Taschenbuch, Bd. 6, Akademie-Verlag, Berlin, 1986, pp. 37
Fahrmeir, L., Tutz, G.: Multivariate Statistical Modelling based on Generalized Linear Models, Springer, Berlin, Heidelberg, New York, 1994
Förster, E., Rönz, B.: Methoden der Korrelations- und Regressionsanalyse, Verlag Die Wirtschaft, Berlin, 1979
Gans, P.: Data Fitting in Chemical Sciences, Wiley, New York, 1992
Mager, H.: Moderne Regressionsanalyse, Salle + Sauerländer, Frankfurt, Berlin, München, Aarau, Salzburg, 1982
Martens, H., Naes, T.: Multivariate Calibration, Wiley, Chichester, New York, Brisbane, Toronto, Singapore, 1989

Experimental Design and Optimization

Bandemer, H., Bellmann, A.: Statistische Planung von Experimenten, B.G. Teubner Verlagsgesellschaft, Leipzig, 1976
Box, G.E.P., Hunter, W.G., Hunter, J.S.: Statistics for Experimenters, Wiley, New York, 1978
Davies, L.: Efficiency in Research, Development and Production: The Statistical Design and Analysis of Chemical Experiments, Royal Society of Chemistry, Cambridge, 1993
Davies, O. L. (Ed.): The Design and Analysis of Industrial Experiments, Oliver and Boyd, London, 1984
Deming, S.N., Morgan, S.L.: Experimental Design: A Chemometric Approach, 2nd Ed., Elsevier, Amsterdam, 1993
Goupy, J.L.: Methods for Experimental Design. Principles and Applications for Physicists and Chemists, Elsevier, Amsterdam, 1993
Massart, D.L., Dijkstra, A., Kaufman, L.: Evaluation and Optimization of Laboratory Methods and Analytical Procedures. A Survey of Statistical and Mathematical Techniques, Elsevier, Amsterdam, 1978
Morgan, E.: Chemometrics: Experimental Design, Wiley, Chichester, 1991

Scheffler, E.: Einführung in die Praxis der statistischen Versuchsplanung, 2. Aufl., Deutscher Verlag für Grundstoffindustrie, Leipzig, 1986

Factor Analysis

Jahn, W., Vahle, H.: Die Faktorenanalyse und ihre Anwendung, Die Wirtschaft, Berlin, 1970
Joliffe, I.T.: Principal Components Analysis, Springer, Berlin, 1986
Malinowski, E.R.: Factor Analysis in Chemistry, 2nd Ed., Wiley, New York, Chichester, Brisbane, Toronto, Singapore, 1991
Malinowski, E.R., Howery, D.G.: Factor Analysis in Chemistry, Wiley, New York, Chichester, Brisbane, Toronto, 1980
Überla, K.: Faktorenanalyse, Springer, Berlin, Heidelberg, New York, 1968
Weber, E.: Einführung in die Faktorenanalyse, Gustav Fischer, Jena, 1974
Weber, E.: Grundriß der biologischen Statistik, 9. Aufl., Gustav Fischer, Jena, 1986

Geostatistical Methods

Akin, H., Siemes, H.: Praktische Geostatistik, Springer, Berlin, Heidelberg, New York, London, Paris, Tokyo, 1988
Cressie, N.A.C.: Statistics for Spatial Data, Wiley, New York, Chichester, Toronto, Brisbane, Singapore, 1991
David, M.: Geostatistical Ore Reserve Estimation, Elsevier, Amsterdam, Oxford, New York, 1977
Journel, A.G., Huigbregts, Ch.: Mining Geostatistics, Academic Press, London, New York, San Francisco, 1978
Ripley, D.B.: Spatial Statistics, Wiley, NewYork, 1981

Graphical Methods of Data Analysis

du Toit, S.H.C., Steyn, A.G.W., Stumpf, R.H.: Graphical Exploratory Data Analysis, Springer, Berlin, 1986
Fleischer, W., Nagel, M.: Datenanalyse mit dem Personalcomputer, Verlag Technik, Berlin, 1989

Multivariate Analysis of Variance and Discriminant Analysis

Ahrens, H., Läuter, J.: Mehrdimensionale Varianzanalyse, 2. Aufl., Akademie-Verlag, Berlin, 1981
McLachlan, G.J.: Discriminant Analysis and Statistical Pattern Recognition, Wiley, Chichester, 1992

Pattern Recognition

Bawden, D., Ioffe, I.T. (Eds.): Application of Pattern Recognition to Catalysis Research, Research Studies Press, Letchworth, 1988
Brereton, R.G. (Ed.): Multivariate Pattern Recognition in Chemometrics, Illustrated by Case Studies, Elsevier, Amsterdam, New York, 1992
Coomans, D., Broeckaert, I.: Potential Pattern Recognition, Research Studies Press, Letchworth, 1986
Jurs, P.C., Isenhour, T.L.: Chemical Applications of Pattern Recognition, Wiley, New York, 1975
Strouf, O.: Chemical Pattern Recognition, Research Studies Press, Letchworth, 1986
Varmuza, K.: Pattern Recognition in Chemistry, Springer, Berlin, 1980
Wolff, D.D., Parsons, M.I.L.: Pattern Recognition Approach to Data Interpretation, Plenum, New York, 1983

Robust Statistics

Hampel, F.R., Ronchetti, E.M., Rousseeuw, P.J., Stahel, W.A.: Robust Statistics: The Approach Based on Influence Functions, Wiley, New York, 1986

Huber, P.J.: Robust Statistics, Wiley, New York, 1981

Rousseeuw, P.J., Leroy, A.M.: Robust Regression and Outlier Detection, Wiley, New York, Chichester, Brisbane, Toronto, Singapore, 1987

Tukey, J.W.: Robust Estimation of Location: Survey and Advances, Princeton University Press, Princeton, 1972

Sampling

Gesellschaft Deutscher Metallhütten- und Bergleute e.V. (Ed.): Probenahme Theorie und Praxis, Verlag Chemie, Weinheim, Deerfield Beach/Florida, Basel, 1980

Green, R.G.: Sampling Design and Statistical Methods for Environmental Biologists, Wiley, New York, Chichester, Brisbane, Toronto, 1979

Gy, P.M.: Heterogeneity-Sampling-Homogenization, Elsevier, Amsterdam, 1991

Gy, P.M.: Sampling of Particulate Materials, Theory and Materials, 2nd Ed., Elsevier, Amsterdam, Oxford, New York, 1982

Gy, P.M. (Ed.): Sampling of Heterogeneous and Dynamic Material Systems. Theories of Heterogeneity, Sampling and Homogenizing Data Handling in Science and Technology, Elsevier, Amsterdam, 1992

Keith, L.H.: Environmental Sampling and Analysis: A Practical Guide, Lewis, Boca Raton, 1991

Keith, L.H. (Ed.): Principles of Environmental Sampling, American Chemical Society, Washington, D.C., 1988

Kraft, G.: Probenahme an festen Stoffen in: Kienitz, H., Bock, R., Fresenius, W., Huber, W., Tölg, G. (Eds.): Analytiker-Taschenbuch, Bd. 1, Akademie-Verlag, Berlin, 1980, pp. 3

Krishnaiah, P.R., Rao, C.R. (Eds.): Handbook of Statistics, Vol. 6, Sampling, Elsevier, Amsterdam, New York, Oxford, 1988

Kutz, D.A. (Ed.): Chemometric Estimators of Sampling, Amount and Error, ACS Symposium Series No. 284, American Chemical Society, Washington, D.C., 1985

Markert, B. (Ed.): Environmental Sampling for Trace Analysis, VCH, Weinheim, New York, Basel, Cambridge, Tokyo, 1994

Nothbaum, N., Scholz, R.W., May, T.W.: Probenplanung und Datenanalyse bei kontaminierten Böden, Erich Schmidt, Berlin, 1994

Pitard, F.F.: Pierre Gy's Sampling Theory and Sampling Practice, Vols. 1 and 2, CRC Press, Boca Raton/Florida, 1989

Quality Control and Assurance

Funk, W., Dammann, V., Donnevert, G.: Qualitätssicherung in der Analytischen Chemie, VCH, Weinheim, New York, Basel, Cambridge, 1992

Funk, W., Dammann, V., Vonderheid, C., Oehlmann, G. (Eds.): Statistische Methoden in der Wasseranalytik, VCH, Weinheim, Deerfield Beach/Florida, 1985

Kateman, G., Buydens, L.: Quality Control in Analytical Chemistry, 2nd Ed., Wiley, Chichester, 1993

Kateman, G., Pijpers, F.W.: Quality Control in Analytical Chemistry, Wiley, New York, Chichester, Brisbane, Toronto, 1981

Kromidas, S.: Qualität im analytischen Labor, VCH, Weinheim, New York, Basel, Cambridge, Tokyo, 1995

Massart, D.L., Dijkstra, A., Kaufman, L.: Evaluation and Optimization of Laboratory Methods and Analytical Procedures, Elsevier, Amsterdam, 1978

Neitzel, V., Middeke, K.: Praktische Qualitätssicherung in der Analytik, VCH, Weinheim, New York, Basel, Cambridge, Tokyo, 1994
Quevauviller, P. (Ed.): Quality Assurance in Environmental Monitoring, VCH, Weinheim, 1995
Subramanian, G. (Ed.): Quality Assurance in Environmental Monitoring, VCH, Weinheim, 1995

Time Series Analysis

Box, G.E.P., Jenkins, G.M.: Time Series Analysis, Forecasting and Control, Holden-Day, San Francisco, 1976
Brockwell, P.J., Davis, R.A.: Time Series: Theory and Methods, Springer, New York, Berlin, Heidelberg, London, Paris, Tokyo, 1987
Chatfield, C.: Analyse von Zeitreihen, BSB B.G. Teubner Verlagsges., Leipzig, 1982
Chatfield, C.: The Analysis of Time Series: An Introduction, 4th Ed., Chapman and Hall, London, 1989
Fomby, T.B., Hill, R.C., Johnson, R.: Advanced Econometric Methods, Springer, New York, Berlin, Heidelberg, London, Paris, Tokyo, 1984
Metzler, P., Nickel, B.: Zeitreihen- und Verlaufsanalysen, S. Hirzel, Leipzig, 1986
Priestley, M.B.: Spectral Analysis and Time Series, Vol. 1, Univariate Series, Academic Press, London, New York, Toronto, Sydney, San Francisco, 1981
Schlittgen, R., Streitberg, B.H.J.: Zeitreihenanalyse, R. Oldenbourg, München, Wien, 1989

Applications in Environmental Analysis

Breen, J.J., Robinson, P.E. (Eds.): Environmental Applications of Chemometrics, ACS Symposium Series, Vol. 292, American Chemical Society, Washington, D.C., 1985
Devillers, J., Karcher, W. (Eds.): Applied Multivariate Analysis in SAR and Environmental Studies, Kluwer Academic Publishers, Dordrecht, Boston, London, 1991
Einax, J. (Ed.): Chemometrics in Environmental Chemistry. Statistical Methods, in the series: Hutzinger, O. (Ed.): Handbook of Environmental Chemistry, Vol. 2, Part G, Springer, Berlin, 1995
Einax, J. (Ed.): Chemometrics in Environmental Chemistry. Applications, in the series: Hutzinger, O. (Ed.): Handbook of Environmental Chemistry, Vol. 2, Part H, Springer, Berlin, 1995
Haccon, P., Mellis, E.: Statistical Analysis of Behavioural Data, Oxford University Press, Oxford, New York, Tokyo, 1994
Karcher, W., Devillers, J. (Eds.): Practical Applications of Quantitative Structure-Activity Relationships (QSAR) in Environmental Chemistry and Toxicology, Kluwer Academic Publishers, Dordrecht, 1990
Lieth, H., Markert, B. (Eds.): Element Concentration Cadasters in Ecosystems – Methods of Assessment and Evaluation, VCH, Weinheim, 1990

References

Baumbach, G.: Luftreinhaltung, Springer, Berlin, Heidelberg, New York, London, Paris, Tokyo, Hong Kong, Barcelona, **1990**, p. 285
Breen, J.J., Robinson, P.E.: Environmental Applications of Chemometrics, ACS Symposium Series, Vol. 292, American Chemical Society, Washington, D.C., **1985**
Brereton, R.G.: Chemometrics. Applications of Mathematics and Statistics to Laboratory Systems, Ellis Horwood, New York, London, Toronto, Sydney, Tokyo, Singapore, **1990**, p. 15
Brereton, R.G. in: Einax, J. (Ed.): Chemometrics in Environmental Chemistry. Statistical Methods, in the series: Hutzinger, O. (Ed.): Handbook of Environmental Chemistry, Vol. 2, Part G, Springer, Berlin, **1995**, pp. 49

Danzer, K.: Lecture Manuscript, Friedrich Schiller University, Jena, **1990**

Devillers, J., Karcher, W. (Eds.): Applied Multivariate Analysis in SAR and Environmental Studies, Kluwer Academic Publishers, Dordrecht, Boston, London, **1991**

Einax, J. (Ed.): Chemometrics in Environmental Chemistry. Statistical Methods, in the series: Hutzinger, O. (Ed.): Handbook of Environmental Chemistry, Vol. 2, Part G, Springer, Berlin, **1995a**

Einax, J. (Ed.): Chemometrics in Environmental Chemistry. Applications, in the series: Hutzinger, O. (Ed.): Handbook of Environmental Chemistry, Vol. 2, Part H, Springer, Berlin, **1995b**

Frank, I.E., Kowalski, B.R.: Anal. Chem. 54 (**1982**) 232R

Geladi, P.: analysis europa (**1995**) 4, 34

Horwitz, W., Kamps, L.R., Boyer, K.W.: J. Assoc. Off. Anal. Chem. 63 (**1980**) 1344

Kateman, G.: Lecture, 5th International Conference on Chemometrics, Leipzig, 27. 1. **1988**

Massart, D.L., Vandeginste, B.G.M., Deming, S.N., Michotte, Y., Kaufman, L.: Chemometrics: A Textbook, Elsevier, Amsterdam, **1988**, p. 5

Reeve, R.N.: Environmental Analysis, Wiley, Chichester, New York, Brisbane, Toronto, Singapore, **1994**, pp. 10

Sharaf, M.A., Illman, D.L., Kowalski, B.R.: Chemometrics, Wiley, New York, **1986**

Thiers, R.E. in: Glick, D. (Ed.): Methods of Biochemical Analysis, Vol. 4, Wiley, New York, **1957**, pp. 274

Vandeginste, B.G.M.: Topics in Current Chemistry 141 (**1987**) 1

Wenning, R.J., Erickson, G.A.: Trends Anal. Chem. 13 (**1994**) 446

Whitby, K.T.: Atmos. Environ. 12 (**1987**) 135

Part A

Essential Chemometric Methods

2 Measurements and Basic Statistics

2.1 Basic Statistics

2.1.1 Introductory Remarks

In general, results from investigations based on measurements may be falsified by three principal types of errors: gross, systematic, and random errors. In most cases **gross errors** are easily detected and avoidable. **Systematic errors** (so-called determinate errors) affect the **accuracy** and therefore the proximity of an empirical (experimental) result to the "true" result, which difference is called **bias**. **Random errors** (so-called indeterminate errors) influence the **precision** of analytical results. Sometimes precision is used synonymously with **reproducibility** and **repeatability**. Note that these are different measures of precision, which, in turn, is not related to the true value.

Remember that British Standard BS 5532 [CAULCUTT and BODDY, 1983] provides qualitative and quantitative definitions of both reproducibility and repeatability. The determination of repeatability and reproducibility for a standard test method by interlaboratory tests is given in [ISO 5725].

As a consequence the **reliability** of results, and hence decisions derived therefrom, is determined by both the accuracy and the precision of the measurements.

2.1.2 Frequency Distribution of Observations

In general statistics there is a difference between the parent population of a random variable, e.g. x (sometimes also characterized by capital letters) and a single realization of the parent population expressed, e.g., as single measurements, x_i, of the variable x. The parent population means an infinity of values which follow a certain distribution function. In the reality of experimental sciences one always has single realizations, x_i, of the random variable x.

In the terminology of statistics analytically measured quantities (properties, features, variables) are random variables x. Such a variable may, e.g., be density, absorbance, concentration, or toxicity. Hence, repeated measurements (observations) using the same sample, or measurements of "comparable" samples, do not result in identical values, x_i, but are single realizations of the random variable x. Using the **frequency distributions**

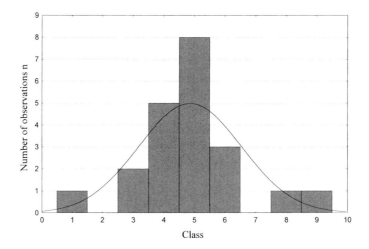

Fig. 2-1. Frequency distribution of 21 measured values

of these single realizations (represented graphically as a histogram) one can estimate the distribution function of the random variable x. An example of 21 measured values is shown in Fig. 2-1 along with a smoothed **distribution function**.

Each column of the histogram represents a discrete value or a certain range of values. If the number n of observations is $30 < n \leq 400$ it is recommended [DIN 53 804] that the class intervals be of width $w \approx L_n/\sqrt{n}$. If $n > 400$ $w \approx L_n/20$ is appropriate. $L_n = x_{max} - x_{min}$ is the **range** of the measurements.

Starting from such frequency distributions one can easily derive **cumulative frequencies** which can be smoothed by the **integrated distribution function** $F(x)$. Fig. 2-2 represents the cumulative frequencies of the example depicted in Fig. 2-1, again with a smoothed distribution function.

$F(x_i)$ of a specific value x_i, i.e. an observation of the random variable x, then represents the probability P that $x \leq x_i$:

$$F(x_i) = P(x \leq x_i) \tag{2-1}$$

Measurements of discrete variables as indicated in Fig. 2-1 yield a stepwise constant function $F(x)$.

More often in practical measurements x and its distribution function are continuous variables and therefore $F(x)$ may be differentiated to give a (probability) **density function** $f(x)$ the shape of which resembles the frequency distribution. Further details are not of interest here, but we should know that we utilize such density functions via well-known statistical tables.

In the same way as measured values of natural variables, values of derived variables such as quotients are also distributed characteristically and may be described by special distributions.

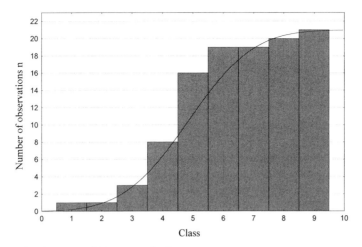

Fig. 2-2. Cumulative frequency of 21 measured values

Examples of distributions are the binomial and the POISSON distribution and the GAUSSian (normal), χ^2-, t-, and F-distributions.

For all types of distribution the following relationship holds:

$$\int_{-\infty}^{\infty} f(x)\,dx = 1 \qquad (2\text{-}2)$$

This means that it is 100% probable that any measured value x_i of variable x will be found within the borders of the integral. Foresee here the contrast between the probability, or certainty, of a result, and the result's usefulness.

2.1.3 The Normal Distribution Model

A continuous random variable x has a **normal distribution** with certain parameters μ (mean, parameter of location) and σ^2 (variance, parameter of spread) if its density function is given by the following equation:

$$f(x;\mu;\sigma^2) = \frac{1}{\sigma\sqrt{2\pi}}\,e^{\left[-\frac{1}{2}\frac{(x-\mu)^2}{\sigma^2}\right]} \qquad (2\text{-}3)$$

where $\sigma > 0$ and $-\infty < x < \infty$, x is said to be $N(\mu, \sigma^2)$ distributed.

To compute parameter independent tables one transforms x into a standardized random variable u:

$$u = \frac{(x - \mu)}{\sigma} \tag{2-4}$$

$$f(u) = \frac{1}{\sqrt{2\pi}}\, e^{-\frac{u^2}{2}} \tag{2-5}$$

where $f(u)$ is called the **normalized** (standardized) **GAUSSian distribution**, where u is $N(0, 1)$ distributed.

In practice, most real random variables which are empirically sampled to provide a limited number of observations do not normally represent exactly normally distributed variables.

The analyst who does not usually worry about the real distribution of a set of results should at least be sure that succeeding computations and tests are not significantly affected by violations of the assumption of a special distribution. Otherwise he may

- prefer robust statistics
- use less sensitive tests
- collect more results hoping they will follow the desired (normal) distribution, or
- sometimes apply an appropriate transformation of the data to produce a normal distribution

2.1.4 Parameters of Distributions

Normally the true parameters (so-called parent population parameters) of distributions are not known. For empirical distributions they have to be estimated (symbol: ^) on the basis of a limited number, n, of observations (so-called sample parameters). Estimates of the most important parameters are:

the **mean** $\bar{x} = \hat{\mu}$ (estimate of μ)
the **variance** $s^2 = \hat{\sigma}^2$ (estimate of σ^2)
the **standard deviation** $s = \hat{\sigma}$

In analytical chemistry, for comparison purposes one sometimes uses the **coefficient of variation** defined as $v = s/\bar{x}$ (estimate of σ/μ). In percent it gives a measure of the **relative standard deviation** s_{rel}. Sometimes the reciprocal of the relative standard deviation is used to express the signal/noise ratio.

Since 1976 it has been recommended by IUPAC that the notion "coefficient of variation" should no longer be used instead of relative standard deviation. In statistics it is still known as a feature of a distribution derived from the moments.

From these basic parameters others are derived which characterize the real shape of a normal distribution: **skewness** and **excess (kurtosis)** (see Section 2.1.5).

Empirical normal distributions often exhibit left (positive) or right (negative) skewness, in other words they are not symmetrical around the arithmetic mean. The skewness increases with increasing deviation of the mean from the median or the mode (modal value). (If a series of measurements of a continuous variable is grouped into equal class intervals, the modal value is a value from the modal class, the class that contains the greatest number of values.)

Positive or negative kurtosis occurs if the convexity of the empirical distribution is more acute or smoother than the curvature of the ideal normal distribution.

Statistical tests (see Section 2.2) exist for both skewness and kurtosis. From the result of such tests one can decide if the deviation of a distribution function based on measurements from an ideal (test) function may be tolerated.

2.1.5 Computation of some Basic Parameters

Typical normal distribution parameters are computed using all measurement values (a certain number of single realizations):

arithmetic mean

$$\bar{x} = \frac{1}{n} \sum_{i=1}^{n} x_i \tag{2-6}$$

variance

$$s^2 = \frac{\sum_{i=1}^{n}(x_i - \bar{x})^2}{n-1} \quad \text{or}$$

$$s^2 = \frac{1}{n-1}\left[\sum_{i=1}^{n} x_i^2 - \frac{1}{n}\left(\sum_{i=1}^{n} x_i\right)^2\right] \tag{2-7}$$

standard deviation

$$s = \sqrt{s^2} \tag{2-8}$$

From the ranked series of n measurements of a continuous variable x

$$x_{(1)}, x_{(2)}, \ldots, x_{(n)} \tag{2-9}$$

one can very quickly read (if n is odd) or calculate (if n is even) the **median** as

$$\tilde{x} = x_m \quad \text{with} \quad m = \frac{n+1}{2} \quad \text{(if } n \text{ is odd)} \tag{2-10}$$

or

$$\tilde{x} = \frac{x_m + x_{m+1}}{2} \quad \text{with} \quad m = \frac{n}{2} \quad \text{(if } n \text{ is even)} \tag{2-11}$$

The median and the **range**

$$L_n = x_{\max} - x_{\min} = x_{(n)} - x_{(1)} \tag{2-12}$$

are also called **robust estimates** of location and spread parameters.

Example 2-1

Assume that by atomic absorption spectroscopy of $n = 10$ samples of sewage water the following concentrations of Zn were found:

195 182 204 202 190 186 192 198 193 188 µg L^{-1}.

The *ranked* series is: 182 186 188 190 192 193 195 198 202 204 µg L^{-1}.

The *median* is: $\tilde{x}$ = (192 + 193)/2 = 192.5 µg L^{-1},
and the *range* is: L_{10} = (204 − 182) = 22 µg L^{-1}.

The *arithmetic mean* is: $\bar{x}$ = 1930/10 = 193 µg L^{-1},
and the *variance* is: s^2 = 1/9 (372 926 − 1930^2/10) = 48.4 µg^2 L^{-2},
and the *standard deviation* is: s = $\sqrt{48.3}$ = 6.967 µg L^{-1}.

Note that the mean does not differ very much from the median.

Statisticians define skewness and other parameters using the parameters of the distribution. With μ_j, the so-called jth central moment, **skewness** is:

$$\gamma_3 = \frac{\mu_3}{\sigma^3} \quad \text{with} \quad \mu_3 = \int_{-\infty}^{\infty} (x - \mu)^3 f(x)\, dx \tag{2-13}$$

where μ is the mean and σ is the standard deviation.

The skewness, sk, of a continous random variable may be calculated as:

$$sk = \frac{\bar{x} - x_{mode}}{s} \tag{2-14}$$

where x_{mode} is that value for which the probability density function has a maximum [CLARKE and COOKE, 1988].

With the skewness estimated by

$$sk = \sqrt{n} \frac{\sum_{i=1}^{n}(x_i - \bar{x})^3}{\left(\sum_{i=1}^{n}(x_i - \bar{x})^2\right)^3} \tag{2-15}$$

the symmetry of the distribution is then classified as:

$sk = 0$ symmetric
$sk > 0$ left sided asymmetry (i. e. left side maximum)
$sk < 0$ right sided asymmetry (i. e. right side maximum)

How plain a distribution is can be evaluated in terms of **excess** or **kurtosis**. Kurtosis is defined as:

$$\gamma_4 = \frac{\mu_4}{\sigma^4} \quad \text{with} \quad \mu_4 = \int_{-\infty}^{\infty}(x - \mu)^4 f(x)\,dx \tag{2-16}$$

Excess is defined as:

$$\gamma_4 = \frac{\mu_4}{\sigma^4} - 3 \tag{2-17}$$

With the excess calculated as:

$$ex = n\frac{\sum(x_i - \bar{x})^4}{\left[\sum(x_i - \bar{x})^2\right]^2} - 3 \tag{2-18}$$

the classification is as follows:

$ex = 0$ normal distribution
$ex > 0$ the convexity of the density function is more acute than for the normal distribution
$ex < 0$ the convexity of the density function is smoother than for the normal distribution

2.1.6 Confidence Intervals

Statisticians have introduced the notion of **fractile** (**quantile**), which is a special argument of the density function. The quantile, x_q, of order q ($0 < q < 1$) is determined by the equation $q = P(x < x_q)$, which means that the probability of finding a value of x below x_q is equal to q (Fig. 2-3).

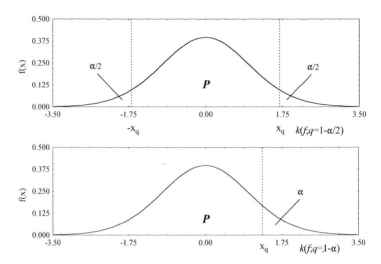

Fig. 2-3. Quantiles of probability density functions of a normal- or t-distributed random variable x. Upper part: Two-sided case. Lower part: One-sided case

In this respect x_q is equal to a certain value k which replaces infinity as the upper integral limit in Eq. 2-2. So we can realize that k values depend on probability P. Again the GAUSSian (normal distribution) function is the simplest model function, because of its sole dependence on P which makes $k = k(P)$.

> ☞ See Appendix A for some selected $k(P)$ of the GAUSSian (Normal) Distribution

Using these k values we are now able to compute (to predict) so-called **confidence intervals**, *cnf*, of measured or calculated (estimated) numerical values x_i.

If, e.g., the true parameter σ (standard deviation) of the distribution of a variable x is known, one is able to calculate

– for a single measurement:

$$cnf(x_i) = x_i \pm \Delta x_i \quad \text{with} \quad \Delta x_i = k(P)\sigma \qquad (2\text{-}19)$$

– for the calculated mean:

$$cnf(\bar{x}) = \bar{x} \pm \Delta\bar{x} \quad \text{with} \quad \Delta\bar{x} = \frac{k(P)\sigma}{\sqrt{n}} \tag{2-20}$$

If we report an analytical result in such a way, we are confident that, with a defined probability P, the true result of, e.g., the mean will lie within the range defined by $\Delta\bar{x}$ around the reported mean value.

More knowledge is necessary to compute confidence intervals in cases where σ is estimated by the sample standard deviation s. Then, $k(P)$ has to be replaced by $k(f = n - 1; P)$, where k now also depends on the so-called **degree of freedom** f. This takes into account that the reliability of an estimate of σ decreases with decreasing n.

Statistical tables list values of $k(f; P)$, called **t-values** in memory of W.S. GOSSET who first published applications under the pseudonym STUDENT.

> ☞ See Appendix B for selected quantiles $k(f; P)$ of the STUDENT t-distribution

Example 2-1 (continued)

If we take the readings of the previous example, tentatively postulate s to be a proper estimate of σ, and fix the confidence level at $P = 0.95$ (95%), according to Eqs. 2-19 and 2-20, with **k(0.95)**, we can write:

$x_6 \pm \Delta x_6 = 193 \pm 1.96 \cdot 6.96 = 193 \pm 14 \; \mu g \; L^{-1}$, and
$\bar{x} \pm \Delta\bar{x} = 193 \pm 1.96 \cdot 6.96/\sqrt{10} = 193 \pm 4 \; \mu g \; L^{-1}$

Note that the confidence range of the mean is much smaller than the confidence range of a single value. So the result is more useful, but as uncertain as before.

The more realistic confidence region of the mean computed using **k(9; 0.975)** = 2.26 now becomes:

$\bar{x} \pm \Delta\bar{x} = 193 \pm 2.26 \cdot 6.96/\sqrt{10} = 193 \pm 5 \; \mu g \; L^{-1}$

Note that the confidence range of the mean is now slightly wider, i.e. using the sample parameters $\bar{x}$ and s we get less optimistic results.

Remark

If no assumption of the type of distribution can be made the values of k may be calculated from CHEBYCHEV's inequality [DOERFFEL, 1990]. It provides an absolute lower bound of probability that P % of the values of variable x are in the interval from $\mu - k\sigma$ to $\mu + k\sigma$:

$$P(|x - \mu| < k\sigma) \geq 1 - \frac{1}{k^2} \tag{2-21}$$

Example 2-1 (continued)

For $P = 0.95$ we calculate from Eq. 2-21 $k = 4.472$ and $\Delta\bar{x} = 9.843$ and hence $\bar{x} \pm \Delta\bar{x} = 193 \pm 10$ µg L^{-1}.

Again, the confidence range of the mean is broadened.

The indication at the end of Section 2.1.2 now can be extended: Until now we only spoke of the **confidence level** P. According to the relationship $P = 1 - \alpha$ this probability is complementary to the **significance level**, α, which represents a certain **risk**, α, that the confidence interval does not encompass the true value of the parameter.

Sometimes it is valuable to be aware of this aspect, namely that statements or decisions are not free from error. Look at the tendency in the tables recommended above: the higher the probability wanted (low risk α), the higher are the values of k. In other words: if you want an absolutely certain analytical result try "to sell" or to get on with a confidence region from $-\infty$ to $+\infty$ ...

Let us have another look at the table of Appendix B and understand why we may need two k-values with the same confidence level P and the same total risk α. If we compare columns 2 and 3 we note that one can use a fractile of the order $q = 0.975$, which means that a risk $\alpha/2$ remains that *both* sides of the symmetrical region around the estimated parameter do not include the true value. On the other hand the fractile of the order $q = 0.95$ may be useful for constructing a one-sided interval hence concentrating the risk α on one side.

As an example, imagine you want to buy a volleyball to pass through a basket ring. Probably you will consider only balls with a diameter less than that of the basket.

Another example may be even more interesting: one-sided intervals are especially useful in comparing heavy metal contamination found in an environmental compartment with a legally fixed threshold value. Because of the experimental random error it is *not* recommended to make this comparison on the basis of a mean value alone. (See next section for the respective statistical test.)

Detailed values of distribution functions can be computed using appropriate statistical software.

Useful approximations or even complete sub-programs which can be used in home-made programs are given, e.g., by HASTINGS [1955], McCORMICK and ROACH [1987], NOACK and SCHULZE [1980], THIJSSEN and VAN DER WIEL [1984].

Remark

According to the general relationship for any estimated parameter we can also calculate confidence regions of standard deviations. The only difference is that these intervals are nonsymmetric.

☞ **See Appendix C for the formula and respective values of k**

Example 2-1 (continued)

The confidence interval for s is: $0.69 \cdot 6.96 = \mathbf{4.8} \leq \sigma \leq \mathbf{12.7} = 1.83 \cdot 6.96$

Use the power of the k values to calculate confidence regions of variances.

2.2 Hypotheses and Tests

2.2.1 Hypotheses, Tests, Decisions, and their Risks

People formulate more or less explicit **hypotheses** in order to derive information from measurements performed on objects, or simply from observation of reality.

In science one may distinguish between working hypotheses, scientific hypotheses, research hypotheses, and statistical hypotheses. In this ranking the degree of generality of a hypothesis decreases whereas its precision increases.

In the following we will therefore focus on statistical hypotheses. Every time one formulates a hypothesis an alternative hypothesis is possible. How can it be determined which hypothesis is true, or at least more likely? Normally a **statistical test** can serve to perform this task. It is a sample (set of measurements) based **decision** procedure which enables decision between the two hypotheses.

Because possible wrong decisions usually have different consequences (weights) one hypothesis is called the **null hypothesis**, H_0, and the **alternative hypothesis** is designated H_1 or H_A.

What about the risks of decisions? As we already know from the construction of confidence intervals, the probability, P, is connected with the complementary **error probability**, α. In a similar manner here the probability, α, characterizes the risk (of the first kind) of making a wrong decision as a result of using a statistical test. But the notion "wrong" may be seen from two different sides.

Remember a confidence limit of a mean: one mistake can be to exclude a value which in fact belongs to the interval around the mean, i.e. to exclude a correct value, another mistake would be to include a "wrong" value. Hence we have two kinds of error: a **type I error** associated with a probability, α, of an error of the first kind, and a **type II error** with a probability, β, of an error of the second kind. The relationship between H_0 and these errors are explained in Tab. 2-1.

Tab. 2-1. Relationships between H_0 and the different types of error

The decision is:	H_0 is retained	H_0 is rejected
If H_0 is true	the decision is OK	the decision is made with type I error (risk α of an error of the 1st kind)
If H_0 is false	the decision is made with type II error (risk β of an error of the 2nd kind)	the decision is OK

Hypotheses can be formulated as **one-sided** or **two-sided** hypotheses. Both hypotheses allow for the risk α but the decisions are made with distinct quantiles (tabulated values):

one-sided tests with: $\alpha = 0.05$ use $q = 1 - \alpha$
two-sided tests with: $\alpha = 0.05$ use $q = 1 - \alpha/2$

Because of the relationship $k(1 - \alpha) < k(1 - \alpha/2)$ one-sided tests are "sharper", i.e. they will indicate "significance" earlier. Therefore a null hypothesis is rejected earlier than it would be by use of a two-sided test.

How do we perform a test? The *first step* is to calculate an empirical test value using the experimentally obtained parameters. Usually one can use typical predefined test values. So, in order to decide on the validity of a hypothesis on the basis of two variances the so-called F-value (to remind us of FISHER) is adequate, and a decision on validity of hypotheses on the basis of the difference between two means or the difference between a mean and a threshold is based on so-called *t*-values (reminding us of STUDENT). According to the operation (mathematical equation) used to obtain the respective value and under the assumption that the attempted hypothesis is true, the calculated value belongs to a certain distribution of values, i.e. the STUDENT or FISHER distribution. Consequently the *second step* is to compare the empirical test value with the theoretical value, the **critical value** from the appropriate table. In the final *third step* the decision is made, where the convention is

to retain H_0 if

> empirical test value < critical value or

to reject H_0 if

> empirical test value ≥ critical value

Note that in contrast to confidence regions, where a value equal to a limit belongs to the region, in test procedures values equal to a critical value are rejected. The decision in the third step critically depends on the proper hypothesis because its precise wording leads to a certain critical test value.

As an *example* think of the comparison of the precision of two methods. Each method can be characterized by its respective variance.

The comparison is based on a typical test value called F. F derived from **exp**erimental values is defined as:

$$F_{exp} = \frac{s_1^2}{s_2^2} \tag{2-22}$$

For practical purposes the convention is to take the larger variance as the numerator and the smaller variance as the denominator. The respective variances s^2 are calculated from n_1 readings of method 1, laboratory 1, ... and n_2 readings of method 2, laboratory 2, ... Consequently there are two versions of wording of the F-test.

The **two-sided test** is based on:

H_0: "The variances s_1 and s_2 are equal"
H_1: "The variances are different"

Then there is no evidence against H_0 if the experimental (calculated) F-value is less than a critical F-value obtained from statistical tables:

$$F_{exp} < F_{crit}\left(n_1 - 1;\ n_2 - 1;\ q = 1 - \frac{\alpha}{2}\right) \tag{2-23}$$

The **one-sided test** is based on:

H_0: "Variance s_1^2 does not exceed variance s_2^2"
H_1: "Variance s_1^2 exceeds variance s_2^2"

Here we accept H_0 if

$$F_{exp} < F_{crit}\,(n_1 - 1;\ n_2 - 1;\ q = 1 - \alpha) \tag{2-24}$$

> ☞ For the critical F-values see Appendix D

Example 2-1 (continued)

Assume that as an alternative to the previous method (AAS with $n = 10$, $\bar{x} = 193$ µg L^{-1} and $s \approx 7$ µg L^{-1}) the Zn content of $n = 4$ samples has been determined by a voltammetric method (196, 204, 197, 203 µg L^{-1}) resulting in a mean $\bar{x} = 200$ µg L^{-1} and a standard deviation $s \approx 4$ µg L^{-1}.
We decide if the performance of both methods is comparable at a confidence level $P = 95\%$ by using a two-sided F-test:

$F_{exp} = 49/16 = 3.06 \quad F_{crit}\,(10 - 1;\ 4 - 1;\ 0.975) = 14.5$

Because $F_{exp} < F_{crit}$ both variances are considered to be equal.

Only if the variances are found equal in the statistical sense it is allowed to proceed with the common t-test for the two corresponding means $\bar{x}_1$ and $\bar{x}_2$. If the F-test signals significant differences one is forced to use special definitions of both the empirical t-value and the degrees of freedom.

The standard t-value for the comparison of two means may be calculated as follows:

$$t_{exp} = \frac{|\bar{x}_1 - \bar{x}_2|}{s_{pooled}}\sqrt{\frac{n_1 \cdot n_2}{n_1 + n_2}} \tag{2-25}$$

with the pooled standard deviation computed from the variance:

$$s_{pooled}^2 = \frac{(n_1 - 1)\,s_1^2 + (n_2 - 1)\,s_2^2}{n_1 + n_2 - 2} \tag{2-26}$$

Example 2-1 (continued)

Now we check if both means are equal by comparison of

$t_{exp} = |200 - 193| \sqrt{[10 \cdot 4/(10+4)]}/s_{pooled} = 1.85$
with $s^2_{pooled} = [(10-1) \cdot 49 + (4-1) \cdot 16]/(10+4-2) = 40.75$
against $t_{crit}(10+4-2; 0.975) = 2.18$

Again, there is no evidence that the two methods give different results.

The tests described above are related to hypotheses based on two or more statistical samples, i.e. several grouped measurements.

Having used the described test to ensure the reliability of results from one's own laboratory, in environmental analysis it is often necessary to compare the mean from a sample of n members against a certain target value given by law or by environmental regulations.

The appropriate t-test if a mean value is below a specified target value is based on:

$$t_{exp} = \frac{(\bar{x} - \mu_0)\sqrt{n}}{s} \tag{2-27}$$

A detailed description of related tests is given in [ISO 2854]. Such tests may be needed to compare a mean with a given value when the variance is known, or to compare two means when the variances are known, or for other purposes.

Example 2-1 (continued)

Let us now compare our experimental result found by AAS with a given critical level for the maximum Zn content in sewage water, e.g. $\mu_0 = 200$ µg L^{-1}. Because this is a one-sided constraint we have to use a one-sided test and keep the critical error level of $\alpha = 0.05$ ($q = 0.95$). We then have to formulate the statistical hypothesis according to:

$H_0 : \bar{x}_0 < \mu_0$ (Null hypothesis with regard to the comparison being made)
$H_1 : \bar{x}_0 \geq \mu_0$ (Alternative hypothesis)

We accept H_0 if $t_{exp} < -t_{crit}(n-1; q = 1 - \alpha)$.

With our experimental results we calculate

$t_{exp} = (193 - 200)\sqrt{10/6.96} = -3.18$.

This value stands against the table value t_{crit} (9; 0.95) = 1.83.

For t_{exp} = −3.18 < t_{crit} (9; 0.95) = −1.83 we retain the null hypothesis, i.e. we can say that the Zn content found does not exceed its recommended limit.
Other countries may have fixed a lower limit for Zn, e.g. μ_0 = 197 µg L^{-1}.
Check your decision under this condition!

Other interesting tests may be undertaken to decide whether the empirical distribution of the measurements obtained from samples follows a certain theoretical distribution, e.g. the normal distribution. In such cases it is quite common to perform the χ^2-**test of goodness of fit** or the **KOLMOGOROV-SMIRNOV test**. Both tests are based on the evaluation of the cumulative frequency of measured data and are described in detail in [MILLER and MILLER, 1993].

2.2.2 How to Estimate Correct Sample Sizes

Those readers who do not want to be frustrated by the possibility that their past decisions might be based on a number of measurements which was not theoretically correct should skip this section.

Obviously tests and decisions drawn from them are dependent on the estimated (calculated) parameters. It is worth mentioning that they also depend on the number, n, of measurements or **replicates**. In addition we know that we do not have one risk alone. Therefore we have to accept that the sensitivity of tests is determined by the proper choice of n, it increases with increasing n. It is, however, not usually possible to increase n without additional costs. On the other hand one can often say, e.g. from experience or from legal requirement which difference between two means may be tolerated and which difference must be assured. Taking into account this and both tolerable risks of wrong decisions, associated with type I and type II errors, one can compute the necessary number of replicates in advance (see textbooks on statistics or ZWANZIGER et al. [1986]):

$$n = \frac{s^2}{(\Delta x)^2} \left[t\left(n-1;\ 1-\frac{\alpha}{2}\right) + t\left(n-1;\ 1-\beta\right) \right]^2 \tag{2-28}$$

where Δx is a certain tolerance. It might, for example, have the following values:

$$\Delta x = (\bar{x}_1 - \bar{x}_2) \quad \text{or} \quad \Delta x = \frac{s}{2} \quad \text{or even} \quad \Delta x = s \tag{2-29}$$

whatever difference or tolerance one assumes to be most important. Statisticians give advice about the magnitude of β: use large values of about 0.04 or 0.05 if the null hypothesis accepted confirms a current state and there is no likelihood of dramatic consequences; use smaller values of about 0.02 or 0.03 if incorrect acceptance of a null hypothesis can lead to severe consequences.

Example 2-2

Assume that a waterworks is able to maintain water quality without problems if the actual Zn concentration is within a tolerance $\Delta x = 10$ µg L^{-1} around $\mu_0 = 190$ µg L^{-1}. Is it acceptable to continue monitoring water quality by drawing $n = 10$ samples at each test interval? Because we further assume that incorrect decisions about Zn have no dramatic effects we can set $\alpha = \beta = 5\%$.

With $\Delta x = 190 - 180$, or $\Delta x = |190 - 200|$, $s = 7$, and $n = 10$ as a first guess (note that in Eq. 2-28 n occurs on both sides!) we calculate the next guess, n':

$n' = 49/100\ (2.26 + 1.83)^2 \approx 8$

Hence we decide that $n = 10$ samples will be sufficient.

2.2.3 Parameter Tests or Outlier Tests?

All the tests previously mentioned are sensitive to so-called outliers. **Outliers** ("stragglers") are defined as single "wrong" values within a series of values which seem to be "right". As already mentioned outliers may influence decisions. There are, therefore, numerous tests for identifying and eliminating possible outliers [RECHENBERG, 1982].

Measuring Values

In the case of continuous variables, one hitherto used the DIXON test for $n \leq 29$ and the GRUBBS test if $n \geq 30$.

Nowadays the DIXON test alone is recommended by standardizing organisations [ISO 5725]. The general formula for the DIXON outlier test is:

$$dix = \frac{x_A - x_B}{x_C - x_D} \quad (2\text{-}30)$$

The indices A, B, C, and D refer to the ranked series and their actual value is to be taken from the DIXON table. These values, the respective indices, and the test in the following example are for the situation when it is not clear at which side of the series the outlier occurs.

> ☞ **See Appendix E for selected DIXON values**

In this table we report the index numbers A to D for the left-sided test in the first line and the index numbers for the right-sided test in the second line. So we can compute two test values: one for the leftmost (*min*) and one for the rightmost (*max*) value of the ranked series. The maximum of both values should be compared against the critical table value and the decision made in the usual way.

In this text we refer to the ISO recommended DIXON test, but note that other tests are available, along with tables, in the statistics literature (see, e.g., [MÜLLER et al., 1979]).

Example 2-1 (continued)

Let us now check the series of Zn concentrations for outliers:
The *ranked* series is:

position	(1)	(2)	(3)	(4)	(5)	(6)	(7)	(8)	(9)	(10)
value (µg L^{-1})	182	186	188	190	192	193	195	198	202	204

For $n = 10$ we have to compute

$dix_{left} = (x_{(2)} - x_{(1)})/(x_{(9)} - x_{(1)})$
$dix_{right} = (x_{(10)} - x_{(9)})/(x_{(10)} - x_{(2)})$

$dix_{left} = (186 - 182)/(202 - 182) = 0.200$
$dix_{right} = (204 - 202)/(204 - 186) = 0.111$

$\max(dix_{left}, dix_{right}) = 0.200$

This means that the lowest value is most suspect, which is interesting because its distance from the mean value is similar to that of the highest value; it is, however, located at the end of the longer tail of the distribution. This value has to be compared with dix_{crit} (10; 0.05) = 0.530.
Because 0.200 < 0.530 we conclude with a risk of 5% that there is no indication of outliers in the series.

Parameters

In a series of standard deviations, variances, or ranges the COCHRAN test [ISO 5725] is used for evaluating the largest value. It is primarily used in (planned) method comparisons, cooperative tests, and analysis of variance where the number of the measurements and the levels of the means should be the same.

Finally we should mention here that any distribution tends to produce outliers. Hence, our advice is to **perform parameter tests before outlier tests and deletion**. For normal distribution we can perform simple tests of skewness and excess using the *t*-test:

Skewness is tested against the critical value:

$$t_{crit} = t(n-2; \; q = 1-\alpha) \sqrt{\frac{6}{n}} \tag{2-31}$$

Excess is tested against:

$$t_{crit} = t(n-2; \; q = 1-\alpha) \sqrt{\frac{24}{n}} \tag{2-32}$$

☞ See Appendix B for selected values of the STUDENT *t*-distribution

General

Valuable advice from KAISER and GOTTSCHALK [1972] is to perform two outlier tests. The first test should use $P = 99\%$, which ensures a low type I error, and the significance of which clearly indicates an outlier. The second test with $P = 0.95$, if nonsignificant, guarantees a low type II error and that the distribution is free from outliers. If the test value is between, *n* should be increased and the tests repeated.

As for most outlier tests, the above procedures **delete** single outliers.

Other ways are **censoration**, where a fixed number of maximum and minimum values are removed, or **winsorization**, where a previously fixed number of values is substituted by their next neighbors.

2.2.4 Repeatability, Reproducibility, and Cooperative Tests

Without going too deeply into details we will refer to **cooperative tests** ("round robins", **interlaboratory comparisons**) in which two basic aims are:

– to compare the performance of two (or more) laboratories relying on well defined and "true" analytical procedures or to compare two procedures relying on "true" values of certified reference material
– to certify a certain material relying on well defined and "true" procedures

Participation in cooperative tests is an important means of **external quality assurance**. In such comparisons two measures are of interest:

(1) **Repeatability** (x_{repeat}) is a quantity such that the probability is P that two test results obtained in the *same* laboratory (on the same material) will not differ by more than x_{repeat} [MANDEL and LASHOF, 1987; ISO 5725].
(2) **Reproducibility** (x_{reprod}) is a quantity such that the probability is P that two test results obtained in *different* laboratories (on the same material) will not differ by more than x_{reprod}.

With the respective degrees of freedom f_{repeat} and f_{reprod} both quantities are defined as repeatability:

$$x_{repeat} = t\left(f_{repeat};\ q = 1 - \frac{\alpha}{2}\right) s_{repeat} \sqrt{2} \tag{2-33}$$

and reproducibility:

$$x_{reprod} = t\left(f_{reprod};\ q = 1 - \frac{\alpha}{2}\right) s_{reprod} \sqrt{2} \tag{2-34}$$

In the case of m laboratories with n_j ($j = 1, \ldots, m$) measurements each the variances are computed according to:

$$s^2_{repeat} = \frac{\sum_{j=1}^{m} \sum_{i=1}^{n_i} (x_{ij} - \bar{x}_j)^2}{(n - m)} \tag{2-35}$$

where n is the total number of measurements and

$$s^2_{reprod} = \frac{\sum_{j=1}^{m} n_j (\bar{x}_j - \bar{x}_{total})^2}{(m - 1)} \tag{2-36}$$

where the denominators define the respective degrees of freedom $f_{repeat} = n - m$ and $f_{reprod} = m - 1$, and the overall mean is:

$$\bar{x}_{total} = \frac{\sum_{j=1}^{m} \sum_{i=1}^{n_i} x_{ij}}{n} \tag{2-37}$$

Example 2-1 (continued)

Let us again use our previous case. In principle this example is void because of the widely different numbers of replicates.
With two labs, $m = 2$, and the numbers of measurements and the respective means, $n_1 = 10$, $\bar{x}_1 = 193$, $n_2 = 4$, $\bar{x}_2 = 200$, $n = 14$, $\bar{x}_{total} = 2735/14 = 195.357$ µg L^{-1} we calculate

$s^2_{repeat} = (315 + 50)/12 = 30.417$ µg^2 L^{-2} and
$s^2_{reprod} = (4 \cdot 2.357^2 + 10 \cdot 4.643^2)/1 = 237.796$ µg^2 L^{-2}

With $f_{repeat} = 14 - 2 = 12$ and $f_{reprod} = 2 - 1 = 1$ we read from the t-table the values $t(12; 0.975) = 2.18$ and $t(1; 0.975) = 12.71$ and obtain

repeatability $x_{repeat} = 2.18 \sqrt{30.417} \sqrt{2} = 17$ µg L^{-1}
reproducibility $x_{reprod} = 12.71 \sqrt{237.796} \sqrt{2} = 277$ µg L^{-1}

Hence, two single values in a series of results provided by one lab may differ by 17 µg L^{-1}, and two single values from the pooled values may differ by 277 µg L^{-1}.
Although the example is not adequate, the results exhibit typical behavior: reproducibility is normally (much) greater than repeatability. (That the example fails may be deduced from the fact that the range of series 1 exceeds x_{repeat}.)

Using the two respective variances a multiple ($m \geq 2$) mean value comparison is possible with the following F-ratio similar to Eq. 2-22 under the assumption that all single laboratory means belong to one "true" mean (to the same distribution):

$$F = \frac{s^2_{reprod}}{s^2_{repeat}} \qquad (2\text{-}38)$$

with $f_{repeat} = n - m$
and $f_{reprod} = m - 1$

This is the common form of the F-ratio because reproducibility is usually (much) greater than repeatability. Following the conventions of analysis of variance, which assume that repeatability is influenced by random errors whereas systematic errors might influence reproducibility and therefore yield larger values of s^2_{reprod}, we have to use *one-sided* critical values F (f_{reprod}; f_{repeat}; $q = 1 - \alpha$). In this case we are allowed to assume that variation of the reproducibility (over all laboratories) is random in nature.

2.3 The Principle of Analysis of Variance

Both types of variances defined and used in the previous section play a central role in the **analysis of variance** (ANOVA). This is because of the assumption that s_{reprod}^2 measures the scatter of data caused by systematic factors (in the above section: the factor "laboratory") and s_{repeat}^2 measures the random scatter of data. Hence we will need this principle in the interpretation of the results from experimental design.

The underlying philosophy is based on the fact that the total sum of squares of the deviations of all single measurement values from the total mean can be split into a within-series sum of squares (used in s_{repeat}^2) and a between-series sum of squares (used in s_{reprod}^2). If there is an influence of a systematic factor (leading to differences in laboratory means or expressed in levels of factors in experimental designs) the above written F-ratio will differ significantly from the critical F-value. In our case the factor "laboratory" may be a possible source of variation of the measurements. In environmental analyses factors like pollution, sampling conditions, particle size, etc. may be considered. If the F-test signaled the significance of a factor, it makes sense to estimate the **variance component** arising from experimental uncertainties and that arising from the factor. The value of s_{repeat}^2 can then be used as an estimate of the random (experimental) component. The systematic (factor caused) variance component, in the simplest case where n_j is constant, may be estimated from $(s_{reprod}^2 - s_{repeat}^2)/n_j$.

Example 2-1 (continued)

Let us determine whether this concept is applicable to the previous results of two methods instead of two laboratories.

We have $m = 2$, $n_1 = 10$, $\bar{x}_1 = 193$, $n_2 = 4$, $\bar{x}_2 = 200$, $n = 14$, $\bar{x}_{total} = 2735/14 = 195.357$

$s_{reprod}^2 = (4 \cdot 2.357^2 + 10 \cdot 4.643^2)/1 = 237.796$
$s_{repeat}^2 = (315 + 50)/12 = 30.417$
$F_{exp} = 7.82$
F_{crit} (12; 1; $q = 0.95$) ≈ 244
(We do not give this value in the Appendix but it can be extrapolated.)

Because F_{exp} is less than F_{crit} we decide that the factor "method" has no influence on the reported results. This is in good agreement with the former finding that the mean values found by both methods are comparable, i.e. they are equal in the statistical sense.

2.4 Causal Modeling, Regression, and Calibration

2.4.1 Correlation or Regression?

By **correlation** we understand the existence of a linear relationship between two variables, the extent of which is described by the correlation coefficient. The aim of determining this coefficient is to assist in assessing whether or not there is a linear dependence of one variable on another. To establish the form of the dependence (or the functional relationship) of the one variable on the other, we perform **regression** analysis, obtaining values for the regression parameters.

2.4.2 Other Notions, their Symbols and Meanings

If we have two experimental variables x and y, with mean values $\bar{x}$ and $\bar{y}$, we should distinguish between two cases **A** and **B**.

Case A: If we have a data set of n pairs of measurements which are essentially the same, although subject to random error (the same measurement repeated), then $\bar{x}$ and $\bar{y}$ are the mean, or average, values of the sets.

Case B: If we have a data set of n different pairs of measurements (as in a calibration set, with concentration and signal) $\bar{x}$ and $\bar{y}$ are called the coordinates of the centre of gravity of the set.

With the following **sums of the squares of deviations** calculated from all n pairs of measurements:

$$SS_{xx} = \sum_{i=1}^{n} (x_i - \bar{x})^2 \qquad (2\text{-}39)$$

and

$$SS_{yy} = \sum_{i=1}^{n} (y_i - \bar{y})^2 \qquad (2\text{-}40)$$

we can use each sum as one possible measure of scatter.
The **sum of the products of the deviations**:

$$SS_{xy} = \sum_{i=1}^{n} (x_i - \bar{x})(y_i - \bar{y}) \qquad (2\text{-}41)$$

then defines a measure of the spread of the cloud of points relative to the centre of gravity.

With these sums one can easily derive the **variances** of both variables:

$$s_x^2 = \frac{SS_{xx}}{n-1} \tag{2-42}$$

$$s_y^2 = \frac{SS_{yy}}{n-1} \tag{2-43}$$

Variance is a measure of the scatter of a group of values about a mean value. It has a useful physical meaning for data in a set of case A. In case B s_x^2 may be used, e.g. as a measure of the spread of the calibration range. As above, the square roots of the variances are known as **standard deviations**.

The **covariance**:

$$s_{xy}^2 = \frac{SS_{xy}}{n-1} \tag{2-44}$$

is used to define the correlation coefficient, which is actually a standardized covariance. It has a useful physical meaning for data in a set of type B.

With the help of the defined sums we can now easily calculate the **linear correlation coefficient**:

$$r_{xy} = \frac{s_{xy}^2}{s_x s_y} = \frac{SS_{xy}}{\sqrt{SS_{xx} SS_{yy}}} \tag{2-45}$$

Test of the Correlation Coefficient

In **case A** it may be convenient to test the significance of the correlation coefficient. In this case the null hypothesis states that the random variables x and y are *independent*. Remember that the correlation coefficient here is a measure of the dependency of y and x. If the hypothesis is valid then the experimental test values:

$$t_{exp} = \frac{r_{xy}\sqrt{n-2}}{\sqrt{1-r_{xy}^2}} \tag{2-46}$$

are *t*-distributed with $f = n - 2$ degrees of freedom and can be tested against t ($f = n - 2$; $q = 1 - \alpha/2$).

Particularly in **case B** where a variable, say y, is assumed to be *dependent* on the other variable, say x, it is rather interesting to test the square of the correlation coefficient, r_{xy}^2, which at least in the standard regression model is a measure of the **coefficient of determination**, i.e. which fraction of the total data variation of y is declared by the mathematical model function of its dependency on x. ($1 - r_{xy}^2$ is called coefficient of nondetermination.)

Here the values:

$$F_{exp} = t_{exp}^2 = \frac{(n-2) \cdot r_{xy}^2}{1 - r_{xy}^2} \tag{2-47}$$

are F-distributed with $f_1 = 1$ and $f_2 = n - 2$ degrees of freedom and may be tested against F_{crit} ($f_1; f_2; q = 1 - \alpha$) or, because of the special case of $f_1 = 1$, by t_{crit}^2 ($f_2; q = 1 - \alpha/2$).

Example 2-3

For **case A** assume that you have a correlation coefficient of $r_{xy} = 0.50$ for $n = 22$ samples of soil and the measured variables Zn and Cd concentration. We obtain

$t_{exp} = 0.5 \sqrt{20}/\sqrt{0.75} = 2.58$

which is to be compared with t_{crit} (20; 0.975) = 2.09. The decision is that changes in the Zn concentration are not independent of changes in the Cd concentration. A significant correlation, at a significance level of 5%, exists between both heavy metals. Even in cases of lower correlation coefficients one can detect significant correlations if the number of samples, measurements, etc. is high enough. It is, nevertheless, hopeless, if one is interested at all, to try linear modeling of, e.g., the dependence of Zn on Cd: $r_{xy}^2 = 0.25$ promises only 25% explanation of the Zn scatter, or vice versa.

Let us consider another situation of type **case A**. Assume that you have a set of bivariate correlation coefficients because in the $n = 22$ samples you have determined Hg, Pb, Cr, and other elements. Obviously it is better not to calculate a lot of t_{exp} values and to compare them against t_{crit} = 2.09. Instead you will calculate r_{crit} from t_{crit} once and quickly compare this value with all the coefficients obtained:

$r_{crit} = t_{crit}/\sqrt{(f + t_{crit}^2)} = 2.09/\sqrt{20 + 2.09^2} = 0.423$

Thus, every calculated correlation coefficient (of your correlation matrix) exceeding a value of 0.423 indicates a significant relationship between the amounts of the two heavy metals.

Example 2-4

As a basic example of **case B** let us take the numerical values of a calibration experiment given by MILLER [1991]. y is the signal and x the concentration variable with the concentration unit cu and the signal unit su.

i =	1	2	3	4	5	6	7
x_i =	0	1	2	3	4	5	6 cu
y_i =	0.1	3.8	10.0	14.4	20.7	26.9	29.1 su

From these two series we compute the following measures:

variances $s_x^2 = 28/6\ cu^2$ and $s_y^2 = 754.72/6\ su^2$
covariance $s_{xy}^2 = 143.9/6\ cu \cdot su$
correlation coefficient $r_{xy} = 143.9/\sqrt{(28 \cdot 754.72)} = 0.996$
coefficient of determination $COD = 0.992$

With $f = 7 - 2 = 5$ and $\alpha = 0.05$ we read a table value t_{crit} (5; 0.975) = 2.57, derive $r_{crit} = 0.754$ or $COD_{crit} = 0.569$ and find that there must be a significant relationship between the variables y and x. Well, this is nothing other than we ought to expect and rely on in calibration.

2.4.3 The Model and the Prediction

In environmental modeling, perhaps after starting from **case A**, or in the extremely important analytical calibration process, **case B**, one prefers linear models for reasons of conceptual simplicity.

The linear **regression model**:

$$y = a_0 + a_1 x \tag{2-48}$$

states that the two variables are related, in this way, because of a causal relationship.

Measurements of y (or even of x also) are subject to random experimental error, so we cannot replace y and x by the values of y_i and x_i to write a simple equation for the line. If, however, we assume that the values of x_i are not subject to measurable error, then we can calculate the values which y_i ought to have if the equation of the line is valid.

These values are denoted by $\hat{y}_i$:

$$\hat{y}_i = a_0 + a_1 x_i \tag{2-49}$$

Because the true values of the regression parameter a_1, called the **regression coefficient**, **slope**, or **sensitivity** (in calibration), and of a_0, called the **intercept**, **offset**, or **blank value** (in calibration) are not known, both have to be "estimated" on the basis of the set of values of n experimental pairs of measurements. These estimates are normally written as $\hat{a}_0$ and $\hat{a}_1$ and in the following, if not stated clearly, the reader should assume that both a_0 and a_1 are used with this meaning.

With the help of the above defined sums the coefficient $\hat{a}_1$ may easily be calculated according to:

$$\hat{a}_1 = \frac{SS_{xy}}{SS_{xx}} \tag{2-50}$$

For a set of measurement values of type B, when we calculate the coordinates of the center of gravity of the set, $(\bar{x}, \bar{y})$ this point must also lie on the line defined by Eq. 2-49 and from:

$$\bar{y} = \hat{a}_0 + \hat{a}_1 \bar{x} \qquad (2\text{-}51)$$

the value of $\hat{a}_0$ can be computed.

Example 2-4 (continued)

For above reported data of **case B** we obtain, in signal units (*su*) or concentration units (*cu*):

$\hat{a}_1 = 143.9/28 = 5.139$ *su/cu*

and using $(\bar{x}, \bar{y})$ which is (3, 15)

$\hat{a}_0 = 15 - 3 \cdot 5.139 = -0.418$ *cu*

Now, from the calibration model

$\hat{y}_i = -0.418 + 5.139\, x_i$

we can calculate the predicted values $\hat{y}_i$:

−0.418, 4.721, 9.861, 15.0, 20.139, 25.279, 30.418 *su*

(The values have not been rounded to the allowed figures to avoid loss of numerical precision in subsequent calculations.)

2.4.4 Conditions to be met for Linear Regression Analysis

The mathematical and statistical assumptions on which the standard regression model is based are the following:

(1) Only if assuming a linear dependence between the variables should a linear regression model be applied.
(2) Random errors associated with x should be negligible. The quantity x can be adjusted at the points x_i without error or with an error much less than the error made in measuring y. If this condition is not met, see Section 2.4.5 for alternative models.
(3) Errors associated with y at each point x_i should fit a normal (GAUSSian) distribution and should be free from outliers. The latter property can be tested by the DIXON

test [ISO 5725]. If this condition is not met, robust regression methods may be a useful alternative.

(4) The variance of y at each point x_i should be equal, i.e. constant over the whole working range of x, or, in other words, the errors in measuring y are independent of the values of x. This property is called **homoscedasticity** and can be tested by the COCHRAN test or by other tests (see [ISO 5725, clause 12]). If this condition is not met, weighted regression models may be considered.

(5) The residuals $e_i = (y_i - \hat{y}_i)$ should also fit a normal distribution, i.e. they should not correlate with x_i. An easy way to check this is to plot e_i versus x_i. If the residuals do not scatter randomly around zero, the linear model may not be adequate for the data. An indication of a wrongly specified model may be the occurrence of autocorrelated residuals, which can be checked by the DURBIN-WATSON test [MAGER, 1982].

Example 2-4 (continued)

Because we do not have, at the moment, replicate measurements of y at x_i we are not able to check conditions (2) to (4). With the above calculated values of $\hat{y}_i$ we obtain the residual e_i:

+0.518, −0.921, +0.139, −0.600, +0.561, +1.621, −1.318 su

which seem to scatter randomly around zero.

In contrast with usual calibration (**case B**), modeling in trace analysis, and also of environmental relationships (**case A**), will probably fail to fulfil condition (2) because x is usually also subject to errors. Alternative linear models must then be considered.

2.4.5 Alternative Linear Regression Models

If both y and x are subject to measurable random errors, one can derive parameters starting from the following equation to obtain a_1 first:

$$\hat{a}_1^2 SS_{xy} + \hat{a}_1 \left[\left(\frac{\sigma_y}{\sigma_x}\right)^2 SS_{xx} - SS_{yy}\right] - \left(\frac{\sigma_y}{\sigma_x}\right)^2 SS_{xy} = 0 \tag{2-52}$$

Again, using Eq. 2-51 the intercept is computable.

From Eq. 2-52 the following special cases may be defined:

(1) $\sigma_y/\sigma_x \to \infty$

This is the standard situation for the ordinary least squares estimation (**OLS**) of a_1.

$$\hat{a}_1 = \frac{SS_{xy}}{SS_{xx}} \tag{2-53}$$

(2) $\sigma_y/\sigma_x \to 0$

The model situation requires the regression of y on x.

$$\hat{a}_1 = \frac{SS_{yy}}{SS_{xy}} \tag{2-54}$$

(3) $\sigma_y/\sigma_x = a_1$

In this situation the error behavior of the two variables is reflected in the slope. The model is called the **reduced major axis** (**RMA**) model

$$\hat{a}_1 = sgn\,(r_{xy})\sqrt{\frac{SS_{yy}}{SS_{xx}}} \tag{2-55}$$

where $sgn\,(r_{xy})$ is the sign of the correlation coefficient.

(4) $\sigma_y/\sigma_x = 1$

Here the errors made in measuring y and x are assumed to be equal in magnitude. Then a_1 is obtained by solving Eq. 2-52. The procedure is called **orth**ogonal (**ORTH**) regression).

$$\hat{a}_1 = \frac{C}{2} + sgn\,(r_{xy})\sqrt{\left(\frac{C}{2}\right)^2 + 1} \quad \text{with}$$
$$C = \frac{SS_{yy} - SS_{xx}}{SS_{xy}} \tag{2-56}$$

(Note, that in this case both variables must have equal units.)

(5) If the measurement error, σ_x^2 or σ_y^2, of one of the variables is known from previous work then a better estimate can be obtained by including it:

$$\hat{a}_1 = \frac{SS_{xy}}{SS_{xx} - n\sigma_x^2} \tag{2-57}$$

or:

$$\hat{a}_1 = \frac{SS_{yy} - n\sigma_y^2}{SS_{xy}} \quad (2\text{-}58)$$

Other models and results of simulation studies may be appropriate in method comparisons [HARTMANN et al., 1993].

Example 2-4 (continued)

With the data used as Example 2-4 we found (with signal and concentration units, *su* and *cu*, respectively):

$\hat{a}_1$ (**OLS**) = 143.9/28 = 5.139 *su/cu* and $\hat{a}_0$ = −0.418 *su*

which may now be contrasted with

$\hat{a}_1$ (**RMA**) = $\sqrt{(745.72/28)}$ = 5.161 *su/cu* and $\hat{a}_0$ = −0.482 *su*

Until now we have no measures of goodness of fit, so we can only see that there is little variation in the coefficients, but that they give slightly different slopes (sensitivities) and hence different values for the intercept (blank value).

2.4.6 Regression Errors and Tests of the Coefficients

If we start from Eq. 2-49 and set $i = a$ for the **a**nalysis step and $i = c$ for the **c**alibration step we obtain:

$$y_a = \bar{y}_c + a_1(x_a - \bar{x}_c) \quad (2\text{-}59)$$

y_a – signal measured in analysis step
$\bar{y}_c$ – mean signal in calibration step
x_a – concentration in analysis step
$\bar{x}_c$ – mean concentration in calibration step

Following the laws of error propagation and using the **residual variance of regression** $s^2(e)$ defined by:

$$s^2(e) = \frac{\sum_{i=1}^{n_c} e_i^2}{n_c - n_p} = \frac{\sum_{i=1}^{n_c} (y_i - \hat{y}_i)^2}{n_c - n_p} \quad (2\text{-}60)$$

where n_c is the total number of calibration experiments and n_p is the number of parameters in the regression equation, i.e. $n_p = 2$ for the model defined with Eq. 2-48, the **error of the predicted values** is:

$$s^2(\hat{y}_i) = s^2(e)\left[\frac{1}{n_c} + \frac{(x_i - \bar{x}_c)^2}{SS_{xx}}\right] \tag{2-61}$$

To replace the concentration term by the respective signal term one can use Eq. 2-59. From this result both variances for the special points $x_i = 0$ and $x_i = \bar{x}$ follow immediately. The first variance then characterizes the **error of the calibration offset** $s^2(a_0)$. The second term, in brackets in Eq. 2-61, shows that we can expect minimum errors of the calibration process around the middle of the calibration (concentration) range, $x_i \approx \bar{x}_c$:

$$s^2(\bar{y}) = \frac{s^2(e)}{n_c} \tag{2-62}$$

The **error of the slope** is defined by:

$$s^2(\hat{a}_1) = \frac{s^2(e)}{SS_{xx}} \tag{2-63}$$

If we generally assume a_i to be a true but unknown parameter of the regression model we have the following possible hypotheses:

H_0: $\hat{a}_i = a_i$
H_1: (in two-sided cases) $\hat{a}_i \neq a_i$
      (in one-sided cases) $\hat{a}_i < a_i$ or $\hat{a}_i > a_i$

The statistical assumption is that $a_i = 0$ and, therefore, the expression

$$t_{exp} = \frac{|0 - \hat{a}_i|}{s(\hat{a}_i)} \tag{2-64}$$

defines the test value which is to be compared with $t_{crit}\,(n_c - n_p;\, 1 - \alpha/2)$ in the two-sided case.

Example 2-4 (continued)

From the above results we can continue calculating

$s^2(e) = 6.175/5 = 1.235\ su^2$, $s(e) = 1.111\ su$
$s^2(\hat{a}_1) = 1.235/28\ su^2/cu^2$, $s(\hat{a}_1) = 0.210\ su/cu$
$s^2(\hat{a}_0) = 1.235(1/7 + 3^2/28) = 0.573\ su^2$, $s(\hat{a}_0) = 0.757\ su$

As in previous sections, with these error estimates it is also possible to determine the confidence interval *cnf* of the parameters:

$cnf(\hat{a}_1) = \hat{a}_1 \pm \Delta\hat{a}_1 = 5.139 \pm 2.57 \cdot 0.210 = 5.139 \pm 0.540\ su/cu$, and
$cnf(\hat{a}_0) = \hat{a}_0 \pm \Delta\hat{a}_0 = -0.418 \pm 1.946\ su$

From the latter it is obviously possible that the true intercept, blank value,... may be zero. The statistically based decision, however, depends on the following.
The values

$t_{exp}(\hat{a}_1) = |0 - 5.139|/0.210 = 24.47$
$t_{exp}(\hat{a}_0) = |0 - 0.418|/0.757 = 0.552$

are to be compared with $t_{crit}(5; 0.975) = 2.57$.
Consequently we reach the decisions that in the case of

$\hat{a}_1$: The hypothesis (assumption) of zero slope is rejected
$\hat{a}_0$: The hypothesis (assumption) of zero intercept is accepted

In the case of the slope we are now sure that the value used in our regression model is justified, but in the case of the offset we run into the dilemma of statistical significance versus practical relevance: should we use the model equation $y = a_1 \cdot x$ for statistical reasons or still continue with model Eq. 2-48 because of possibly smaller residuals. In most cases people will prefer the second method.

2.4.7 Weighted and Robust Regression

If the variance of the dependent variable (signal) is not constant over the range of the independent variable (concentration) the data are said to be **heteroscedastic**. To test this situation, e.g., by the COCHRAN test, at each x_i or selected x_i replicate measurements of y are necessary. If the data are heteroscedastic weighted regression methods will sometimes give reasonable results for parameters of the regression model. Formulae are available, e.g. that given by HWANG and WINEFORDNER [1988]. Another possible treatment is maximum likelihood regression, which utilizes an iteratively weighted procedure.

Sometimes **weighted regression** is a way of preserving the conceptual simplicity of linear models.

In principle, weighing factors w_i can be based upon the information of the measurement error of y at position x_i. If the error has been estimated from n_i measurements at each calibration position x_i,

$$w_i = \frac{1}{s^2(y_i)} \tag{2-65}$$

In analytical chemistry the highest variance often is associated with the highest values of the concentration. Then it is possible to obtain reasonable results by weighing the least squares of the residuals in the following way:

$$\sum_{i=1}^{n_i} \frac{e_i^2}{x_i^2} \rightarrow \text{Minimum!} \tag{2-66}$$

In a previous section we mentioned that outliers and highly deviating values in a series of measurements are known to have a severe effect on most tests. In regression models also, the parameters are most sensitive to the response values near the borders of the calibration range. In order to moderate the influence of possible outliers one should try **robust techniques**. These so-called **nonparametric regression statistics** start from the common model:

$$y_i = a_0 + a_1 x_i + e_i \text{ with } i = 1, \ldots, n \tag{2-67}$$

where a_0 and a_1 are the true coefficients in the robust model equation, and treats the ranked values:

$$x_{(1)} \leq x_{(2)} \leq \cdots \leq x_{(n)} \tag{2-68}$$

as known constants.

If all x_i are different they can form $m = n(n-1)/2$ pairs of values from which the following constants can be computed:

$$\hat{a}_{1ij} = \frac{y_j - y_i}{x_j - x_i} \text{ with } x_i \neq x_j \text{ and } i < j \tag{2-69}$$

One estimate of the slope most often used in the regression equation written above is the median of all the constants obtained from Eq. 2-69:

$$\tilde{a}_1 = med(\hat{a}_{1ij}) \tag{2-70}$$

From all computed:

$$\hat{a}_{0ij} = \frac{x_j y_i - x_i y_j}{x_j - x_i} \tag{2-71}$$

in a similar way one finds the estimate of the offset from:

$$\tilde{a}_0 = med\,(\hat{a}_{0ij}) \tag{2-72}$$

On the other hand, each pair of regression coefficients defined by Eqs. 2-69 and 2-71 yields residuals:

$$e_i = y_i - \hat{y}_i \quad \text{with} \quad i = 1, \ldots, n \tag{2-73}$$

with which classical minimization is possible:

$$\sum_{i=1}^{n} e_i^2 \rightarrow \text{Minimum!} \tag{2-74}$$

Some methods, therefore, take as the best pair of regression coefficients that which gives the "least median of squares":

$$\min\left[med(e_i^2)\right] \tag{2-75}$$

Details may be found in the references [PHILLIPS and EYRING, 1983; ROUSSEEUW and LEROY, 1987; DANZER, 1989]. Robust estimates in linear regression analysis are compared by DIETZ [1987] in a simulation study. An application of robust statistics in the field of environmental analysis is described in detail in Section 9.3.

Example 2-5

To show the performance of the concept of robust regression let us assume three cases with three pairs of (x_i, y_i) each, where the outliers are underlined:

	(A)			(B)			(C)		
x_i:	1	2	3	1	2	3	1	<u>0</u>	3
y_i:	4	6	8	4	<u>9</u>	8	4	6	8

(The reader may quickly design a draft plot to compare the cases. Note that outliers in x produce the most critical problems.)

From the usual regression calculation we obtain:

(A): $\hat{a}_0 = 2$, $\hat{a}_1 = 2$ (B): $\hat{a}_0 = 3$, $\hat{a}_1 = 2$ (C): $\hat{a}_0 = 4.857$, $\hat{a}_1 = 0.857$

Robust estimation for case B with the above given formulae yields the following three values for the coefficients:

pair	1,2	2,3	1,3	
$\hat{a}_1$	5	−1	2	$med(\hat{a}_1) = 2$
$\hat{a}_0$	−1	11	2	$med(\hat{a}_0) = 2$

The model $y = 2 + 2 \cdot x$ is in accordance with the model for the outlier free case A.

2.4.8 Nonlinear Models

Here it is convenient to distinguish two kinds of nonlinear model. Model functions of the **first kind** still contain linear coefficients as weights of the degree of the dependence of the variable y on the variable x which now may occur in a **polynome** of type:

$$y = a_0 + a_1 x + a_2 x^2 + \ldots \tag{2-76}$$

Such models are called **quasi linear** because the variable(s) raised to a higher power may be simply substituted by a new variable.

If, for example, x^2 in Eq. 2-76 is substituted by x_2, and for consistency x is written as x_1 the model

$$y = a_0 + a_1 x_1 + a_2 x_2 + \ldots \tag{2-77}$$

is now a multivariate one, but is still linear with respect to the coefficients, which are accessible by the usual estimation procedure. Such replacements or transformations and nonlinear models are usually implemented in instrumental software.

In nonlinear functions of the **second kind** at least one coefficient is not linear, e.g.:

$$y = ax^b \tag{2-78}$$

Then one has to utilize special algorithms (software) to calculate reliable coefficients. The appropriate nonlinear parameter (coefficient) estimation should be preferred. Sometimes again it is possible to linearize a nonlinear model. For the above function, e.g., using a logarithmic transformation will give:

$$\log(y) = \log(a) + b \log(x) \tag{2-79}$$

which can be handled like:

$$y' = a'_0 + a'_1 x'_1 \tag{2-80}$$

where $y' = \log(y)$, $a'_0 = \log(a)$, $a'_1 = b$, and $x'_1 = \log(x)$.

Such transformations should, however, be carried out with caution, because one has to ensure that the statistical properties of the variables are not changed. The creation of heteroscedastic behavior may be mentioned as one possible problem. Proper nonlinear parameter (coefficient) estimation is, therefore, preferable.

Example 2-4 (continued)

In Example 2-4, despite the significant linear correlation coefficient one may feel that a nonlinear model according to Eq. 2-77 should be tried. The data are to be extended by the values of x_i^2 in this way:

i	=	1	2	3	4	5	6	7
x_i	=	0	1	2	3	4	5	6 cu
x_i^2	=	0	1	4	9	16	25	36 cu^2
y_i	=	0.1	3.8	10.0	14.4	20.7	26.9	29.1 su

Using the software package STATISTICA [1995] we obtain the following results:

$\hat{a}_0 = -0.638$ su, $s(\hat{a}_0) = 1.070$ su, $t_{exp} = 0.596$, which equals $\alpha \approx 58\%$ in F_{crit}
$\hat{a}_1 = -5.404$ su/cu, $s(\hat{a}_1) = 0.835$ su/cu, $t_{exp} = 6.468$, which equals $\alpha \approx 0.3\%$ in F_{crit}
$\hat{a}_2 = -0.044$ su^2/cu^2, $s(\hat{a}_2) = 0.134$ su^2/cu^2, $t_{exp} = 0.329$, which equals $\alpha \approx 76\%$ in F_{crit}

From these results we derive the same conclusions as above: neither the intercept $\hat{a}_0$ nor the coefficient of the nonlinear term is signficant and we can rely on the simple linear dependence of y on x.

2.4.9 Measures of Goodness of Fit

Regardless of the model, one can use certain measures to estimate appropriate ratios of fractions of variation encountered in regression analysis. A first but (as we will see later) rough measure is the **coefficient of determination**, *COD*. In any case it should be used instead of the correlation coefficient! It is the ratio of the data variation explained by the model (via the predicted values) to the total data variation

$$COD_{xy} = \frac{\sum_{i=1}^{n}(\hat{y}_i - \bar{y})^2}{\sum_{i=1}^{n}(y_i - \bar{y})^2} \qquad (2\text{-}81)$$

For simple linear regression it holds that COD is related to the square of the correlation coefficient by $COD_{xy} = r_{xy}^2$.

Some other coefficients of determination are used for **multivariate regression** models.

Because they may be of interest after having performed nonlinear regression of the first kind (see Section 2.4.8) they are mentioned here. These coefficients lead to information according to:

- **multiple determination**: if all variables x are considered simultaneously
- **partial determination**: if the relationship between variables y and x_k are considered excluding the influences of the remaining variables ($j = 1$, ..., m; where m is the total number of variables and $j \neq k$)
- **inner determination**: if the dependence of one x on another x is of interest

The last mentioned coefficient is a very important indicator of the so-called **multicollinearity** which can cause severe numerical problems or at least affect the reliability of the derived regression coefficients.

No coefficients of determination should be used as sole criteria for the evaluation of regression functions. Problems may, e.g., arise with autocorrelated measurements (see also Section 6.6).

All coefficients of determination can be evaluated by one-sided F-tests. The (simple) COD of Eq. 2-81 is tested by:

$$F_{exp} = \frac{COD(n-2)}{1 - COD} \quad \text{against} \quad F_{crit}(f_1 = 1; f_2 = n-2; q = 1-\alpha) \qquad (2\text{-}82)$$

Another important measure is called the **goodness of fit**, GOF. It characterizes the fraction of the residual variance declared by the model. The test is based on the GOF value written as an F-distributed variable:

$$F_{exp}(GOF) = \frac{\sum_{i=1}^{n}(\hat{y}_i - \bar{y})^2}{\sum_{i=1}^{n}(y_i - \hat{y}_i)^2} \cdot \frac{n - n_p}{n_p - 1} \qquad (2\text{-}83)$$

in which n_p is the number of estimated regression coefficients, normally equal to 2 in standard linear regression models.

The test against one-sided critical values $F_{crit}(n_p - 1; n - n_p; q = 1 - \alpha)$ always should be significant.

Sometimes the GOF test is called "test of significance of the regression".

Example 2-4 (continued)

For Example 2-4 we calculate the following values:

Model	OLS	RMA	
COD_{xy}	0.992	1.000	
$F_{exp}(COD)$	620	∞	$F_{crit}(1; 5; 0.95) = 6.61$
$F_{exp}(GOF)$	598.8	599.6	$F_{crit}(1; 5; 0.95) = 6.61$

Obviously all the models explain the data variation significantly. Usually the OLS model will be tried and tested first, so there is no need for an alternative model. It is, however, interesting that the RMA model explains 100% of the data variation whereas to minor degrees the OLS model underestimates the data variation.

Note that the residual deviations (denominator in *GOF*) both arise from modeling and experimental errors. It would, therefore, be appropriate to perform an analysis of variance to test both sources of error.

2.4.10 Analysis of Variance for Regression Models

An excellent paper on the subject of variance tests in regression modeling has been written by DEMING and MORGAN [1979].

Here, we want to emphasize that one is able to calculate the fraction of the experimental error only if replicate measurements (at least at one point x_i) have been taken. It is then possible to compare model and experimental errors and to test the sources of residual errors. Then, in addition to the *GOF* test one can perform the **test of lack of fit**, *LOF*, and the **test of adequacy**, *ADE*, (commonly used in experimental design). In the lack of fit test the model error is tested against the experimental error and in the adequacy test the residual error is compared with the experimental error.

With
n – the total number of calibration measurements
n_p – the number of regression coefficients estimated
m – the number of calibration points with replicates ($j = 1, \ldots, m$)
n_j – the number of replicates at point j

the lack of fit value, *LOF*, is defined as:

$$F_{exp}(LOF) = \frac{\sum_{j=1}^{m}(\bar{y}_j - \hat{y}_j)^2}{\sum_{j=1}^{m}\sum_{i=1}^{n_j}(y_{ij} - \bar{y}_j)^2} \cdot \frac{n-m}{m-n_p} \qquad (2\text{-}84)$$

This value can be tested against $F_{crit}(m - n_p; n - m; q = 1 - \alpha)$ and should normally lead to the decision that the model error is greater than the experimental error.

The adequacy of the model, *ADE*, can be tested by comparing:

$$F_{exp}(ADE) = \frac{\sum_{i=1}^{n}(y_i - \hat{y}_i)^2}{\sum_{j=1}^{m}\sum_{i=1}^{n_j}(y_{ij} - \bar{y}_j)^2} \cdot \frac{n-m}{n-n_p} \qquad (2\text{-}85)$$

against

$$F_{crit}(n - n_p; n - m; q = 1 - \alpha)$$

Example 2-6

We extend Example 2-4 by simulating $n_j = 3$ replicates at each of the $m = 7$ calibration points in such a way that the means of each 3 replicates are equal to the original values given in Example 2-4. The $n = 3 \cdot 7 = 21$ *y* values are given below:

m =	1	2	3	4	5	6	7
x =	0	1	2	3	4	5	6 cu
y =	0.05	3.85	10.0	14.4	20.6	26.7	28.7 su
	0.1	3.75	10.05	14.3	20.8	27.1	29.1 su
	0.15	3.8	9.95	14.5	20.7	26.9	29.5 su

(In the following ratios in each paranthesis the numerator is the respective sum of squares defined in Eqs. 2-83 to 2-85 and the denominator is the corresponding degree of freedom.)

$F_{exp}(GOF) = (2218.6/1) / (18.98/19) = 2221 \leftrightarrow F_{crit}(1; 19; 0.95) = 4.38$
$F_{exp}(LOF) = (18.53/5) / (0.451/14) = 115 \leftrightarrow F_{crit}(5; 14; 0.95) = 2.96$
$F_{exp}(ADE) = (18.98/19) / (0.451/14) = 31 \leftrightarrow F_{crit}(19; 14; 0.95) = 2.40$

As with single measurements in the calibration range, the *GOF* test indicates that the regression model is appropriate for description of variation in the data. The *LOF* confirms that highly precise measurements are at hand and that the model error exceeds the experimental error. The analyst testing calibration performance will be satisfied with this level of agreement. The analyst performing experimental designs will possibly look at the *ADE* test result. In the light of the precise measurements the *ADE* test even indicates that another, maybe nonlinear, model could be tried.

2.4.11 Regression Functions and Confidence Regions

By analogy with the **confidence range** of a single value or a mean, we can calculate a confidence interval for a single predicted value according to:

$$\hat{y}_i \pm \Delta \hat{y}_i = \hat{y}_i \pm t(f;\ q) \cdot s(\hat{y}_i) \qquad (2\text{-}86)$$

with

$$s^2(\hat{y}_i) = s^2(e) \left[\frac{1}{n_c} + \frac{(x_i - \bar{x}_c)^2}{SS_{xx}} \right] \qquad (2\text{-}87)$$

n_c – number of measured values in calibration step

and $s^2(e)$ as defined in Eq. 2-60. The degree of freedom is $f = n_c - 2$, with n_c the number of measurements from which the regression model has been estimated, and $q = 1 - \alpha/2$.

With one measurement in the analysis step at x_j, which is not necessarily identical to one of the previous x_i in the calibration process, and, from the mathematical point of view, not even within the regression range of x, $s^2(\hat{y}_j)$ then becomes:

$$s^2(\hat{y}_j) = s^2(e) \left[1 + \frac{1}{n_c} + \frac{(x_j - \bar{x}_c)^2}{SS_{xx}} \right] \qquad (2\text{-}88)$$

Then the confidence interval can be made smaller by replicate measurements, n_a, of a sample at x_j and $s^2(\hat{y}_j)$ becomes:

$$s^2(\hat{y}_j) = s^2(e) \left[\frac{1}{n_a} + \frac{1}{n_c} + \frac{(x_j - \bar{x}_c)^2}{SS_{xx}} \right] \qquad (2\text{-}89)$$

On each side of the calibration line the border points of the confidence regions may be connected or smoothed to give a **confidence band** around the calibration line, see Fig. 2-4.

We can only note here that, in order to be statistically correct, the procedure used for calculating the confidence band has a very sophisticated basis (see, e.g., [MAGER, 1982; LUTHARDT et al., 1987].

In the course of an analytical procedure after the **calibration step**, modeled in the same manner as described above, it is even more important to have an estimate of the confi-

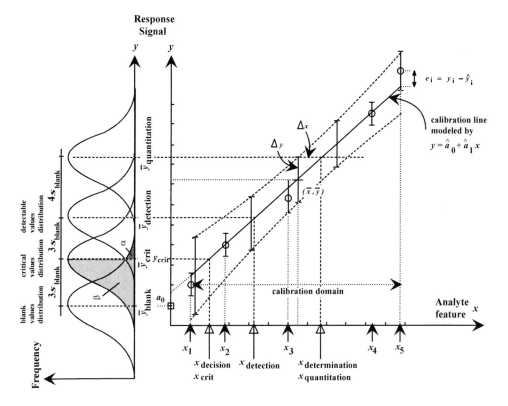

Fig. 2-4. Blank statistics and confidence band statistics in calibration (according to [MÜLLER et al., 1994])

dence interval of an analyte concentration, x_a, found in an **analyzing step**; this utilizes the inverted regression function.

For discussion of factors influencing a certain mean value of analyte concentration, $\bar{x}_a$, after repeated signal measurements, n_a, one can use the following expression for the confidence interval of $\bar{x}_a$, assuming that $n_a \approx n_c$:

$$cnf(\bar{x}_a) = \bar{x}_a \pm \Delta\bar{x}_a = \bar{x}_a \pm \frac{s(e)\, t(f;\, q)}{a_1} \sqrt{\frac{1}{n_a} + \frac{1}{n_c} + \frac{(\bar{x}_a - \bar{x}_c)^2}{SS_{xx}}} \qquad (2\text{-}90)$$

Several analytical papers deal with the construction of such intervals (see, e.g., [EBEL, 1983; EBEL et al., 1983; 1989]. Details of exact predictions, especially for limits of decision, detection, and determination (see Section 2.4.12) are considered by LUTHARDT et al. [1987].

According to MILLER [1992] after replacing $1/n_a$ by 1, Eq. 2-90 can even be used for the estimation of an analyte value x_a after a single reading of y_a. In that case a special value

$$g = \frac{t^2(f; q) \, s^2(e)}{a_1^2 SS_{xx}} \tag{2-91}$$

should be less than 0.05 [MILLER, 1992].

It is important to note, that the symmetrical confidence interval stated in Eq. 2-90 is only obtained if $\bar{x}_a$ is not too far from $\bar{x}_c$, the centre of the calibration domain. In principle, the confidence region of $\bar{x}_a$ is nonsymmetric. For details of computation and for examples, see BONATE [1990].

2.4.12 Limits of Decision, of Detection, and of Determination

In calibration, the simplest model of the dependence of signal values on concentration values is given by Eq. 2-48:

$$y = a_0 + a_1 x \tag{2-92}$$

where y denotes the signal feature and x stands for the analyte feature.

In principle, all performance measures of an analytical procedure mentioned in the title of this section can be derived from a certain **critical signal value**, y_{crit}. These performance measures are of special interest in **trace analysis**. The approaches to estimation of these measures may be subdivided into **'methods of blank statistics'**, which use only blank measurement statistics, and **'methods of calibration statistics'**, which in addition take into account calibration confidence band statistics.

Methods of Blank Statistics and the Limit of Decision

The critical signal value may then be best obtained from repeated analysis of several blank analytical samples. From the mean blank value of n_b repeated measurements:

$$\bar{y}_{blank} = \frac{1}{n_b} \sum_{i=1}^{n_b} y_i \tag{2-93}$$

and the standard deviation of the blanks:

$$s_{blank} = \sqrt{\frac{\sum_{i=1}^{n_b} (y_i - \bar{y}_{blank})^2}{n_b - 1}} \tag{2-94}$$

the critical signal value y_{crit} is then regarded as the upper limit of the confidence interval of the mean blank signal:

2.4 Causal Modeling, Regression, and Calibration

$$y_{crit} = \bar{y}_{blank} + s_{blank}\, k(f;\, q)\, \sqrt{\frac{1}{n_b}} \qquad (2\text{-}95)$$

where the degree of freedom $f = n_b - 1$ and $q = 1 - \alpha$ are used in the quantile k of the t-distribution. In this equation, via the t-value we assume a symmetrical distribution of the measured blank signal. For trace determinations, however, this assumption may not be valid. Then adequate transformations are necessary to get an appropriate distribution, e.g. a logarithmic normal distribution.

If the signal of any test sample is above this critical value, then, with a certain error risk of, say, $\alpha = 0.05$, one can decide that the signal comes from an analyte present in the sample. According to this **qualitative decision** the corresponding critical concentration $x_{crit} = x_{decision}$, the **limit of decision**, is easily calculated by combination of the above given calibration equation, Eq. 2-92, for $y = y_{crit}$ and Eq. 2-95 assuming that $\bar{y}_{blank} = a_0$:

$$x_{decision} = \frac{s_{blank}}{a_1} k(f;\, q)\, \sqrt{\frac{1}{n_b}} \qquad (2\text{-}96)$$

If the quantile $k(q)$ of the normal distribution is used instead of $k(f;\, q)$ the definition of KAISER results, see e.g. [KAISER, 1973].

From Eq. 2-96 optimization is thus possible by reducing the standard deviation of the blank value, by raising the number of blank measurements n_b, and by using a method with a high sensitivity, a_1.

The numerical value of the limit of decision can also be reduced by taking into account the future analytical effort involved in n_a. If for the analyzing step the same standard deviation as for the blank determinations can be ensured, the limit of decision is:

$$x_{decision} = \frac{s_{blank}}{a_1} k(f;\, q)\, \sqrt{\frac{1}{n_b} + \frac{1}{n_a}} \qquad (2\text{-}97)$$

Basically, this is one of the proposals of DIN [DIN 32 645] for the German 'Nachweisgrenze' ($f = n_b - 1$).

Methods of Calibration Statistics and the Limit of Decision

This approach uses the upper confidence limit of the calibration intercept a_0 derived from n_c measurements of calibration samples [DIN 32 645]:

$$y_{crit} = a_0 + s(e)\, k(f;\, q)\, \sqrt{\frac{1}{n_c} + \frac{1}{n_a} + \frac{\bar{x}_c^2}{SS_{xx}}} \qquad (2\text{-}98)$$

The residual standard deviation of the regression line $s(e)$ is defined in Eqs. 2-60 and 2-61, and the sum of squares SS_{xx} is defined in Eq. 2-39, $\bar{x}_c^2$ is the squared mean value of the concentrations used in the calibration step ($f = n_b - 2$). Again, by combination of Eq. 2-98 with the calibration model a limit of decision is accessible:

$$x_{decision} = \frac{s(e)}{a_1} k(f; q) \sqrt{\frac{1}{n_c} + \frac{1}{n_a} + \frac{\bar{x}_c^2}{SS_{xx}}} \quad (2\text{-}99)$$

Here one recognizes another way of optimizing, namely to set up a calibration experiment with $\bar{x}_c$ close to the limit of decision. Consequently is has been recommended that the highest concentration should be below $10 \cdot x_{decision}$.

Limit of Detection
From Fig. 2-4 one can see that at and near the decision limit the distribution is such that there is a 50% probability that the measured values lie within the spread of the blank value. In other words, at this point the risk of wrongly accepting a blank signal is low (probability of α) but there is a high probability ($\beta = 50\%$) of a risk of the second kind, namely that of mistaking an analyte signal for a blank. Hence, an analyte signal is reliably detected when the risk β is as low as possible. For $\alpha = \beta$ the **limit of detection** (in German: 'Erfassungsgrenze') is twice the limit of decision (in German: 'Nachweisgrenze').

In cases where $\alpha \neq \beta$ the DIN recommendation [DIN 32645] is:

$$x_{detection} = x_{decision} + x'_{decision} \quad (2\text{-}100)$$

($x'_{decision}$ is calculated from Eq. 2-96 or Eq. 2-97 with $q = 1 - \beta$.)

Limit of Quantitation (Limit of Determination)
From Fig. 2-4 it can be derived that for a particular observed value $y_{quantitation}$ an acceptable analyte interval $\pm \Delta x$ for the respective $x_{quantitation}$ (in German: 'Bestimmungsgrenze') can only be reported if the ratio $s(e)/a_1 \sim \Delta x$ is not too large.

"This quotient is the critical quantity for the practical application of an analytical method. In addition, the relative standard deviation of an analyte result sometimes must not exceed a specified maximum. These two requirements lead to the ... pragmatic setting of so-called limits of quantitation. Since such limits depend on the particular value observed, they cannot be regarded as meaningful characteristics of the method" [MÜLLER et al., 1994]. In brief conclusion, the analyst is well advised not to report determinations below this limit, because they will probably be biased owing to unacceptably high random errors.

History
Historically, the simple so-called "$k\sigma$"-criteria are based on the standard deviation ("σ") of blank measurements. The limit of decision ('Nachweisgrenze') has been defined at $k = 3$, the limit of detection ('Erfassungsgrenze') at $k = 6$, and, among other possibilities,

the quantitation limit ('Bestimmungsgrenze') has been used for $k = 10$ [LONG and WINEFORDNER, 1983].

On the other hand it has to pointed out that the k-values offer the advantage of taking into account statistical (α) and hence analytical risks of wrong decisions. The trace analyst then may adjust k and α to meet the purposes of the analytical procedure at hand. Basic considerations can be traced back, e.g., to the early work of EHRLICH [1967].

In analytical and in environmental sciences there is a constant challenge to enhance instrumental performance and to detect lower and lower trace amounts. It seems also to be a challenge for statistical refinement of the definition of the limits briefly considered above. Until now the soundest approach to this field is probably that given by LUTHARDT et al. [1987]. The reader who is not forced to follow statutory regulations (DIN, ISO, BS, ...) may try his own literature search.

References

Bonate, P.L.: J. Chromatogr. Sci. 28 (**1990**) 559
Caulcutt, R., Boddy, R.: Statistics for Analytical Chemists, Chapman and Hall, London, **1983** p. 29
Clarke, G.M., Cooke, D.: A Basic Course in Statistics, 2nd Ed., Edward Arnold, London, **1988**, p. 193
Danzer, K.: Fresenius' Z. Anal. Chem. 335 (**1989**) 869
Deming, S.N., Morgan, S.L.: Clin. Chem. 25 (**1979**) 840
Dietz, E.J.: Commun. Statist. – Simula. 16 (**1987**) 1209
DIN 32645, part 1: Chemische Analytik; Nachweis-, Erfassungs- und Bestimmungsgrenze; Ermittlung unter Wiederholbedingungen; Begriffe, Verfahren, Auswertung 5, **1994**
DIN 53804, part 1: Statistische Auswertungen; Meßbare (kontinuierliche) Merkmale, Beuth, Berlin, **1981**, p. 3
Doerffel, K.: Statistik in der Analytischen Chemie, Deutscher Verlag für Grundstoffindustrie, Leipzig, **1990**, p. 45
Ebel, S.: Comp. Anwend. Lab. 1 (**1983**) 55
Ebel, S., Alert, D., Schaefer, U.: Comp. Anwend. Lab. 1 (**1983**) 172
Ebel, S., Lorz, M., Weyandt-Spangenberg, M.: Fresenius' Z. Anal. Chem. 335 (**1989**) 960
Ehrlich, G.: Fresenius' Z. Anal. Chem. 232 (**1967**) 1
Hartmann, C., Smeyers-Verbeke, J., Massart, D.L.: Analusis 21 (**1993**) 125
Hastings, C.: Approximation for Digital Computers, Princeton University Press, Princeton, New Jersey, **1955**
Hwang, J.D., Winefordner, J.D.: Progr. Anal. Spectrosc. 11 (**1988**) 209
ISO 2854 in: ISO Standards Handbook, 3rd Ed., ISO, Geneva, **1989**, pp. 6
ISO 5725 in: ISO Standards Handbook, 3rd Ed., ISO, Geneva, **1989**, pp. 412
Kaiser, H.: Pure Appl. Chem. 34 (**1973**) 35
Kaiser, R., Gottschalk, G.: Elementare Tests zur Beurteilung von Meßdaten, Bibliographisches Institut, Mannheim, **1972**
Long, G.L., Winefordner, J.D.: Anal. Chem. 55 (**1983**) 712A
Luthardt, M., Than, E., Heckendorff, H.: Fresenius' Z. Anal. Chem. 326 (**1987**) 331
Mager, H.: Moderne Regressionsanalyse, Salle und Sauerländer, Frankfurt, Aarau, **1982**
Mandel, J., Lashof, T.W.: J. Qual. Technol. 19 (**1987**) 29
McCormick, D., Roach, A.: Measurement, Statistics and Computation, Wiley, **1987**
Miller, J.N.: Spectrosc. Int. 3 (**1991**) no. 4, p. 41
Miller, J.N.: Spectrosc. Int. 4 (**1992**) no.1, p. 41

Miller, J.C., Miller, J.N.: Statistics for Analytical Chemistry, 3rd Ed., Ellis Horwood, Chichester, **1993**, p. 74 and pp. 165
Müller, H., Zwanziger, H.W., Flachowsky, J.: Trace Analysis, in: Ullmann's Encyclopedia of Industrial Chemistry, Vol. B5, VCH, Weinheim, **1994**, p. 101
Noack, S., Schulze, G.: Fresenius' Z. Anal. Chem. 304 (**1980**) 250
Phillips, G.R., Eyring, E.M.: Anal. Chem. 55 (**1983**) 1134
Rechenberg, W.: Fresenius' Z. Anal. Chem. 311 (**1982**) 590
Rousseeuw, P.J., Leroy, A.M.: Robust Regression and Outlier Detection, Wiley, New York, **1987**
STATISTICA 5.0 for Windows, StatSoft, Tulsa, **1995**
Thijssen, P.C., van der Wiel, P.F.A.: Trends Anal. Chem. 3 (**1984**) 11
Zwanziger, H., Rohland, U., Werner, W.: Fresenius' Z. Anal. Chem. 323 (**1986**) 371

3 Remarks on Experimental Design and Optimization

3.1 Basic Notions and Ideas

Methods of optimization are useful in all steps of the analytical process. Whenever optimization starts from a sound, not necessarily statistical basis, one can speak of a design or of designed experiments. Besides optimization in the laboratory itself, for example with the aim of obtaining the highest signal responses during method development or the most precise results, in environmental analysis results are tremendously influenced by the sampling step (see Section 4). A number of sampling plans exists and most of them are definitely experimental designs.

In the widest sense in an **experiment** people want to investigate or to influence their surroundings. In most cases their activity is aimed at the recording and the interpreting the result of the experiment. Interpretation also may be important to the conclusions drawn about the current state of an academic or a practical situation. In any case the result will be affected by chance. In a narrower sense, especially for environmental investigations where complex systems or matrices may influence the concentration or behavior of analytes, the value of experiments is twofold. Firstly, planned investigations of the environment ('experiments') can, in a systematic manner, consider possible influencing features. Secondly, experimental design offers the advantage of the opportunity to model complex systems (in the laboratory) and to investigate the influence of selected conditions on analytes or other system features.

In such **statistically designed experiments** one wants to exclude the random effects of a limited number of features by varying them systematically, i.e. by variation of the so-called factors. At the same time the order in which the experiments are performed should be randomized to avoid systematic errors in experimentation. In another basic type of experiment, **sequential experiments**, the set-up of an experiment depends on the results obtained from previous experiments. For help in deciding which design is preferable, see Section 3.6. In principle, **statistical design** is one recommendation of how to perform the experiments. The design should always be based on an exact question or on a working hypothesis. These in turn are often based on models.

According to the number n of included **independent variables** $x_1, ..., x_n$ (influencing quantities, features, factors, ...) we distinguish **one-factorial designs** ($n = 1$) and **multi-factorial designs** ($n > 1$). According to the number m of recorded **response features** $y_1, ..., y_m$ we will get **univariate results** ($m = 1$) or **multivariate results** ($m > 1$). Therefore we can arrange our numerical results in data matrices with k lines, the experiments, and m

columns, which are associated with the measured quantities. An example of such a table is given in Tab. 3-1. Eq. 3-1 shows the corresponding **matrix of results**:

$$Y = \begin{pmatrix} y_{11} & y_{12} & y_{13} & y_{14} \\ y_{21} & y_{22} & y_{23} & y_{24} \\ y_{31} & y_{32} & y_{33} & y_{34} \\ y_{41} & y_{42} & y_{43} & y_{44} \end{pmatrix} \qquad (3\text{-}1)$$

Tab. 3-1. Example of designed sampling and multivariate results

Water Sample	Influencing features		Responses			
	Origin	Temp.	Pb	Zn	TOC	Conductivity
	x_1	x_2	y_1	y_2	y_3	y_4
1	Tap	Summer	y_{11}	y_{12}	y_{13}	y_{14}
2	Tap	Winter	y_{21}	y_{22}	y_{23}	y_{24}
3	Sewage	Summer	y_{31}	y_{32}	y_{33}	y_{34}
4	Sewage	Winter	y_{41}	y_{42}	y_{43}	y_{44}

The **design matrix** in a preliminary form is then given by:

$$X = \begin{pmatrix} \text{tap} & \text{summer} \\ \text{tap} & \text{winter} \\ \text{sewage} & \text{summer} \\ \text{sewage} & \text{winter} \end{pmatrix} \qquad (3\text{-}2)$$

Each experiment can also be called a treatment. It refers to a defined combination of fixed levels of the factors. Repeated (multiple) measurements of such a combination are said to be members of one and the same class, sometimes called a **cell**. In our above example, we may then have performed four replicate analyses, so that in each cell we have four members (objects, measurements), e.g. in 'class 11', $\{y_{11,1}, y_{11,2}, y_{11,3}, y_{11,4}\}$. Analysis of variance (see also Section 3.3.9) then can be used to evaluate the results.

As mentioned above, in analysis the response variables are mostly optimization criteria, e.g. precision, signal height, limit of detection, certain performance criteria, and others.

There are several ways to find the optimum:
– via models derived from statistical designs (optimum factorial designs)
– via the results of sequential experiments based on defined algorithms (simplex method, ...)
– via unplanned, stochastic (but throughout systematic) methods (grids, screening, ...)

The methods of analytical chemistry also offer the possibility of monitoring the effects of environmental redevelopment; analysis thus provides the basis for regression modeling of the mutual interrelationships of environmental parameters or pollution. Sampling plans should, therefore, be clever strategies encompassing all possible sources of influence on experimental results. Especially when using statistical designs the investigators are able to conduct the necessary sampling experiments in an optimum way, at least in saving time. A deeper discussion of environmental sampling is given in Section 4.

In the following it is our aim to encourage the reader to use methods of (statistical) experimental design rather than to give a complete introduction to all the variants possible.

For deeper study of the matter the reader can be directed to monographs on experimental design and optimization, e.g. [MORGAN, 1991; DEMING and MORGAN, 1993]. More references on this subject are given in Section 1.8.

3.2 Factorial Designs

In factorial designs we have n variable factors which we are able to adjust at fixed **levels**. Sometimes the factors only 'exist' in discrete levels but sometimes we are interested in or are only able to set up these discrete levels. The assignable causes, the factors which we assume will affect the recorded responses, may be **quantitative or qualitative features**. The quantitative character of temperature, pH, stirring velocity, etc. is obvious. Examples of qualitative factors are the age of membranes in ion selective electrodes or filter devices (with the levels old and new) or a certain medication or noxious agent (present or absent). In this situation the notion 'version' of a factor makes sense (ISO 3534/3 in [ISO STANDARDS HANDBOOK, 1989]).

If each factor is represented by the same number of levels we have **symmetrical designs**.

If the number of repeated experiments (if any) according to each level combination (cell) is constant throughout the design it is a **balanced design**.

Very often it is sufficient to start with plans where only the lowest and the highest levels of each factor are considered. In this case we have two levels and n factors, which gives rise to a **complete design** of type 2^n. If all possible combinations of the two factors in two levels each are to be performed, a total of $k = 2^n$ experiments results.

Using the results of these k experiments we are then able to estimate the coefficients a_n of appropriate polynomial models for each response variable y:

$$y = f(x_1, \ldots, x_n) \tag{3-3}$$

Mathematically, this equation models the response surface of the recorded response variable y.

From mathematics we know that k unknown coefficients from k equations can be computed (estimated) via regression analysis; this has been described in Section 2.4. In Section 3.3 we shall show how straightforward and simple it is to calculate the coefficients of certain regression models on the basis of an (orthogonal) experimental design.

A possible practical situation for a 2^n factorial design is depicted in Fig. 3-1; the situation is treated in Example 3-1.

For the moment, as a general consideration, let us assume $k = 4$ experiments performed within type of design mentioned above. Then we can calculate a maximum of four coefficients, i.e. all the coefficients of:

$$y = a_0 + a_1 x_1 + a_2 x_2 + a_{12} x_1 x_2 \tag{3-4}$$

For k coefficients the underlying design is called a **saturated design**.

The coefficients (associated with variables) are proportional to the so-called **main effects** and the **interaction effects** if more than one variable is involved. The identification of such effects is important in evironmental research. Of special interest may be the interaction of influencing factors in the sense of synergistic or antagonistic effects. An illustrative example from the field of environmental research is given in Section 10.2.

Statistical tests on the **significance** of the above coefficients are possible if we have estimates of the experimental variance from the past or if we can repeat some experiments or even the total design. Alternatively, variance can be computed if the so-called **central point** of a design is (sampled and) measured repeatedly. This is a point between all factor levels. In Fig. 3-1 this is the point between location 1 and location 2 and between depth 1 and depth 2. The meaning of another term which is used, **zero level**, will be clear after we have learned how to construct general designs.

Statisticians have defined several optimum characteristics of designs with regard to the estimated properties. For our purposes it should be sufficient to speak of an **optimum design** with regard to the number of experiments to be undertaken.

One way of achieving a smaller number of experiments is the use of **fractional factorial designs**. Such designs are possible if interactions of factors can be neglected. Then

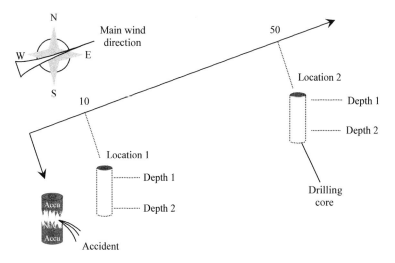

Fig. 3-1. Typical situation for experimental design

the corresponding column of the design matrix (see below) of the complete design will be used for the computation of the main effect of an additional factor.

Eq. 3-4 then becomes:

$$y = a_0 + a_1 x_1 + a_2 x_2 + a_{12} x_3 \tag{3-5}$$

The respective coefficient a_{12} then is a **confounded estimate** of the effect of the new factor x_3 and the possible interaction effect $x_1 x_2$.

A further reduction of experimental effort may be achieved by the selection of special designs developed, for example, by HARTLEY [1959], BOX and BEHNKEN [1960], WESTLAKE [1965], and others. In these designs the ratio of experiments to the number of coefficients necessary is reduced almost to unity. (This situation is somewhat different from *regression analysis* or random selection of experiments where, in principle, k experiments or measurements are sufficient to estimate k parameters of a model. In *experimental design* the optimized number of experiments is derived from statistical consideration to encompass as much variation of the factors as possible.)

Certain circumstances may force us to follow the opposite direction and to burden ourselves with additional experimental expense. Models of higher order may be unavoidable if the response variable follows a nonlinear function of the primary variables, or factor variables. Then one can utilize designs which take into account more than two levels of each factor. As an example, in 3^n designs one has n factors with three levels each.

Sometimes it is sufficient to use **central composite designs** or **noncentral composite designs** to enable the calculation of additional coefficients of effects with a reasonable amount of extra work. In the upper part of Fig. 3-2 a 2^2 design is extended to a central composite 3^2 design.

As already mentioned, it is recommended that **the order** of the experiments be **randomized** to avoid systematic errors (bias), which may happen if, e.g., all experiments on the highest level of a concentration are made with a faulty pipet on the same day. As an example from environmental analysis, a systematic error could be introduced if in a certain contaminated region the surface is sampled on one (rainy) day and a deeper layer is sampled on a different (sunny) day.

In the last mentioned case, where a time factor may introduce bias, or in cases where inhomogeneities of the sampled material may influence the responses, block designs can be chosen. To eliminate several block effects one can select a latin square design. With regard to the arrangement of experiments we refer to appropriate text books, e.g. [BANDEMER et al., 1973; MORGAN, 1991]. There the reader may find special types of design which fit his requirements.

Finally, there are two families of factorial designs which depend on the combination mode of the factors. If it is possible to combine each level of one factor with each level of every other factor this is denoted **cross-classified design**. The second family consists of **nested designs** or hierarchical experiments. In those designs the levels of one factor may not be combined with all the levels of another factor. As an example consider an investigation performed by two laboratories (factor no. 1) which unfortunately cannot

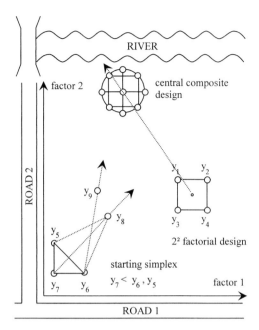

Fig. 3-2. Factorial experimental designs and sequential designs
y_8 – next step in modified simplex algorithm
y_9 – next step in weighted simplex algorithm

exchange personnel (factor no. 2). So, in laboratory no. 1 technicians A and B run the analyses and in laboratory no. 2 technicians C and D perform the analyses.

An excellent review on experimental design which is followed by a valuable discussion of numerous scientific and educational aspects is given by STEINBERG and HUNTER [1984]. In analytical chemistry experimental design has been used to optimize almost all types of analytical procedure because of its ease of use.

3.3 Example: Set-up and Interpretation Steps of Factorial Designs

3.3.1 Fixing the Levels of Factors

This is sometimes a crucial point. In environmental analysis in particular it is not always obvious which factors to consider as relevant. The inclusion of too many factors in the first step may on the other hand, substantially increase the number of experiments to be performed or samples to be taken, even in the simplest 2^n design. In the following discussion we will refrain from this type of factorial design.

Example 3-1

Imagine the gross modeling of the distribution of a contaminant, say Cd concentration (mg kg^{-1}), in a certain region (direction) using samples which originate from different sampling locations and depths (see Fig. 3-1). Then the simplest way would be to follow a 2^2 design. In that case one reckons with two factors, say "depth" and "location". Let us fix "depth" at 10 cm and 20 cm below the surface and "location" at distances of 10 m and 50 m from a suspected point of contamination.

3.3.2 Performing the Measurements

The next step is to perform the experiments according to the adjusted factor levels and their possible combinations and to record the response variable in which one is interested.

Example 3-1 (continued)

Measurement no.	Depth in cm	Location in m	Cd concentration (response) found in mg kg^{-1}
1	10	10	12
2	10	50	16
3	20	10	60
4	20	50	58

3.3.3 Generalizing the Design

As one may expect from 'optimum designs' there is only one scheme for similar situations. As for the box above the next will show other examples of 2^2 designs.

Example 3-2

Assume, e.g., the treatment of a contaminated soil. Attempts are made to decompose the pollutant y, a particular organic compound, by application of an acid or a base at low or high temperature. Hence for every soil sample (possibly a third factor ...) we have four treatments:

Treatment no.	pH	T in °C	Concentration, y, of the organic component in mg kg^{-1}
1	2	20	...
2	2	70	...
3	8	20	...
4	8	70	...

Example 3-3

Another example may be checking of the blank response of an ammonia ion selective electrode under different experimental conditions, namely the age of the sensor membrane and the speed at which the measuring solution is stirred.

Blank no.	Age	Speed in r.p.m.	Response in mV
1	old	400	...
2	old	800	...
3	new	400	...
4	new	800	...

The general scheme of the 2^2 design, and of similar ones, is based on a simple transformation procedure which uses the **half-range**, hr, and the **central value**, cv, of the values (levels) of each factor, x:

$$hr = \frac{x_{\max} - x_{\min}}{2} \qquad (3\text{-}6)$$

$$cv = \frac{x_{max} + x_{min}}{2} \tag{3-7}$$

Then, all the levels, x_i, of each factor can be transformed according to:

$$z_i = \frac{x_i - cv}{hr} \tag{3-8}$$

into values of a standardized variable (or factor), z.

This generally leads (for 2^n designs) to low levels at "−1" and high levels at "+1". Therefore in the future one can use one and the same scheme of coded measurements independent of the real values of the n factors, x, under study.

The regression model then becomes:

$$y = a_0 + a_1 z_1 + a_2 z_2 + a_{12} z_1 z_2 \tag{3-9}$$

With regard to the number of standardized variables we shall speak of the **core of a design** which means that the combinations of the values in the respective columns, z, determine the set-up of the particular experiments.

Example 3-1 (continued)

For factor no. 1, 'depth' we calculate $hr = (20 − 10)/2 = 5$ and $cv = (20 + 10)/2 = 15$ and hence $x_1 = 10$ is transformed into $z_1 = (10 − 15)/5 = −1$ and $x_2 = 20$ is transformed into $z_2 = (20 − 15)/5 = +1$. The same procedure applied to factor no. 2, 'location', the core of the design for gross modeling of Cd distribution then becomes:

Measurement no.	Factor z_1 ('depth')	Factor z_2 ('location')
1	−1	−1
2	−1	+1
3	+1	−1
4	+1	+1

So, with the general scheme in the following example we describe the core design.

3.3.4 Constructing the General Computation Scheme

From the core design we are now able to construct a **general computation scheme** in which we omit all the "1" and introduce a **common factor**, z_0, as well as the interaction of pairs of factors, e.g. $z_1 z_2$. The factor z_0 is assumed to be present independently of the variation of the 'planned' factors z_1 and z_2. It may, e.g., be considered as a blank value or a basic level of contamination. The interaction of factors is expected to be important in antagonistic or synergistic effects.

Example 3-1 (continued)

For z_0 we simply set +1 for each measurement; this means that the overall factor gives each measurement the constant "weight" "+1". To obtain the interaction column we multiply the values (+1 or −1) of the respective factor columns:

Measurement no.	z_0 ('common')	z_1 ('depth')	z_2 ('location')	$z_1 z_2$ ('interaction')	y (response)
1	+	−	−	+	y_1
2	+	−	+	−	y_2
3	+	+	−	−	y_3
4	+	+	+	+	y_4

3.3.5 Extending for More Factors

If necessary one can easily extend the scheme by an additional factor. In principle, all the previous designed measurements must now be repeated with each level of the new factor. But one should not forget to randomize the order of experiments. In this way one quickly constructs a complete 2^{n+1} factorial design.

When interaction of factors z_1 and z_2 is negligible it is even possible to create an incomplete or fractional factorial design 2^{n-1}. Obviously this saves $2^2 = 4$ measurements or experiments, the "drawback" is, as already mentioned, that the estimate of the major effect of z_3 is confounded with the (minor) effect of interaction of z_1 and z_2.

Example 3-1 (continued)

If there are significant interactions between z_1 and z_2 the core of the design is extended as follows:

Measurement no.	z_1	z_2	z_3
1	−	−	−
2	−	+	−
3	+	−	−
4	+	+	−
5	−	−	+
6	−	+	+
7	+	−	+
8	+	+	+

If the interaction $z_1 z_2$ can be assumed to be, or is tested to be, negligible, the core of the design is identical to the core in the example above:

Measurement no.	z_1	z_2	$z_3 \, (= z_1 z_2)$
1	−	−	+
2	−	+	−
3	+	−	−
4	+	+	+

3.3.6 Computing the Main Effects

The next step is the very easy computation of the main effects of each factor, the two-factor interaction, and the model parameters of the possible regression function.

In the case of 2^n this is extremely simple: For each column, z, in the computational scheme one has to multiply the signs with the respective response, to sum up these values and to average the sums. The average values are the estimates of the wanted **regression coefficients**.

$$\hat{a}_j = \frac{\sum_{i=1}^{k} z_{ji} y_i}{k} \tag{3-10}$$

j is the index of the respective entry, z, in the computational scheme.

Leaving out the z_0 column, these coefficients multiplied by 2 immediately yield the **effects** in the statistical or variance analytical sense. This difference in the computation of the regression coefficients and the effects comes from the different models used in regression analysis and in analysis of variance. The first refers to the zero or medium level, the second refers to the lowest level.

Without jumping into regression analysis the main and interaction effects are also easily computed directly from responses. The main effect of each factor is the difference between the mean values of the responses obtained at the highest and lowest levels of the factor. The interaction effect is the mean of the difference between the effects of one factor at the highest level and the other factor at the lowest level.

Example 3-1 (continued)

Regression coefficients and effects

Measurement no.	$z_0 y$	$z_1 y$	$z_2 y$	$z_1 z_2 y$	Response y
1	$+y_1$	$-y_1$	$-y_1$	$+y_1$	$y_1 = 12$
2	$+y_2$	$-y_2$	$+y_2$	$-y_2$	$y_2 = 16$
3	$+y_3$	$+y_3$	$-y_3$	$-y_3$	$y_3 = 60$
4	$+y_4$	$+y_4$	$+y_4$	$+y_4$	$y_4 = 58$
column sums	146	90	2	-6	146
column sums/4	36.5	22.5	0.5	-1.5	

estimates the coefficients
of the model: $y = \hat{a}_0 \ + \ \hat{a}_1 z_1 \ + \ \hat{a}_2 z_2 \ + \ \hat{a}_{12} z_1 z_2$,
so we obtain: $y = 36.5 \ + \ 22.5 z_1 \ + \ 0.5 z_2 \ - \ 1.5 z_1 z_2$

which describes the dependence of soil Cd content as a function of depth and location.

The effects in the sense of analysis of variance are:

36.5   45   1 -3

(N.B.: The effect of the overall factor, z_0, is equal to the coefficient in the regression model.)

Effects directly from the responses

main effect of z_1:	$(60 + 58)/2 - (12 + 16)/2 = 45$
main effect of z_2:	$(16 + 58)/2 - (12 + 60)/2 = 1$
interaction effect of $z_1 z_2$:	$[(58 - 60) - (16 - 12)]/2 = -3$ or
	$[(58 - 16) - (60 - 12)]/2 = -3$

3.3.7 Testing the Model Adequacy and the Coefficients

If one wants to test the adequacy of the model and the effects or coefficients, replicates of the entire design or measurements at a central point of the design are necessary. The real position of the central point is easily obtained by the use of Eq. 3-8 which in our simple case immediately leads to x_i = central value.

Model Adequacy

There is only one way to circumvent additional experiments, namely if the experimental variance is known *a priori* as σ^2 and the number of experiments k exceeds the number of estimated coefficients n_a in the model.

In that case the test of adequacy is performed with a χ^2-test:

$$\chi^2_{exp} = \frac{(k - n_a) s^2_{model}}{\sigma^2} \quad \text{versus} \quad \chi^2_{crit}(k - n_a; q = 1 - \alpha) \tag{3-11}$$

n_a – number of estimated coefficients a_j (see Eq. 3-10)
k – number of experiments

with

$$s^2_{model} = \frac{\sum_{i=1}^{k} (\hat{y}_i - y_i)^2}{k - n_a} \tag{3-12}$$

To test adequacy of the model we can utilize the test procedure outlined in Section 2.4.10; for a practical example, see Section 10.2.4.

In every case again the decision is: if the experimental test value is greater than or equal to the critical value one is advised to use models with, e.g., quadratic terms, i.e. higher order polynomials.

In **simple linear models** (first order polynomials) the adequacy can be **tested by** another **simple procedure**. A *t*-test is performed between the mean result, $\bar{y}_0$, from the experiments on the central point and the mean, $\bar{y}$, of response values from factorial design:

$$t_{exp} = \frac{|\bar{y} - \bar{y}_0|}{s(\bar{y}_0)} \quad \text{versus} \quad t_{crit}\left(n_0 - 1; \ q = 1 - \frac{\alpha}{2}\right) \tag{3-13}$$

Instead of performing central point experiments the general recommendation is, however, to replicate the total design.

Coefficients

The last mentioned test is also useful for testing each coefficient:

$$t_{exp} = \frac{|\hat{a}_j|}{s(\bar{y}_0)} \quad \text{versus} \quad t_{crit}\left(n_0 - 1; \ q = 1 - \frac{\alpha}{2}\right) \tag{3-14}$$

Here the t-test is performed to test the difference between a_j and zero.

Example 3-1 (continued)

Adequacy

Assume that we have made $n_0 = 4$ additional measurements at the central point of the design, at level zero (0) of each factor with $y_0 = \{34, 38, 32, 36\}$. Then $\bar{y}_0 = 35$ results ($s(\bar{y}_0) = 2.58/2$), and with $\bar{y} = 36.5$ the difference of 1.5 in $t_{exp} = 1.5/1.29 = 1.16$ is to be tested against

$t_{crit}(4 - 1; 0.975) = 3.18$ for the two-sided test

Because the experimental t is less than the critical t at the 5% level of significance there is no need for quadratic fitting of the results.

Coefficients

Using $s(\bar{y}_0) = 1.29$ as an estimate of each $s(\hat{a}_j)$ and coefficients from the model

$y = 36.5 + 22.5 \cdot z_1 + 0.5 \cdot z_2 - 1.5 \cdot z_1 z_2$
and $t_{crit}(3; 0.975) = 3.18$ we see for
coefficient $\hat{a}_1$: $t_{exp} = 22.5/1.29 = 17.43$ > than t_{crit}
coefficient $\hat{a}_2$: $t_{exp} = 0.5/1.29 = 0.39$ < than t_{crit}
coefficient $\hat{a}_{12}$: $t_{exp} = -1.5/1.29 = -1.16$ < than t_{crit}

and we decide at the 5% level of significance that only z_1, the sampling depth, is a significant factor.

Effects

Because there are no replicates of the entire design we skip the tests.

3.3.8 Interpreting the Results

After testing the coefficients the remaining model can be interpreted in terms of the influence or significance of the respective factors, z_j, seen as the independent variables in the regression model. If one wants to see the factors as effects on the dependent variable, the coefficients must be multiplied by 2.

Example 3-1 (continued)

From the test of the coefficients in the model function

$$y = 36.5 + 22.5 \cdot z_1 + 0.5 \cdot z_2 - 1.5 \cdot z_1 z_2$$

all that remains is

$$y = 36.5 + 22.5 \cdot z_1$$

which shows that there was no horizontal spread of the contamination ('location factor', z_2). There is probably no substantial interaction between depth and location.
The effect of z_1 is to double 22.5. That gives 45. An enrichment in depth can be assumed: Cd content increases by 45 (%) irrespective of the sampling location.

If interpolation of the values of the target variable y is necessary, for example to find better estimates of the model error (for tests of adequacy), the variables, z, in Eq. 3-9 must be retransformed using Eqs. 3-6 to 3-8.

Again, notice the possible conflict between the **statistical significance** and the **practical relevance** of the models. In general, the models derived so far should only be thought of as rough approximations of reality. In any case it is better to acquire as many data as possible, or at least as many as necessary, and to perform a complete regression analysis or analysis of variance (see Section 3.3.9). In the first case it is, nevertheless, advantageous to follow a statistical design, for example to avoid multicollinearity, i.e. the linear dependence of the independent variables.

Because introducing the reader to actual optimization techniques is beyond the scope of this book, let us only indicate here that with the model obtained the analyst or technician is able to find the optimum value of y by partially differentiating the regression function. Setting each differential to zero he or she he may find the optimum values of the single variables. Substitution of these values into the model equation will yield the optimum value of y.

Example 3-1 (continued)

Although it does not make much sense to 'optimize' the Cd content in the current example and to find the 'optimum' location and depth let us briefly demontrate the procedure.

The first derivative of the complete model function

$$y = 36.5 + 22.5 \cdot z_1 + 0.5 \cdot z_2 - 1.5 \cdot z_1 z_2$$

must be set to zero. With $\partial y/\partial z_1 = 22.5 - 1.5 z_2$ one calculates $z_{2,opt} = 1/15$ and similarly $z_{1,opt} = 1/3$; inserting both values into the model yields $y(z_{1,opt}, z_{2,opt}) = 44$ mg kg^{-1} Cd.
In a real experiment, where 'location' and 'depth' could have been tunable features for optimizing the 'yield of Cd'; this may be the value of Cd expected at the settings $x_1 = 16.7$ cm and $x_2 = 42$ m.

For linear functions of the first order (without interaction terms) the vector of the coefficients forms the gradient of the estimated model. Following the steps indicated by the coefficients one will reach the optimum in the steepest ascent mode.

3.3.9 Analysis of Variance

Finally keep in mind that analysis of variance (ANOVA) is the most powerful statistical technique for evaluating the results of factorial designs with replications if the significance of factors is of interest, rather than the models of their relationship.

To look at the experimental design from the point of view of ANOVA we invert the scheme used so far. For the two-factorial case we can build Tab. 3-2.

Tab. 3-2. Data scheme for use in ANOVA

		Factor 2 ('location') Level **1**	Factor 2 ('location') Level **2**
Factor 1 ('depth')	Level **1**	*results of* *cell* (**1, 1**)	*results of* *cell* (**1, 2**)
Factor 1 ('depth')	Level **2**	*results of* *cell* (**2, 1**)	*results of* *cell* (**2, 2**)

3.3 Example: Set-up and Interpretation Steps of Factorial Designs

As we may remember from Sections 2.3 and 2.4.10, the ANOVA technique is useful in cases where the number of results in each cell is *different* (but see below!). This may happen sometimes when single experiments fail or, in environmental analysis, when some samples are exhausted more quickly than others or when sampling fails. We also recognize ANOVA to be a valuable technique for the evaluation of data from planned (designed) environmental analysis. In this context the principle of ANOVA is to subdivide the total variation of the data of all cells, or factor combinations, into "meaningful component parts associated with specific sources of variation for the purpose of testing some hypothesis on the parameters of the model or estimating variance components" (ISO 3534/3 in [ISO STANDARDS HANDBOOK, 1989]).

In addition it is now time to think about the two assumption models, or types of analysis of variance. **ANOVA type 1** assumes that all levels of the factors are included in the analysis and are fixed (**fixed effect model**). Then the analysis is essentially interested in comparing mean values, i.e. to test the significance of an effect. **ANOVA type 2** assumes that the included levels of the factors are selected at random from the distribution of levels (**random effect model**). Here the final aim is to estimate the variance components, i.e. the variance fractions with respect to total variance caused by the samples taken or the measurements made. In that case one is well advised to ensure balanced designs, i.e. equally occupied cells in the above scheme, because only then is the estimation process straightforward.

Example 3-1 (continued)

Assume that we have sampled soil four times at each field location given by the factor combinations, with the following result:

		Factor 2 ('location')	
		Level 1	Level 2
	Level 1	10	16
		14	17
		12	17
		12	18
Factor 1 ('depth')	Level 2	60	59
		60	59
		60	58
		59	60

At this stage we shall use appropriate software on a PC, rather than a pocket calculator. From the results above one easily obtains the effects: 44.9, 2.1, and −2.9.

In the following ANOVA table we report rounded values, it is, therefore, not possible to reproduce the F-values exactly as ratios of the mean sums of the respective line and the residual mean sum of squares (according to Section 2.3):

Line no.	Source of variation	Sum of squares	f	Mean sum*	F_{exp}	α_{calc} in %
1	'depth'	8055	1	8055	7581	0.0
2	'location'	18.1	1	18.1	17.0	0.1
3	'interaction'	33.1	1	33.1	31.1	0.0
4	'residual'	12.8	12	1.1		

* Sum of squares divided by f.

The surprising outcome is in contrast with the model (Example 3-1) derived from four samples (one run of the design)! From the computed levels of significance, which are always less than 5% (our usual threshold) we have to conclude that all main effects and the interaction effect are significant sources of variation of the Cd content.

Because all sources of variation are significant one could now proceed to calculate their quantitative contribution to the total variation.

3.3.10 Analysis of Covariance

Analysis of covariance is a technique for estimating and testing the effects of certain variables when so-called **covariables** are present; the technique has, unfortunately not yet attracted the interest it merits.

In environmental studies in particular, very often one is not able to adjust these concomitant variables but only to measure them. Typical situations occur during sampling. In principle all sampling circumstances are concomitant variables, for example pH, temperature, salinity in water sampling, or humidity in air sampling, redox potential, and particle size distribution in soil sampling.

If the undesigned effect of these covariables is not taken into account, the results of analysis of variance may be biased and serious misinterpretation is possible.

In analysis of covariance the influence of the covariable(s) is basically corrected for by means of a regression model with the covariable(s) as the independent variable(s). Hence analysis of covariance appears as a combination of both regression analysis and analysis of variance.

3.3 Example: Set-up and Interpretation Steps of Factorial Designs

It is beyond the scope of this introduction to give further mathematical details; see, e.g., [HOCHSTÄDTER and KAISER, 1988]. Let us rely on the developers of dedicated software ...

Again, let us stress our previous example, which now includes a possible covariable.

Example 3-1 (continued)

Assume that together with the Cd content we have determined the moisture content by weighing the samples before and after infrared drying. The % moisture of the samples is given in parentheses:

		Factor 2 ('location')	
		Level 1	Level 2
	Level 1	10 (**38**)	16 (**42**)
		14 (**40**)	17 (**43**)
		12 (**38**)	17 (**43**)
		12 (**37**)	18 (**43**)
Factor 1: ('depth')	Level 2	60 (**80**)	59 (**79**)
		60 (**82**)	59 (**80**)
		60 (**80**)	58 (**78**)
		59 (**80**)	60 (**79**)

The results from analysis of covariance are summarized in a pattern identical to the ANOVA output:

Line no.	Source of variation	Sum of squares	f	Mean sum	F_{exp}	α_{calc} in %
1	'depth'	2.88	1	2.881	3.63	8.3
2	'location'	3.08	1	3.083	3.89	7.4
3	'interaction'	0.94	1	0.935	1.18	30.1
4	'residual'	8.73	12	0.728		

Suprisingly enough, we now obtain the result that all main effects and the interaction effect may be neglected (all error probabilities P computed for the respective F-values are greater than 5%)! The result is not too surprising if one reviews the numerical values: the moisture must have an influence, because all samples near the surface are 'dry' and the samples from the deeper layer are 'wet'.

The analysis clearly reveals that the covariable dominates the influence of depth and location.

3.4 From Factorial to Sequential Designs

In environmental analysis we shall extremely seldom expect questions to be answered by methods of sequential design. It could, however, be possible that one is, for example, asked to locate the maximum contamination within a certain area, so that the decontamination process can start there. In the following discussion we will, therefore, provide only a brief summary of some optimization methods which could be useful.

In order to link the section on exactly planned experiments and the section on 'pure' sequential methods let us roughly describe the method of BOX and WILSON [1951] with two factors under study.

Imagine that the two factors are, for example, spatial directions along and perpendicular to a road, see Fig. 3-2. In addition, assume that the origin of the area spanned by the two factors is approximately where the road crosses another road. In a 2^2 factorial design the lowest levels of both factors are determined by the location of the crossroads. Let us 'place' the four points of the starting 2^2 design at the corners of a square region near this crossing and measure BTEX (benzene, toluene, ethylbenzene, xylenes) concentrations there.

According to BOX and WILSON the next steps could be:
(1) From the measurement results (in Fig. 3-2: y_1, y_2, y_3, and y_4) of the complete design derive the regression coefficients for a first order polynomial.
(2) Starting from the central point of the design (the center of the square) **sequentially** step forward in proportion to these coefficients until a so-called quasi- or near-stationary region is reached, i.e. no further significant effects occur.
(3) At that location it may be necessary to fix the positions of another design and to acquire missing data to find a higher order polynomial and to build the final model of the near-optimum region.

(Usually the 'optimum' lies on a convex or concave region of the response surface which can be approximated by functions with quadratic or cubic terms.)

If the borders of the experimental space are reached, for example a river parallel to road no. 1, it is normally sufficient to set-up a first order polynomial there and to follow the possibly changed direction of its gradient. Remember that the gradient of a first order polynomial is simply constructed from the linear coefficients of the polynomial.

The conceptual transition to fully sequential methods is now easy to understand: why step forward in a fixed direction for possibly a long time? One possibility could be, in our example, to start from only three points of a triangle. After measuring the desired quantity (in Fig. 3-2: y_5, y_6, and y_7) and comparing the results one could discard the worst result (in Fig. 3-2: y_7) and step forward in the direction opposite to that of the respective point (i.e. that which gave the worst result). The new result (in Fig. 3-2: y_8) may then be compared with the two former results (in Fig. 3-2: y_5 and y_6). Then again the worst result can be discarded (in Fig. 3-2: y_6) and one can proceed forward in the other direction. This is, in short, the principle of **simplex methods**, see next Section. Step by step the simplex will move in the direction of the optimum. The name 'simplex' comes from geometry where any figure is called a simplex if it has one point more than the dimension of the space in which it is defined.

3.5 Sequential Methods for Optimum Search

Let us restrict ourselves to putting the numerous variants of sequential optimum search into some sort of order. Most of the methods mentioned in the following review are described by BUNDAY [1984a], who also gives BASIC programs; details of the simplex method and its programming may be found in BUNDAY [1984b].

If we expect optima as functions of *one variable*, i.e. one independent variable or one factor one can utilize:

- the FIBONACCI search
  Here it is advantageous that the number of experiments can be determined before the work starts. The method is adequate for unimodal response functions.
- the golden section search
  The number of experiments may be undetermined.
- the uniplex method
  This is a useful method for open intervals.
- the method of POWELL
  The method uses quadratic interpolation.
- the method of DAVIDON
  This method uses cubic interpolation.

If we expect optima as functions of *several variables* we can use:

- direct search methods
  - searches parallel to the axes
  - the HOOKE-JEEVES method
  - simplex methods (regular or modified simplex, e.g. according to NELDER and MEAD [1965])

- gradient methods
  - steepest descent/ascent (remember the BOX-WILSON method as an example)
  - quadratic functions (NEWTON and POWELL)
  - the DAVIDON-FLETCHER-POWELL method
  - the FLETCHER-REEVES method

If we can or must impose certain restrictions during optimization we may select:

- search methods
  - the modified HOOKE-JEEVES method
  - the BOX complex-method

On the other hand, conversion of restricted into free methods is possible, for example by use of

- penalty functions
- barrier functions (SUMT method, FIACCO and McCORMICK)

By far the most popular technique is based on **simplex methods**. Since its development around 1940 by DANTZIG [1951] the simplex method has been widely used and continually modified. BOX and WILSON [1951] introduced the method in experimental optimization. Currently the 'modified simplex method' by NELDER and MEAD [1965], based on the simplex method of SPENDLEY et al. [1962], is recognized as a standard technique. In analytical chemistry other modifications are known, e.g. the 'super modified simplex' [ROUTH et al., 1977], the 'controlled weighted centroid', the 'orthogonal jump weighted centroid' [RYAN et al., 1980], and the 'modified super modified simplex' [VAN DER WIEL et al., 1983]. CAVE [1986] dealt with boundary conditions which may, in practice, limit optimization procedures.

Both the development and the optimization of simplex methods are still continuing. Several functions have been designed to test the performance of the simplex algorithms, one example is the famous ROSENBROCK valley. Other test functions have been reported by ABERG and GUSTAVSSON [1982]. Most analytical applications of simplex optimization are found in atomic spectroscopy [SNEDDON, 1990] and chromatography [BERRIDGE, 1990].

3.6 Factorial or Sequential Methods?

The primary decision on which family of methods to consider can be based on the answers to two questions:

(1) Do I really need the functional relationship, the modeling of the dependency of the signal, target, response variable, ... on influencing variables, factors, ...?
(2) Am I primarily interested in finding an optimum response and the respective conditions?

In the *first case* (1) and when the user wants to ensure optimum conditions for regression modeling and analysis of variance, the family of methods comprising statistical, experimental, and factorial designs should be consulted. In the *second case* (2) an appropriate procedure from the sequential method family should be selected.

Some general hints may further assist the decision on which family to select. Certainly each single method will have its own special usefulness or its own prerequisites or conditions of application.

Factorial designs

1 the number of experiments is known in advance
2 the techniques are well suited for discontinuous (discretely adjustable) variables
3 an empirical (polynomial) model can be derived
4 the statistical significance of the parameters (variables) can be tested
5 the optimum may be calculated by differentiation of the derived model functions

Simplex methods and other search algorithms

1 the number of experiments is usually not fixed or known in advance
2 the techniques are only suited for continuously adjustable variables
3 no explicit functional relationship is available
4 the methods give the optimum without providing information about the importance of the factors
5 the global optimum may sometimes not be found (holds true even for explicit functions subjected to search algorithms)

Drawback 3 may be overcome by setting up a factorial design around the optimum value.

References

Aberg, E.R., Gustavsson, A.G.T.: Anal. Chim. Acta 144 (**1982**) 39
Bandemer, H., Bellmann, A., Jung, W., Richter, K.: Optimale Versuchsplanung, Akademie-Verlag, Berlin, **1973**
Berridge, J.C.: in: Glajch, J.L., Snyder, L.R. (Eds.): Computer-Assisted Method Development for HPLC, Elsevier, Amsterdam, **1990**, p. 3
Box, G.E.P., Behnken, D.W.: Ann. Math. Stat. 31 (**1960**) 838
Box, G.E.P., Wilson, K.B.: Roy. Stat. Soc. Ser. B 13 (**1951**) 1
Bunday, B.D.: Basic Optimization Methods, Edward Arnold, London, **1984 a**
Bunday, B.D.: Basic Linear Programming, Edward Arnold, London, **1984 b**
Cave, M.R.: Anal. Chim. Acta 181 (**1986**) 107
Dantzig, G.B.: Rand Corp. Rept., P-891, **1956**
Deming, S.N., Morgan, S.L.: Experimental design: A Chemometric Approach, 2nd Ed., Elsevier, Amsterdam, **1993**

Hartley, H.O.: Biometrics 15 (**1959**) 611
Hochstädter, D., Kaiser, U.: Varianz- und Kovarianzanalyse, Verlag Harri Deutsch, Frankfurt am Main, Thun, **1988**
ISO Standards Handbook 3, Statistical Methods, 3rd Ed., ISO, Genova, **1989**, p. 23 and p. 292
Morgan, E.: Chemometrics: Experimental Design, Wiley, Chichester, **1991**
Nelder, J.A., Mead, R.: Comp. J. 7 (**1965**) 308
Routh, M.W., Swartz, P.A., Denton, M.B.: Anal. Chem. 49 (**1977**) 1422
Ryan, P.B., Barr, R.L., Todd, H.D.: Anal. Chem. 52 (**1980**) 1460
Sneddon, J.: Spectr. Int. 2 (**1990**) 42
Spendley, W., Hext, G.R., Himsworth, F.R.: Technometrics 4 (**1962**) 441
Steinberg, D.M., Hunter, W.G.: Technometrics 26 (**1984**) 71
van der Wiel, P.F., Maassen, R., Kateman, G.: Anal. Chim. Acta 153 (**1983**) 83
Westlake, W.I.: Biometrics 21 (**1965**) 324

4 Sampling and Sampling Design

4.1 Introduction

The quality and utility of analytical data are usually highly dependent on the suitability of the sample and the adequacy of the sampling program. Sampling is not only the first, but also a very important and complex step in the analytical process.

Both the sample and the whole sampling program are the basis for answering the specific analytical question. This means the samples must represent the quality of the whole object under investigation, the so-called parent population defining the specific task or problem.

In other words, the samples have to reflect, without distortion, the piece of information required from the population. Otherwise, the conclusions from the analytical data – the output of the analytical laboratory – about the state of the investigated object are definitely arbitrary and may cause momentous errors in the interpretation of results. Samples of a population must, therefore, be representative according to the specific query. This means that they must be both accurate and reproducible. This implies that the sampling process is affected by errors in each case. The question of representativeness is a question which has to be answered for each individual case in relation to the heterogeneity of the population to be sampled, the accuracy required, and the reproducibility of results. The extent of representativeness is, therefore, highly dependent on the expenditure on sampling and analysis and the time needed for the investigation.

The sources of sampling error are constitutional error, distributional error, and error caused by use of the incorrect sampling technique [GARFIELD, 1989; MINKINNEN, 1987].

The environment and its compartments are often characterized by a high degree of inhomogeneity and movement. Some examples which demonstrate the large spatial and time-dependent variation in environmental media are represented by the case studies discussed in Chapters 7–9.

In each case, the investigation of a specific environmental problem requires a problem-adapted strategy and a sampling process design. Thereby, aspects of chemical, physical, and biological features of the environment and problems concerning analytical chemistry and statistical aspects both have to be considered. Regulatory aspects and practical considerations have a considerable influence on the sampling process. The various **purposes of sampling** such as:

- quality control in the sense of estimating the extent of pollution,
- prognosis in the sense of forecasting, and
- the estimation of damage

render the sampling process very complex and difficult.

Because it is obvious that optimum sampling must always be a compromise between statistical requirements, economic aspects, and empirical knowledge, only the constructive cooperation of the environmental specialist, the analytical chemist, and the chemometrician can substantially improve the results of sampling and analytical programs.

The **planning of representative sampling** is based on the methodology of GREEN [1979]:

- Samples must be replicated within each combination of time, location, or other variables of interest.
- An equal number of randomly allocated replicate samples should be taken.
- The samples should be collected in the presence and absence of conditions of interest in order to test the possible effect of the conditions.
- Preliminary sampling provides the basis for the evaluation of sampling design and options for statistical analysis.
- The efficiency and adequacy of the sampling device or method over the range of conditions must be verified.
- Proportional focusing of homogeneous subareas or subspaces is necessary if the whole sampling area has high variability as a result of the environment within the features of interest.
- The sample unit size (representative of the size of the population), densities, and spatial distribution of the objects being sampled must be verified.
- The data must be tested in order to establish the nature of error variation to enable decision on whether the transformation of the data, the utilization of distribution-free statistical analysis procedures, or the test against simulated zero-hypothesis data is necessary.

Considering these aspects, an **optimized sampling program** has to take the following aspects into account [HOFFMANN, 1992]:

- place, location, and position of sampling
- size, quantity, and volume of the sample
- number of samples
- date, duration, and frequency of sampling
- homogeneity of the sample
- contamination of the sample
- decontamination of the sample
- sample conservation and storage

The Committee on Environmental Improvement of the American Chemical Society [1980] suggests the following minimum requirements for an **acceptable sampling program** [BARCELONA, 1988]:

- proper statistical design that takes into account the goals of the study and its certainties and uncertainties
- instructions regarding sample collection, preservation of labeling, and transport to the analytical laboratory
- training of personnel in the sampling techniques and the procedures specified

To an extent in accordance with the title and goal of this book, some fundamentals of the sampling process are discussed in more detail below. The reader should not, however, forget that the sampling process is very complex and, therefore, that environmental aspects, the analytical chemistry to be used, and mathematical statistics must be considered simultaneously.

It is remarkable that at the present level of knowledge legally fixed procedures for environmental investigations provide much more information on sampling techniques and preparation than on the statistical description of the sampling process (see for example the different ISO norms referred to at the end of this chapter, or [DIN 4021, 1990]). In the scientific literature also description of the technical and analytical handling of the sampling process predominates (see for example [STOEPPLER, 1994; MARKERT, 1994]). A good overview on sampling in chemical analysis is given in the literature [KRATOCHVIL and TAYLOR, 1981; KRATOCHVIL et al., 1984; GARFIELD, 1989] and an overview on sampling in environmental analysis is available from KRATOCHVIL et al. [1986], KEITH [1988], and BARNARD [1995]. Statistical aspects of the sampling process are described in more detail by KATEMAN and PIJPERS [1981; 1988], KATEMAN [1987], KRISHNAIAH and RAO [1988], GY [1982; 1986; 1990; 1991a; 1991b; 1992], and PITARD [1989]. Aspects of quality concepts and practices applied to sampling are discussed by THOMPSON and RAMSEY [1995]. An overview on statistical aspects of the planning of the sampling process for polluted soils is given by NOTHBAUM et al. [1994]. A collection of both practical and theoretical aspects of sampling for environmental trace analysis has been published by MARKERT [1994].

The most important norms dealing with the fundamentals of sampling and sampling design have been fixed by the International Organization of Standardization in Geneva (see for example [ISO 1213, Part 2, 1992; ISO/DIS 5667, Part 1–13, 1991; ISO 9359, 1989; ISO/DIS 10381, Part 1–6, 1995; ISO 11074, Part 2, in prep.; ISO/DIS 11464, 1995; ISO/DIS 14507, 1995]. Some of these norms are in the process of revision and updating.

A detailed overview on references for sampling by HANNAPPEL [1994] contains not only references for general, statistical, and detailed aspects, but also for norms of the International Organization of Standardization in Geneva.

4.2 Basic Considerations

At the beginning of this passage some definitions concerning the term **sample** are necessary. According to the IUPAC Commission on Analytical Nomenclature [HORWITZ, 1990] the general definition is: "The sample is the actual material investigated, whether diluted or undiluted." The sample is a portion of material, selected in a specialized manner, which represents a larger body of material. The analytical result obtained from the sample should be a distortion-free estimate of the quantity or concentration of a constitutent or property of the parent material. The parent material, also known as the lot, may be homogeneously or – substantially more frequently in the environment – heterogeneously distributed. The quality or the quantity of the parent material may be constant in space and/or time (static conditions) or may change dynamically (dynamic conditions).

Each sample contains an uncertainty – the sampling error – arising from the heterogeneity of the parent material. The potential hazard resulting from extrapolation from the smaller portion (laboratory sample) to the larger portion (parent population) becomes obvious with the aid of the following theoretical example:

The influence of emission of some airborne pollutants on an agricultural area has to be investigated. From an area of 10 000 square meters, ten soil samples were taken from the surface soil horizon. 1 g of each sample was recently analyzed in the laboratory. This means that the results, for instance a pollutant concentration, obtained from ten 1 g samples are used to estimate the level of pollution in the whole area with a medium depth of the surface horizon of 30 cm and a medium soil density of 1.5 g cm^{-3}. A parent population of more than 10^9 g has to be assessed from a total sample mass of 10 g! The necessity of extrapolation over a range of several orders of magnitude indicates the problems connected with sampling process.

This theoretical example may serve to illustrate the importance of preparation and implementation of the sampling to both the accuracy and the reproducibility of the analytical results reflecting the environmental problem. It is, furthermore, obvious that the sampling process must be adapted and optimized for each separate case. This requires a very sensitive balance between environmental knowledge and experience on the one hand and statistical requirements on the other.

This section not only introduces some statistical principles important to sampling, but also considers their connection with our empirical knowledge of the environment. It is, for example, obvious that a container of a few kilograms of arsenic located in an unknown position in a deposit or an old mine dump with a volume of tens of thousands of cubic meters cannot, owing to economic considerations, be found with realistic statistical probability. In this case human experience or empirical knowledge are necessary.

Fig. 4-1 shows the different steps of sample preparation, starting with the parent population, according to the IUPAC, Analytical Chemistry Division, Commission on Analytical Nomenclature [HORWITZ, 1990].

The **sampling plan** is a predetermined procedure for the selection, withdrawal, preservation, transport, and preparation of the portions removed from a parent population as a sample. Each sample contains a **sampling error**. It is defined as a part of total error associated with the use of only a small part of the parent population to extrapolate to the whole population, as distinct from the analytical error. It arises from the degree of inhomogeneity in the population. The inhomogeneity (or heterogeneity) is the extent to which a property or a quality is distributed nonuniformly throughout the quantity of the material or the compartment to be investigated.

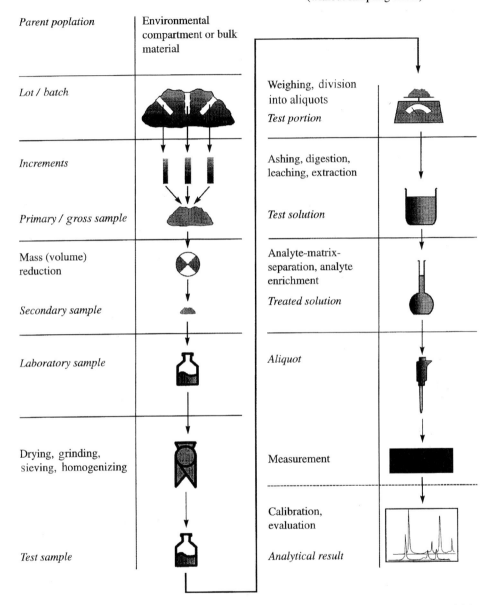

Fig. 4-1. Terms used in sampling operations according to the IUPAC-Analytical Chemistry Division Commission on Analytical Nomenclature [HORWITZ, 1990]

The term **specimen** is of particular importance in environmental investigations. It is a specifically selected portion of a material taken from a dynamic system and assumed (on the basis of *a priori* knowledge or experience of the environment) to be representative of the parent population at the time it is taken and at the site from which it is taken. Even when the sampling error cannot be determined, taking a specimen is sometimes the only way of obtaining information on the state of pollution of the environment. The strongly time-dependent discharge of waste water from small factories is a typical example.

A further difficulty of sampling in the environment must be taken into consideration: in many queries, the parent population, e.g. the volume of polluted air over a town, can be defined only very subjectively.

The main areas of environmental investigations are description, monitoring, and control. These can be achieved by probability, nonprobability, or bulk sampling. On the basis of the models used in their derivations sampling strategies approach the underlying space-time reality of the part of environment investigated in different ways. Monitoring includes investigation of spatial and/or temporal changes in an environmental compartment, or in other words, its dynamic character. This monitoring can be seen as the first step of process control. In controlling the pollution or loading state of an environmental medium (field survey) and comparing the results obtained with thresholds or legally binding limits (threshold infringement), more static and averaging aspects are described. From the viewpoint of common analytical chemistry, it corresponds more to bulk analysis. As introduced by KATEMAN [1987], an object can be described either globally (gross description) or in more detail.

The main types of sample must be described briefly:
– **Random samples** have to be selected in such a manner that any portion of the population has an equal (or known) chance of being chosen. But random sampling is, in reality, quite difficult. A sample selected haphazardly is not a random sample. Thus, random samples have to be obtained by using a random sampling process (for instance with random number generation for specimen selection). The samples must reflect the parent population on the basis of an equal probability distribution.
– **Systematic samples** are taken in a systematic manner to monitor or test a systematic hypothesis, such as the dependence of changes in composition on space or time. The algorithms used for systematic sampling are described in more detail in Section 4.5.3.
– **Stratified samples** consist of portions obtained from identified subparts (strata) of the parent population. Within each stratum, the samples are taken randomly.
– **Composite samples** are, as the name suggests, combinations of individual samples which together represent the average dependence of the composition on the mass or volume stream or the time, i.e. they represent a bulk average. The purpose of such sampling is to produce one representative sample for the whole population investigated or a subpart of it. Sometimes a composite sample is obtained by a division or reduction process (see also Fig. 4-1). A composite sample should contain the same

information as the collection of the individual samples, i.e. the mean values and the variance of the analytes of interest.

In soil pollution studies, the use of composite samples produces an average over a definite area or space. In air or water pollution studies, composite sampling is usually a means of integrating over a longer time period. Flow-proportional composite sampling is often applied, especially in river water analysis.

– A **subsample** may be a portion of the sample obtained by selection or division. It is an individual unit of the lot taken as a part of the sample or the final unit of multistage sampling.

More detailed information on the nomenclature of sample types is given by HORWITZ [1990].

The definition of the term **representative sample** is given in Section 4.1. In each case the representativeness should be regarded as being defined for the specific purpose. In environmental investigations it is impossible in practice to obtain one representative sample, except for the case of a composite sample for bulk analysis. It is, however, possible to obtain representative samples with a properly designed and optimized sampling plan (see Section 4.5.3).

4.3 Theoretical Aspects of Sampling

Because of the economic implications of sampling problems in industry, many investigations have been carried out in order to investigate the variability of the analyte in a material to be sampled. In oil and mining research, in particular, sampling theories have been developed for solving problems which are quite similar to environmental problems [KRATOCHVIL et al., 1986]. The majority of environmental compartments to be investigated must be regarded as heterogeneous masses as, for example, soils and sewage sludges, the water body in rivers, or the atmosphere over a factory. The sampling of heterogeneous masses must, therefore, be discussed.

4.3.1 Number of Individual Samples Required

A normally distributed parent population X is characterized by its expected mean value μ and standard deviation σ. If the samples taken from this population are representative, the sample average $\bar{x}$ is an unbiased estimate of μ:

$$\bar{x} = \frac{\sum_{i=1}^{n} x_i}{n} \qquad (4\text{-}1)$$

Generally, for a sample of size n, the range

$$\bar{x} - u(P)\frac{s}{\sqrt{n}} \leq \mu \leq \bar{x} + u(P)\frac{s}{\sqrt{n}} \quad (4\text{-}2)$$

is an estimate of μ at an assumed confidence level, expressed by $u(P)$. The values of $u(P)$ which are the expression of the standardized normal distribution are given in monographs on mathematical statistics and chemometrics (see for instance [SHARAF et al., 1986; MASSART et al., 1988]).

For a small sample size and a particular confidence level other than GAUSSian distribution, the STUDENT t-distribution has to be assumed. Furthermore, the confidence range for the estimate $\bar{x}$ is:

$$\bar{x} - t(f; P)\frac{s}{\sqrt{n}} \leq \mu \leq \bar{x} + t(f; P)\frac{s}{\sqrt{n}} \quad (4\text{-}3)$$

where the specific value for the t-distribution is determined by the level of confidence required, P, and the degree of freedom, $f = n - 1$.

If the maximum allowed uncertainty is U and the variance of the population is s^2, which includes the variances of all steps of the analytical process from sampling to analysis, the sample size n is given very simply and pragmatically by:

$$n = \left(\frac{t(f; P) \cdot s}{U}\right)^2 \quad (4\text{-}4)$$

with:

$$U = \bar{x} \cdot s_{rel} \quad (4\text{-}5)$$

The total weight w_{total} of the composite sample can be calculated according to:

$$w_{total} = n \cdot w_{increment} \quad (4\text{-}6)$$

$w_{increment}$, which is the mass of the individual sample, can be determined by means of the different methods described in Section 4.3.2 or is given by fixed sampling procedures.

An estimate of $\bar{x}$ and s^2 can be obtained from past experience or preliminary sampling. The application of this equation is described in detail by KRAFT [1978; 1994] and HOLSTEIN [1980].

A simple example may illustrate this method: Ten tons of sewage sludge have to be investigated to determine its average loading by lead, an ecotoxicologically relevant heavy metal. It has been discovered from previous investigations that the average contamination of sewage sludge by lead is $\bar{x} = 200$ mg kg^{-1} and the standard deviation $s = 40$ mg kg^{-1}. The maximum allowed uncertainty shall be $U = 10\%$. Because the

t-value depends on the degree of freedom f, Eq. 4-4 has to be solved by iteration. Eq. 4-4 is satisfied for a level of condidence $P = 0.95$ if 18 samples are taken from the sludge heap. Assuming that the weight of a subsample is $w_{increment} = 1$ kg, which is usual in investigations of soil and sludge pollution, the resulting weight of the composite sample is $w_{total} = 18$ kg.

If the population is not normally distributed and cannot be transformed into a GAUSsian distribution, then the required number of individual samples should be determined on the basis of other distributions [KRATOCHVIL et al., 1986; SHARAF et al., 1986]; if no classical distribution type exists, robust statistics must be applied [ROUSSEEUW and LEROY, 1987]. HÅKANSON [1984] has described the relationship between the pretreatment of river sediment samples (sieving, centrifugation) and the value of information defined by analogy with Eq. 4-4.

It should be noted that the number of samples extracted from a parent population can also be regarded as a problem of homogeneity. If no other information is available the samples can also be tested for their homogeneity or heterogeneity by univariate or multivariate statistical procedures. An overview of the investigation of homogeneity is given in the literature [DANZER and SINGER, 1985; LIEBICH et al., 1989; 1992].

4.3.2 Mass of Individual Samples Required

Many theories deal with the determination of the required mass of a sample from a heterogeneous bulk. Some of the most important of these must, furthermore, be discussed briefly. The dependence of sampling error on the grain size was discussed by REED [1881/82; 1884/85] over 100 years ago.

The question about the mass of sample required, which first appeared in the mining industry and which is widely accepted and also applicable to many environmental problems, is described by TAGGART's nomogram (Fig. 4-2). The dependence of the required total sample mass on the heterogeneity of the material and the concentration of the analyte are also discussed.

A common and exact theory of sampling of inhomogeneous materials with stochastic composition is presented by BRANDS [1983]. Because experimental evaluation of this theory involves some difficulties, it was verified by simulation experiments.

The more detailed theory of GY [GY, 1982; 1991 a; 1991 b; PITARD, 1989] describes the sampling of heterogeneous materials. The sampling variance s_S^2 in a sample of weight w, which depends on the particle shape, the particle size distribution, the composition of the phases comprising the particles, and the degree to which the analyte is liberated from the remainder of the material during particle size reduction by grinding or crushing, can be estimated from:

$$s_S^2 = \frac{f_{shape}\, f_{size\ distribution}\, f_{composition}\, f_{liberation}\, d_l^3}{w} \qquad (4\text{-}7)$$

d_l – linear dimension of the largest particle

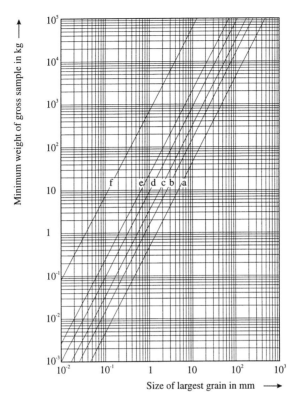

Fig. 4-2. Dependence of sample mass on grain size and analyte concentration in the ore according to TAGGART [1948] and KRAFT [1980]. (a) Very small grain size or very uniformly distributed metal content, (b) small grain size or uniformly distributed metal content, (c, d) medium metal content and normal metal distribution, (e) high metal content or irregular distribution, (f) very high metal content or very irregular distribution

The shape factor, f_{shape}, is the ratio of the average volume of all particles with a maximum linear dimension equal to the applied mesh size of a sieve to that of the cubes which will just pass the same sieve. The value of f_{shape} can be assumed to be 0.5 for the most materials. The particle size distribution factor, $f_{size\ distribution}$, is the ratio of the upper size limit to the lower size limit. The composition factor, $f_{composition}$, is defined as:

$$f_{composition} = \frac{1-c}{c}(1-c)\rho_A + c \cdot \rho_M \qquad (4\text{-}8)$$

c – overall concentration of the component of interest
ρ_A – density of the component of interest
ρ_M – density of the remaining material

The liberation factor, $f_{liberation}$, is defined as:

$$f_{liberation} = \sqrt{\frac{d_A}{d_l}} \qquad (4\text{-}9)$$

d_A – average grain diameter of the sought-for component
d_l – diameter of the largest particle in the mixture

An algorithm for the evaluation of the fundamental sampling error in the sampling of particulate solids, based on GY's theory, is described by MINKINNEN [1987].

A quite similar consideration was published by INGAMELLS and SWITZER [1973] and by INGAMELLS [1974b; 1976]. A sampling constant C_S is proposed which enables estimation of error of subsampling, i.e. the withdrawal of a small portion from a well mixed material. The relationship to GY's equation (Eq. 4-7) is given by:

$$C_S = f_{shape}\, f_{size\ distribution}\, f_{composition}\, f_{liberation}\, (d_l^3 \cdot 10^4) \qquad (4\text{-}10)$$

The required sample weight, w, for a relative standard deviation of sampling of 1% at a confidence level of 68% is given by:

$$w = \frac{C_S}{s_{rel}^2} \qquad (4\text{-}11)$$

s_{rel} – experimentally found relative standard deviation

INGAMELLS' constant, C_S, is well estimated by measurements of sets of samples with different weights [INGAMELLS, 1974a]. If INGAMELLS' theory is applied to several sets of individual samples with different weights, the degree of homogenization can be determined.

The theory of VISMAN [1969] describes the relationship between the sampling variance s_S^2 and the heterogeneity of the population:

$$s_S^2 = \frac{C_{homogeneity}}{w \cdot n} + \frac{C_{segregation}}{n} \qquad (4\text{-}12)$$

w – individual increment size
n – number of increments

$C_{homogeneity}$ is the homogeneity constant and $C_{segregation}$ is the segregation constant. These values can be obtained by collecting a series of increment pairs from sites close to the parent population. The members of the pairs must have the same weight w. The connection to INGAMELLS' constant is given by:

$$C_{homogeneity} = 10^4 \bar{x}^2 C_S \qquad (4\text{-}13)$$

4 Sampling and Sampling Design

An early theory for description of the random sampling error in well mixed particulate materials with particle-size-dependent composition is described by BENEDETTI-PICHLER [1956].

The number of particles required, n, can be calculated from:

$$n = \left[\frac{\rho_1 \cdot \rho_2}{\bar{\rho}^2}\right]^2 \left[\frac{100(c_1 - c_2)}{s_{rel}\bar{c}}\right]^2 p(1-p) \tag{4-14}$$

ρ_1 and ρ_2 – densities of the two kinds of particle
$\bar{\rho}$ – average density of the sample
c_1 and c_2 – percentage analyte composition of the two kinds of particle
$\bar{c}$ – overall analyte concentration in the sample
s_{rel} – relative standard deviation of sampling
p and $1-p$ – mass fractions of the two kinds of particle in the bulk

Assuming spherical particles, the required minimum sample weight, w_S, can be calculated:

$$w_S = \frac{1}{6}\pi d^3 \rho n \tag{4-15}$$

The relationship between the minimum sample weight required and the composition of the two kinds of particle in the bulk is illustrated in Fig. 4-3.

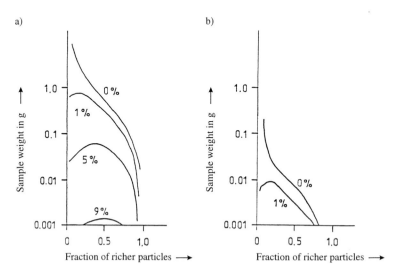

Fig. 4-3. Relationship between the minimum sample weight required and the composition of the two kinds of particle in the bulk for different analyte concentrations (0, 1, 5, 9%) according to BENEDETTI-PICHLER [1956]. (a) Relative standard deviation $s_{rel} = 0.1\%$, (b) relative standard deviation $s_{rel} = 1.0\%$

If the fraction of large particles is small and the small particles contain only a very low concentration of analyte, then a relatively large sample weight is required.

It is recommended as an empirical rule that the samples taken from the bulk should have a mass at least ten times that of the largest grain. If, for example, communal waste deposits have to be investigated, the required sample mass should be 100–250 kg [RUMP and HERKLOTZ, 1990].

Often, all the empirical constants necessary for calculation of the required sample masses according to the above equations are unknown or the costs of determining them are too high. Therefore, the very easy use of Eqs. 4-4 to 4-6 may be recommended for the determination of the required sample mass.

Another method of estimation of the critical weight of the test portion used for the analysis of heavy metals in lake sediments is described by HENRION et al. [1991].

According to the law of error propagation (see Eq. 4-21), the total variance of the analytical result can be expressed as:

$$s^2 = s_S^2 + s_A^2 \tag{4-16}$$

The relationship between the sampling variance estimate, s_S^2, and the sample weight, w_S, is characterized by:

$$s_S^2 \sim \frac{1}{w_S} \tag{4-17}$$

Combining Eqs. 4-16 and 4-17 the total variance can be expressed by:

$$s^2 = \frac{a_1}{w_S} + s_A^2 \tag{4-18}$$

For the determination of the critical sample weight, $w_{S\ crit}$, the condition

$$s_S^2 \leq s_A^2 \tag{4-19}$$

is used [DOERFFEL and ECKSCHLAGER, 1981]. This critical weight can therefore be approximately determined by:

$$w_{S\ crit} = \frac{a_1}{s_A^2} \tag{4-20}$$

An example from soil investigation may illustrate this method: The influence of high-metal emission on the top soil in the surroundings of a metallurgical factory was investigated by KRIEG and EINAX [1994]. The question which had to be answered was: what is the minimum (or critical) mass of the test portion which fulfils the condition expressed in Eq. 4-19? A bulk sample taken from the top soil was dried at 60 °C, pulverized and sieved (particle diameter <63 μm). Different masses (0.1 g to 2.5 g) of these samples were then digested with aqua regia [DIN 38414, 1983]. Eight replicates were prepared for each sam-

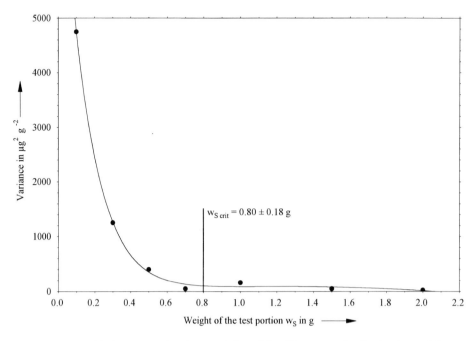

Fig. 4-4. Relationship between total variance and the weight of the test portion for the element Mn

ple weight. The concentrations of several metals (Ca, Cr, Cu, Fe, K, Mg, Mn, Na, Ni, and Zn) in equal volumes of aliquots of these extracts were determined by means of optical emission spectroscopy with an inductively coupled plasma. The determination of the critical sample weight for the analysis of Mn according to Eq. 4-19 is demonstrated in Fig. 4-4. The critical sample weight for Mn is $w_{S\ crit} = 0.80 \pm 0.18$ g. The critical sample weights for all analyzed elements and the corresponding confidence levels are illustrated in Tab. 4-1.

Tab. 4-1. Critical weights, $w_{S\ crit}$, of soil samples for the analyzed elements and their confidence levels, $\Delta w_{S\ crit}$, for $P = 0.95$

Element	$w_{S\ crit}$ in g	$\Delta w_{S\ crit}$ in g
Ca	0.70	0.12
Cr	0.65	0.15
Cu	0.80	0.09
Fe	0.70	0.08
K	1.35	0.21
Mg	0.70	0.15
Mn	0.80	0.18
Na	0.80	0.18
Ni	0.70	0.17
Zn	0.55	0.10

The value is highest for potassium because of the greater microinhomogeneity for this element. For the other elements, 1 g was the minimum sample weight recommended for the test portion taken! GARDNER [1995] proved that a test portion of at least 1 g of particles with a diameter of less than 63 µm ensures that the relative standard deviation of the analysis is not increased by more than approximately 1% as a result of inhomogeneity. The determination of the critical sampling weight from linear regression of the relationship between the total variance and the reciprocal value of the weight of the test portion was, in contrast to HENRION et al. [1991], not possible. Obviously, the cited authors' assumption of a linear relationship between the weight of the test portion and the variance of the analysis was not fulfilled in the case under investigation.

In the field of environmental monitoring the sampling of a relatively high mass is usually recommended, as was described for soil sampling by HOFFMANN and SCHWEIGER [1981] and BERROW [1988]. In the majority of cases under investigation in the field of environmental research, there is enough raw material for a sample mass of sufficient size to be taken. Therefore, the use of a sample weight significantly greater than the critical value should not be the problem in the practice of environmental investigations. If only a small sample mass is available, as, for example, for sampled airborne particulate matter, it must be proved that the sample mass is sufficiently large that sampling error as a result of heterogeneity does not predominate.

4.3.3 Minimization of the Variance of Sampling and Analysis

The inherent error of each sample taken, expressed as its variance, can be described by the law of error propagation:

$$\sigma^2 = \sigma_S^2 + \sigma_{A_1}^2 + \sigma_{A_2}^2 + \cdots \quad (4\text{-}21)$$

The analytical error can be split into suberrors according to the various steps of sample pretreatment and the analytical procedure. A qualitative discussion of the different contributions to the sampling error is given by WEGSCHEIDER [1994].

Minimization of the variance of sampling and analysis shall be discussed using an example from environmental monitoring according to the scheme of SHARAF et al. [1986].

A river is analyzed for its loading of adsorbable organohalogen compounds (AOX). Twenty samples are taken at the same time and location. Each sample has an AOX value of 20 µg L^{-1} with a standard deviation of sampling s_S = 2.0 µg L^{-1}. Each analytical determination has a standard deviation s_A = 1.0 µg L^{-1}.

Now the following analysis plans must be considered in terms of their variance:

Plan 1: All samples are analyzed separately and the mean value is reported. The variance of the mean value s_m^2 is given by:

$$s_m^2 = (s_m^2)_S + (s_m^2)_A \quad (4\text{-}22)$$

The two subvariances in Eq. 4-22 $(s_m^2)_i$ can be expressed by:

$$(s_m^2)_i = \frac{s_i^2}{k} \tag{4-23}$$

k – number of individual units (in this example $k = 20$)

$$s_m^2 = \frac{2.0^2}{20} + \frac{1.0^2}{20} = 0.25$$
$$\underline{\underline{s_m = 0.5 \, \mu g \, L^{-1}}}$$

Plan 2: After mixing all twenty samples, one sample, being one twentieth of the whole, is analyzed. The variance of sampling is definitely unchanged after mixing. Then, the variance of the mean value is given by:

$$s_m^2 = \frac{2.0^2}{20} + \frac{1.0^2}{1}$$
$$s_m^2 = 1.2$$
$$\underline{\underline{s_m = 1.095 \, \mu g \, L^{-1}}}$$

Plan 3: After mixing all twenty samples, the mixture is divided into twenty samples, each of which is analyzed. The variance of the mean is given by:

$$s_m^2 = \frac{\frac{s_S^2}{20}}{20} + \frac{s_A^2}{20}$$
$$s_m^2 = \frac{2.0^2}{400} + \frac{1.0^2}{20} = 0.06$$
$$\underline{\underline{s_m = 0.245 \, \mu g \, L^{-1}}}$$

Plan 4: After mixing all the samples, only four samples are analyzed (each sample again being one twentieth of the mixture). The variance of the mean value is given by:

$$s_m^2 = \frac{2.0^2}{80} + \frac{1.0^2}{4} = 0.3$$
$$\underline{\underline{s_m = 0.548 \, \mu g \, L^{-1}}}$$

Plan 5: If the variance of the analysis is independent of the sample size (see the following implementations), the twenty individual samples can be mixed and the whole mixture can be analyzed once. The variance of the analysis of n mixed samples is equal to:

$$s_{A\,mixed}^2 = \frac{s_A^2}{n^2} \tag{4-24}$$

The variance of the mean value is, furthermore, given by:

$$s_m^2 = \frac{s_S^2}{n} + \frac{s_A^2}{n^2} \qquad (4\text{-}25)$$

$$s_m^2 = \frac{2.0^2}{20} + \frac{1.0^2}{20^2} = 0.202$$

$$\underline{s_m = 0.45 \text{ µg L}^{-1}}$$

The smallest variance of the mean value is obtained by using plan 3, but twenty samples have to be analyzed. The second smallest variance can be obtained by use of plan 5, i.e. by performing only one analysis. This plan should be the most recommended if economy is the limiting factor of the task. It should be guaranteed that the standard deviation of the analysis is independent of the sample size for this plan (see the following implementations).

4.3.4 Investigation of the Origin of Variance

If the resolution of the error according to Eq. 4-21 is of interest, then simple univariate analysis of variance (ANOVA, described in detail in Sections 2.3 and 3.3.9) must be applied.

One example may illustrate the algorithm used for error resolution (for mathematical details see also [DOERFFEL, 1990]): Four samples were taken simultaneously from the same location in a river. These samples were divided into four subsamples and centrifuged in order to separate the suspended material. The subsamples were then analyzed four times and the chromium concentration was determined by means of AAS with electrothermal atomization. By following this procedure, error resolution for these origins of variance is possible:

– inhomogeneity of the river water and sampling error
– error of the separation of the suspended material by centrifugation
– error in the measurement of sample volume and in the analytical measurement

Variance splitting can be performed by applying ANOVA (Tab. 4-2).

Tab. 4-2. Contributions to variance in the analysis of chromium in river water samples using AAS with electrothermal atomization

Variance origin	Total error in %	Relative error* in %
Sampling	9.2	3.3
Centrifugation	73.1	9.6
Dosage and analysis	17.3	4.6

* Related to the mean value of the measured concentration

The error of the AAS determination of chromium in river water with a determined mean concentration of 4.8 µg L^{-1} is acceptable for this particular environmental purpose. The highest variance percentage arises as a result of centrifugation. The sampling error arising from river inhomogeneity is relatively small in this particular environmental situation.

THOMPSON and MAGUIRE [1993] have compared sampling and analytical error for the example of trace metals in soils. They demonstrate that to obtain valuable information on the magnitude of sampling and analytical errors the application of robust nested analysis of variance is to be preferred to classical parametric analysis of variance.

4.3.5 Sampling Location and Frequency

The previous examinations were conducted without consideration of any correlation between the samples taken from the parent population. This is useful for gross description and if there are no correlations between the individual samples, i.e. the samples are stochastically distributed in space or time. In the reality of environmental investigation, the sampling location and frequency may have a very real and strong effect on the quality and usefulness of the data obtained. Data from the environment are often correlated in space and/or time. In this circumstance, the optimum number of locations and frequencies (or better: the location in space and time) of the samples can best be estimated after preliminary sampling experiments or as a result of prior knowledge. At this point it should be noted that sampling location and frequency are only two aspects of the same problem. If the spatial dependence of environmental data is considered, the question of sampling location is essential. If the temporal changes of an environmental compartment are of interest, the question of sampling frequency must be answered. The mathematical tools for solving these two problems are the same.

Objects which are internally correlated, for example volumes sampled from rivers, soils, or ambient air, can be treated by autocorrelation analysis or semivariogram analysis. The range up to a critical level of error probability is an expression of the critical spatial or temporal distance between sampling points.

One-dimensional algorithms for autocorrelation analysis for one or more variables are described in detail in Section 6.6. Some case studies which illustrate the application of autocorrelation analysis both in its univariate and multivariate manner are discussed in Sections 7.1 and 9.1. BARCELONA et al. [1989] offer another example of investigation of sampling frequency for assessment of groundwater quality where autocorrelation analysis was used. Investigation of representativeness in soil sampling (see Section 9.1) illustrates that the one-dimensional technique can also be a useful tool for solving the problem of representativeness in a two-dimensional problem (agricultural area).

A quite different way of monitoring dynamic processes is described by PIJPERS [1986]. Planning sampling for such a system can be conducted advantageously using a KALMAN filter algorithm and concepts borrowed from information theory. The underlying ideas are demonstrated by different examples of dynamic processes, e.g. river sampling.

A further problem concerning the sampling location must be discussed. Many environmental queries include the question of the depth at which the sample must be taken. For example, it is of essential importance to know the depth or, better, the depth interval in the studies of soil, river, or groundwater pollution. The sampling depth must be adapted to the specific purpose of the investigation. Often this aim depends on the planned use of the soil area or water body.

Two approaches must be considered for soil studies [PAETZ and CRÖSSMANN, 1994]: depth- or soil-horizon-related sampling. Depth-related sampling is performed for agricultural purposes and for screening analysis of potentially contaminated areas. For environmental reasons, particularly those concerned with soil and groundwater protection, ISO/DIS 10381-1 [1995] recommends sampling in soil horizons, as this is more representative. For example, soil fertility studies deal primarily with the A horizon whereas studies involving water movement focus on the horizon where permeability would probably be reduced [BARNARD, 1995]. NOTHBAUM et al. [1994] have published suggestions of sampling depth for soil pollution studies according to the different categories of use of the soil. If a precise assessment of the mass input in an area is to be obtained and the depth distribution of the pollutant is unknown, or the depth strata are disturbed, as, for example, for dumps and disposal sites, systematic investigation of the depth distribution of the pollutants in question must be performed (see also Section 9.2.2).

4.4 Geostatistical Methods

Geostatistical approaches may be very useful for the treatment of environmental data, particularly if large numbers of heterogeneous samples are needed to find minimum confidence ranges for sampling results [BARCELONA, 1988]. Geostatistics includes all statistical methods for the analysis of data correlated in space-time-reality. Geostatistical methods are also referred to as the **theory of regionalized variables**.

The development of geostatistical methods was initiated by problems of representative sampling over a large geographic region with significant, but unknown, long- and short-term changes. The first empirical work was performed by KRIGE [1951a; 1951b] and DE WIJS [1951; 1953]; the statistical theory was developed by MATHERON [1957; 1963; 1965]. Initially developed for prospecting for gold deposits in South Africa, geostatistical methods were soon used for investigation in the mining geology. In recent years, these methods have been increasingly applied to the assessment of environmental pollution, particularly in the study of soil (see, for example, [EINAX and SOLDT, 1995]).

Geostatistical methods are also termed random field sampling as opposed to random sampling [BORGMAN and QUIMBY, 1988]. The spatial dependence of data and their mutual correlation can be analyzed by use of semivariograms. Statements on the anisotropy of the spatial distribution are also possible. Kriging, a geostatistical method of spatial estimation, is the tool commonly used for estimating the amount of contaminant present at an unsampled location [SINGH et al., 1993]. Unsampled points can be estimated by means of their neighboring points. The variance of estimation is minimized. The kriging method is useful for describing the concentrations of investigated elements in contour plots. The mathematical fundamentals of both semivariogram analysis and the kriging method are described in detail in the literature [DAVID, 1977; JOURNEL and HUJIBREGTS, 1978; AKIN and SIEMES, 1988; MYERS, 1991].

Only the first part of geostatistical considerations – semivariogram analysis – belongs directly to sampling problems. The second part – kriging estimation – is more related to the problem of the evaluation of environmental data. But because the two steps of the geostatistical approach are part of a single statistical method, they are introduced together in the following text.

4.4.1 Intrinsic Hypothesis

The theoretical basis of the kriging method is the so-called intrinsic hypothesis. A random function $Z(x)$ accomplishes this hypothesis if two assumptions are fulfilled:

– The expectation value of the difference between $Z(x+l)$ and $Z(x)$ equals zero:

$$E[Z(x+l) - Z(x)] = 0 \qquad (4\text{-}26)$$

l – distance
x – coordinate

– The variance of the differences between the two realizations of the random variable depends only on the distance:

$$Var[Z(x+l) - Z(x)] = 2\gamma(l) \qquad (4\text{-}27)$$

$\gamma(l)$ – semivariance

Under these suppositions, the application of linear geostatistical methods, like point kriging, is possible on the basis of the semivariogram.

4.4.2 Semivariogram Analysis

Description of the mutual dependence of sample values for a variable in space is possible with the correlogram, the semivariogram (or the variogram), and the covariogram. The most useful tool in geostatistics is the semivariogram or, in more general terms, the variogram.

Usually the computation of the semivariance is the first step in geostatistical analysis. The **semivariance** $\gamma(l)$ is expressed by:

$$\gamma(l) = \frac{1}{n(l)} \sum_{i=1}^{n-l} [z(x_{i+l}) - z(x_i)]^2 \qquad (4\text{-}28)$$

$n(l)$ — number of sample pairs at each distance l
$z(x_i)$ and $z(x_{i+l})$ — values of the variable x at the location i and $i + l$

The graphical representation of the semivariance $\gamma(l)$ as a function of the distance is called the **semivariogram** (see Fig. 4-5); the representation of $2\gamma(l)$ is referred to as the variogram [GILBERT and SIMPSON, 1985]. These two terms are not, however, consistently distinguished in the literature. The semivariogram function describes the relationship between spatially correlated data. The semivariogram has to be constructed on the basis of experimental data obtained by sampling in the area under investigation.

An important advantage of geostatistical methods is that the sampling points do not necessarily have to be regularly distributed. For environmental investigation, this point is of particular importance because sometimes certain locations in the area cannot be sampled.

If the sampling points are not distributed regularly, the locations of the sampling points have to be transformed into polar coordinates (direction $\Psi \pm \Delta\Psi$, distance of the points $l \pm \Delta l$) for the computation of the semivariogram.

According to Eq. 4-28 the theoretical semivariogram function has the value $\gamma = 0$ for $l = 0$. Semivariograms obtained from experimental data often have a positive value of intersection with the $\gamma(l)$-axis expressed by C_0 (see Fig. 4-5). The point of intersection is named **nugget effect** or nugget variance. This term was coined in the mining industry and indicates an unexplained random variance which characterizes the microinho-

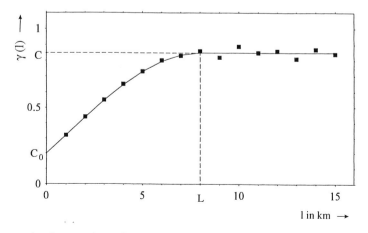

Fig. 4-5. Example of an experimental semivariogram with nugget effect C_o, sill C, and range L

mogeneity at the sampling location. From this point of intersection the semivariance increases until the variance of the data set called **sill** C is reached. Up to this point, the regionalized variables in the sampling locations are correlated. Then they must be considered to be spatially independent. This distance of correlation is called **range** L. An example of an experimental semivariogram with nugget effect, sill, and range is demonstrated in Fig. 4-5.

An experimental semivariogram can be modeled by fitting a simple function to the data points. Linear, spherical, or exponential models are often used [KATEMAN, 1987; AKIN and SIEMES, 1988].

In some cases a linear model may be applied:

$$\begin{aligned} \gamma(l) &= al \quad \text{for} \quad l < L \\ \gamma(l) &= s^2 \quad \text{for} \quad l \geq L \end{aligned} \tag{4-29}$$

L – range of the semivariogram

An exponential model is expressed by:

$$\gamma(l) = s^2 \left(1 - e^{-\frac{l}{L}}\right) \tag{4-30}$$

The "usual" model is a spherical one:

$$\gamma(l) = s^2 \left(\frac{3l}{2L} - \frac{l^3}{2L^3}\right) \tag{4-31}$$

But other types of nonlinear interpolation may also be applied [LEMMER, 1986].

At this point it should be noted that the experimental semivariogram may be influenced by many errors. One way to minimize these errors is the computation of the indicator semivariogram $I(x, Z_{crit})$ [JOURNEL, 1983]. The function $Z(x)$ has to be transformed according to:

$$I(x, Z_{crit}) = \begin{cases} 1 & \text{if} \quad Z(x) \geq Z_{crit} \\ 0 & \text{if} \quad Z(x) < Z_{crit} \end{cases} \tag{4-32}$$

For example the median can be used as threshold Z_{crit}. The advantage of this method is its ruggedness against outliers.

Another way to handle "irregular" raw data, e.g. nonnormally distributed data, is the computation of the semivariogram on ranks of the original data [SINGH et al., 1993]. The resulting semivariogram is more robust than the semivariogram based on the original values.

A comparison between an autocorrelogram and a semivariogram, as exemplified by the design of a sampling plan for aquatic sediments, is given by WEHRENS et al. [1993].

If the sampling locations are not distributed in a regular raster, the following methods are available to determine the semivariogram [AKIN and SIEMES, 1988]:

- If the deviation of the sampling points from the screening raster is small, the points must be shifted to the neighboring raster point.
- The sampling points can be projected to lines of different orientation. The semivariogram can then be computed along these lines with the distance $l \pm \Delta l$.
- In the case of irregularly distributed points a direction can be determined by an angle Ψ with a tolerance of $\pm \Delta\Psi$ (opening angle); the corresponding distance is again $l \pm \Delta l$.

If the semivariogram depends on the direction of the distance vector **l**, it is termed **anisotropic**. Otherwise, the spatial structure of the population to be investigated is **isotropic**. Comparing the semivariograms calculated for different directions it is also possible to investigate the spatial structure of the parent population, for example an agricultural area influenced by particulate emissions. In practice, other types of semivariogram may result from nested or periodic structures; hole effects also [FRÄNZLE, 1994]. The shape of the semivariogram gives useful information concerning the spatial structure of the contamination, but because numerical description is more complicated in such cases, the following geostatistical step, the kriging estimation, can be performed only in restricted form.

ASSMUS et al. [1985] have investigated the spatial variability of soil phosphorus by the use of semivariograms. As described above, the authors can determine the minimum sampling distance. The spatial variability can also be detected by applying a simple strip sampling technique (see also Section 4.5.3.2). Comparison of composite samples taken from vertical strips of the field with composite samples taken from horizontal strips can reveal large differences in the phosphorus levels at different parts of the field. This strategy for detecting lateral differences is qualitatively comparable with the application of the twofold univariate analysis of variance in soil sampling as described in detail in Section 9.1.3.1.

4.4.3 Estimation of New Points in the Sampling Area – Kriging

The semivariogram as a variance function can also be used to estimate the value and the variance for new points not sampled in the investigated area. The method applied for this purpose is termed kriging. **Kriging** is a special regression method for interpolation of spatially or temporally correlated data with minimization of variance. The normal distribution of the data is an important condition. If the original data are not normally distributed, which is often the case for trace components in environmental compartments, the logarithm of the data or otherwise transformed data have to be applied to obtain a normal distribution of the data (see also Section 9.4).

The technique of kriging uses a weighted moving average interpolation method [BUSCHE and BURDEN, 1991] in the sense of a straightforward linear regression [JOURNEL, 1988]. The weights, w_i, are determined on the basis of the experimental semivariogram so that the kriging variance becomes a minimum. Qualitatively expressed, neighboring points give a better estimate of the value at an unknown location than do more distant locations. The estimation function for simple point kriging is:

$$\hat{z}(x_0) = \sum_{i=1}^{n} [w_i z(x_i)] \qquad (4\text{-}33)$$

$\hat{z}(x_0)$ is the estimated value of the random function Z at the unsampled location x_0 and w_i are the weights of the $z(x_i)$. The weights, w_i, can be obtained by means of the kriging system (see also [MYERS, 1991]):

$$\sum_{j=1}^{n} w_i \gamma(x_i, x_j) + \mu = \gamma(x_i, x_0), \qquad i = 1, 2, \ldots, n \qquad (4\text{-}34)$$

(x_i, x_j) – distance between the sampling points x_i and x_j
(x_i, x_0) – distance between the sampling points x_i and the unsampled point x_0 at which the value is being estimated
μ – LAGRANGE multiplier

The empirical weights have to be normalized with the following condition to obtain a distortion-free estimate:

$$\sum_{i=1}^{n} w_i = 1 \qquad (4\text{-}35)$$

The kriging variance is given by:

$$\sigma_k^2 = \sum_{i=1}^{n} w_i \gamma(x_i, x_0) + \mu \qquad (4\text{-}36)$$

Applying the kriging point estimation, isolines can be computed both from regular distributed sampling points and from irregular distributed points. Fig. 4-6 demonstrates the isolines; for ease of visualization, the figure also contains a three-dimensional representation of nickel levels near a large metallurgical factory. The values are obtained as the estimates from an irregular distributed sampling grid plan [EINAX and SOLDT, 1995]. A detailed description of this case study is given in Section 9.4.

The advantage of kriging is that it furnishes not only an estimate of the unsampled point, but also an estimate of the variance at this location. For a sampling location the exact value is estimated with the variance being zero. If the kriging method is applied, an exact and undistorted interpolation is possible.

GILBERT and SIMPSON [1985] describe the estimation of the spatial pattern of radionuclides in soils. They demonstrate the advantage of the kriging method applied to logarithmically transformed data. But they refer to the fact that the validity of kriging

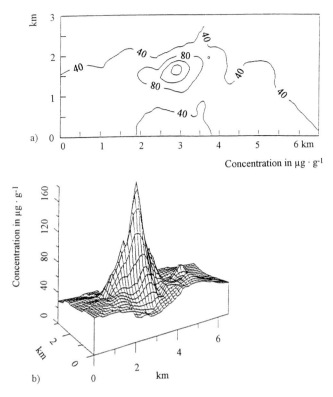

Fig. 4-6. Kriging estimation of nickel near a metallurgical factory [EINAX and SOLDT, 1995].
(a) Representation of the isolines, (b) three-dimensional plot

estimates depends on the accurate estimation and modeling of the spatial correlation structure of the objects under investigation. Their paper illustrates the bias which can result when a changing correlation structure over space is ignored.

4.4.4 Cross-Validation

The quality of a theoretical semivariogram model can be assessed by means of point kriging and cross-validation [MYERS, 1991]. Each point of the data set is deleted one after another and then newly estimated according to Eqs. 4-33 to 4-35 by means of its neighbors. Additionally, the kriging variance is calculated for each estimated point (Eq. 4-36).

n difference values are obtained from n sampling values and their n estimated values:

$$\Delta z = z(x_i) - \hat{z}(x_i) \qquad (4\text{-}37)$$

These values are squared and divided by the corresponding kriging variance:

$$\Delta z_{rel}^2 = \frac{[z(x_i) - \hat{z}(x_i)]^2}{\sigma_k^2(x_i)} \quad (4\text{-}38)$$

The mean of these quotients Δz_{rel}^2 should be situated near zero and their standard deviation near unity, i.e. it should be a standardized normal distribution if the kriging estimation is distortion-free.

The goodness of fit for experimental data to a theoretical semivariogram model can be tested by means of cross-validation. The best model is that with the smallest deviation from the mean zero and with the smallest standard deviation.

4.5 Sampling Plans and Programs

4.5.1 Basic Considerations

In each case the design of the sampling process has to start with problem-adapted planning.

First considerations have to include [KRATOCHVIL, 1981]:
- purpose of sampling
- population to be studied
- substances or groups to be measured
- extent to which speciation is to be determined
- precision required
- extent to which the distribution of the substance within the population is to be obtained

As the most important criteria for planning, the following relationships have to be considered:

- representativeness-sufficiency-indicativity
- objectivity-reliability-validation
- economy-time

These relationships are strongly influenced by the amount of *a priori* knowledge of the subject to be investigated relative to the amount of knowledge required. The accuracy and precision of the sampling data clearly influence the above relationships. The choice of the most suitable sampling strategy depends on many different factors. In general terms, the best strategy combines the highest possible representativeness with the lowest costs, taking into account available information and resources [FORTUNATI et al., 1994].

4.5.2 Purpose of Sampling and the Chemometric Methods Applicable

A brief overview shall be given of the purpose of sampling on the one hand and the chemometric methods applicable on the other hand. In the field of environmental investigation, samples are taken mostly for purposes of:

– quality control
– prognosis statement and forecasting
– damage and risk assessment [GUDERNATSCH, 1982]

To answer these questions, the following purposes of sampling have to be regarded [KATEMAN, 1987; 1988].

Sampling for Gross Description
The description of an object in the sense of environmental investigation may be the determination of the gross composition of an environmental compartment, for example the mean state of a polluted area or particular location. If this is the purpose, the number of individual samples required and the required mass or size of these increments have to be determined. The relationship between the variance of sampling and that of analysis must be known and both have to be optimized. The origin of the variance of the samples can be investigated by the study of variance contribution of the different steps of the analytical process by means of the law of error propagation (Eq. 4-21) according to Section 4.3.4.

Sampling for Detailed Description and Control
If the purpose of sampling is the detailed description of the composition of an object, the character of the internal correlation has to be investigated. The methods of autocorrelation and/or semivariogram analysis, as described in Sections 6.6 and 4.4.2, may be useful for clarification of the internal spatial and/or temporal relationships existing within the parent population to be sampled. Geostatistical methods, e.g. kriging, enable undistorted estimation of the composition of unsampled locations in the area of investigation.

Principles of sampling for threshold control are described by KATEMAN and PIJPERS [1981]. The authors show that the sampling frequency depends on the distance between the last value measured and the threshold level. They explain both the adaptation of the sampling frequency to the existing state of the process or the environmental compartment under control, and the effect of the distance between samples on the cost of the project.

Sampling for Monitoring
In the field of environmental investigation, it is not usually possible to distinguish clearly between detailed description and control on the one hand and monitoring on the other hand. Monitoring is more the investigation of the dynamic character of the state of an environmental compartment. So methods of time series analysis, as described in mathematical detail in Section 6, can be effectively applied.

4.5.3 Sampling Plans

The sampling plan should include a fixed and detailed description of:

- size, number, and location of sample increments or specimens
- extent to which samples may be combined
- procedures for reduction of gross or grab samples to the size of sample required by the laboratory, and then to the test portion
- all steps of the analytical process, beginning with sampling, through sample pretreatment, analytical measurement, and data evaluation

The quantities or relationships required which have to be fixed in the specified sampling plan are [BORGMAN and QUIMBY, 1988]:

- regional mean and variance
- regional trend in the mean
- differences between means and variances obtained in different areas
- estimation of probability laws for various sampled variabilities
- detailed contour maps of variables of interest
- spatial correlation functions, e.g. autocorrelogram, semivariogram, or covariogram
- probabilities of extreme values
- fraction of the investigated area which has values higher than a critical value, for example a legally fixed threshold
- confidence levels and limits on various sampled variables
- test statistics for hypothesis tests related to particular parameters of the random field
- regional location for spatial pattern, for example stack emissions into the atmosphere
- assessment of the future behavior of the processes present within the territory being sampled

4.5.3.1 Basic Types of Sampling

In general, three broad methods are available for planning a sampling procedure [GARFIELD, 1989]: **probability sampling**, **nonprobability sampling**, and **bulk sampling**.

The theoretical fundamentals of these methods are described by KRISHNAIAH and RAO [1988]. These basic types can be realized as:

Random Sampling

- **Simple random sampling.** Simple random sampling is performed directly on the whole population (area or section) under investigation. Any increment taken from the parent population has an equal chance of being selected. In practice, the problem is that the sample has to be taken in space or time after random number generation, not haphazardly.

The main problem of random sampling is that the experience of the person who is sampling is very important and can strongly influence any systematic investigation or input hypothesis [NOTHBAUM et al., 1994]. The most important subjective distortions may be:
- preferred sampling at easily accessible locations
- intuitive selection of obviously heavily or slightly loaded points
- tendency towards an intuitive regular distribution of sampling points

Simple random sampling should, therefore, generally be used either in conjunction with other sampling methods or in cases involving only small study populations [SPRINGER and McCLURE, 1988].

- **Stratified random sampling.** The parent population has first to be subdivided or stratified into more homogeneous subparts of the whole. This can be done by use of previously acquired empirical knowledge or by applying methods for testing the significance of the differences between these *a priori* classes (see also Section 5.6). The samples have then to be taken randomly in the subareas, the so-called **strata**. Reduced variability within each stratum produces stratified sampling estimates which have smaller sampling errors than the corresponding simple random sampling estimates from the same sample size.

In comparison with single random sampling, the procedure of stratified random sampling has the advantage of being more precise, because of the greater resolution, and thus reducing the variability arising from sampling error [PETERSEN and CALVIN, 1986]. The design of different random sampling plans, including the possibility of structuring in layers, clots, degrees, or phases, is described by HILDEBRANDT [1987]. After classification of the total area to be investigated into more homogeneous subunits, a random selection without any restrictions should be preferred.

Important empirical criteria for finding homogeneous subareas may be [WEHRENS et al., 1993]:
- climatic conditions
- morphological conditions
- geomorphological conditions
- human activities

Systematic Sampling

The first sample is taken randomly and the rest are selected on a systematic basis, for example in a definite step width or, e.g. for packaged food, every tenth unit. Systematic sampling is easier, quicker, and cheaper than random sampling. The precision of the results obtained is mainly influenced by the distance between the sampling points. The primary disadvantage of systematic sampling is that no valid estimate of sampling error can be calculated from a single sample [SPRINGER and McCLURE, 1988]. Systematic sampling is, nevertheless, often used in environmental pollution studies because of its simplicity. The different types of grid plan used as the primary tool for systematic sampling are described in Section 4.5.3.2.

Judgmental Sampling

The most widely used method of nonprobabilistic sampling is judgmental or "targeted" sampling. The selection takes place according to the judgment or experience of the sampling specialists. Statistical approaches cannot be applied. The risk of applying judgmental sampling is the very subjective character of this technique; this can lead to systematic errors. Without additional data, the estimates cannot be extrapolated to the whole population. Judgmental sampling may lead to smaller errors than random sampling if the number of observations is small and the place was suitably chosen [FORTUNATI et al., 1994]. In small areas, judgmental sampling can serve as an initial approach (primary sampling) for subsequent systematic mapping. In this way, judgmental sampling enables assessment of the state of pollution, increasing the representativeness of the single analysis while minimizing the total effort [FINNERN, 1988].

The different types of sampling can be realized in **rigid or hypothesis-directed plans**. The various strategies and sampling approaches can be used separately or together, depending on needs and costs.

An empirical, but often practical way of sampling has been presented by EXNER et al. [1985]. They divided an area into suitable clean-up units of practical size to solve a problem of hazardous waste. Inside these strips, samples were taken along transects and some composite samples were formed from different combinations of the individual samples. After homogenization of the composite samples randomly selected aliquots of soil were taken and analyzed. The estimated arithmetic mean and its standard deviation enabled computation of an upper confidence limit for the mean concentration of the clean-up units.

4.5.3.2 Grid Plans

In practice, the plans generally used for environmental sampling are called grid plans. These rigid plans have to be adapted to the specific environmental situation [SCHULIN et al., 1993]. If no previous knowledge exists, the distance between the sampling points of the grid has to be optimized in the sampling process. The criterion is, on the one hand, the minimization of the largest distance or area which is not sampled [SCHOLZ et al., 1994] and, on the other hand, the maximization of the sampling distance while still maintaining the representativeness of sampling (see Sections 8.1.2 and 9.1). A brief description is given for the different grid plans. For a more detailed discussion see [NOTHBAUM et al., 1994; SCHOLZ et al., 1994].

Tab. 4-3 demonstrates the relationship between the assumed pollutant distribution and the grid plan which should be used for sampling.

Fig. 4-7 shows principal examples of the various types of rigid grid plans.

Most legally binding sampling procedures merely demand regular distribution of the sampling points over the area under investigation [ABFKLÄRV, 1992; NETHERLANDS NORMALISATIE INSTITUUT, 1991]. Other fixed procedures suggest a minimum distance between the sampling points of 50 m, or, if an area >100 hectares is to be investigated, 100 m [LANDESANSTALT FÜR ÖKOLOGIE, 1988].

4.5 Sampling Plans and Programs

Tab. 4-3. Relationship between the assumed pollutant distribution and the grid plan to be used

Assumed distribution	Grid plan
No assumption	Rectangular or bottle rack grid
Relatively homogeneous distribution	Rectangular or bottle rack grid
Centered contamination	Polar grid or stratified rectangular or bottle rack grid
Linear contamination	Linear grid, several diagonals (sometimes in zig zag manner, e.g. in the shape of capital letters: N, W, or X)
Special cases	Problem-adapted grid

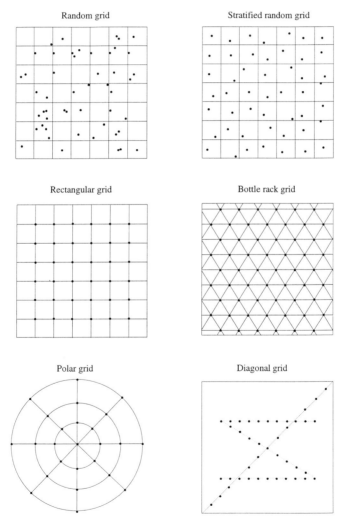

Fig. 4-7. Principal examples of various types of rigid grid plan

If there is no information about the distribution of the pollutant, or it can be assumed that the distribution is relatively homogeneous, **rectangular or bottle rack grids** (Fig. 4-7) should be used. Bottle rack grids are also referred to as regular triangular grids [PARKHURST, 1984]. The probability of finding a limited area of contamination is higher if bottle rack grids rather than rectangular grids are used [NOTHBAUM et al., 1994; SCHOLZ et al., 1994]. In the case of very dense grids, the results obtained from both types are similar. From the statistical point of view, bottle rack grids should be preferred to rectangular grids. Otherwise, from the practical point of view, it has to be taken into consideration that the effort of fixing the sampling points is greater than for rectangular grids. Each second row of the bottle rack plan has to be shifted by half of the distance between the sampling points. If the sampling personnel do not have adequate experience, rectangular grids should be used.

In the case of a centered contamination distribution, the sampling points in the highest loaded area should be closer together than those in areas more distant from the source of pollution. It is obvious that *a priori* information (resulting from record search and/or primary sampling) must exist. Because of easier handling, stratified rectangular grid plans seem more suitable than **polar grids**. An excellent example of the application of a problem-adapted stratified rectangular grid plan to the assessment of dioxin pollution in the surroundings of Seveso (Italy) is described by FORTUNATI et al. [1994].

If the contamination took place in one dimension only, as for example in rivers or groundwater streams, a simple **linear or diagonal grid** should be selected.

A practical method for the investigation of two-dimensional contamination, which includes the use of grid plans and easy-to-handle composite samples, is described in the literature [ASSMUS et al., 1985] under the name **strip sampling**.

The area under investigation has to be divided into rectangular subunits. Composite samples are then taken from both horizontal and vertical strips. The results obtained from the horizontal strips must be compared with those from the vertical strips with regard to their variability. The characterizing parameter of the variability is defined as:

$$\text{Parameter of variability} = \frac{x_{\max} - x_{\min}}{n} \qquad (4\text{-}39)$$

$x_{\max}$ – maximum value in the strip
$x_{\min}$ – minimum value in the strip
n – number of strips

Fig. 4-8 illustrates an example of strip sampling for determination of lead pollution in an agricultural area [KRIEG, 1995].

The mean values from the strips (numbers on the right margin and on the bottom of Fig. 4-8) give a good overview of the areal distribution of the lead content of the soil. The parameter of variability according to Eq. 4-39 is 45 mg kg^{-1} in the direction of the x-axis and 23.8 mg kg^{-1} in the direction of the y-axis. This indicates that the variability in the horizontal direction is higher, and so easy-to-handle strip sampling gives a first impression of the areal distribution of contamination. The method can serve as a useful

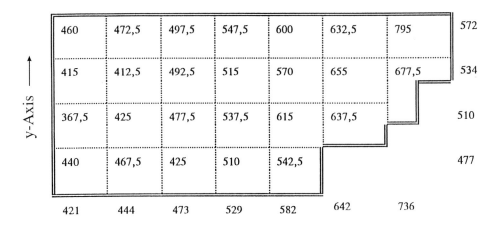

Fig. 4-8. Strip sampling for assessment of lead pollution in an agricultural area (values in mg kg^{-1})

primary sampling procedure. A more quantitative study of the heterogeneity can be conducted by applying univariate analysis of variance, as described in detail in Section 9.1.3.1.

4.5.3.3 Primary Sampling

The whole sampling process may be subdivided into two practical steps: primary and secondary sampling. This division is described by FLATMAN et al. [1988] for the application of geostatistical methods in particular. These authors differentiate between primary sampling to obtain the internal correlation of the sampling locations in space and the secondary step for map-making.

If no *a priori* information is available, sampling in a rigid plan should be the first step in the investigation of an environmental compartment with two different goals:

– determining the extent of the polluted areas
– determining the internal correlation between the sampling locations by computing the empirical semivariogram

Approaches to this problem include regular grids (square, rectangular, or radial), transects, and combinations of them. Some authors [McBRATNEY et al., 1981] suggest transect sampling. Other workers [FLATMAN et al., 1988] have reported that in pollution monitoring, transects alone give very unclear semivariograms. Yet others [SANDERS, 1982] report that for river sampling a lateral transect is the optimum strategy for obtaining representative samples. The different zones of relatively complete mixing in a river are tested with a criterion based upon a two-way analysis of variance. Combinations of transects with additional sampling points in the major and minor axes of the plume may also be useful [FLATMAN et al., 1988].

The orientation and extension of the sampling field, particularly in air and soil pollution monitoring, is, furthermore, highly dependent on meteorological parameters such as wind direction and magnitude and the geographical situation, for example hills as a natural barrier. The specific design should be adapted to the individual case study.

Empirical guidelines for determining the density of air monitoring networks have been published in the literature [BRYAN, 1976] and in governmental regulations [USEPA, 1971a; 1971b]. Some criteria concerning the minimum number of air monitoring sites on the basis of a priority classification have been published as USEPA regulations [1971a; 1971b].

The following procedures should be a practical means of **primary sampling** from a chemometric point of view: If no previous information on the extent of the polluted area is available, the transect samples should be spaced closer together at the grid center (for example near a dust-emitting stack) and further apart at the boundaries of the grid. The grid should be large enough to sample unpolluted locations at the outer boundary of the investigated region so that the background concentration in the specific environmental compartment can be assessed.

Sometimes, it may be very useful to investigate the microinhomogeneity (or microvariability) in the studied area by sampling a few points in a very narrow screen. An example of a study on sampling representativity and homogeneity for river sediments has been published by TRUCKENBRODT and EINAX [1995]. They investigated the inhomogeneity of river sediments in a very narrow rectangular grid with a sampling distance of only 1 meter.

In each case, initial sampling should be combined with field duplicates for quality analysis and control.

The optimum distance for initial sampling may be determined by use of previous knowledge of the investigated area or – if this does not exist or if this is not available – by environmental experience. In reality, however, the limiting factor is often the money available for sampling and analysis. Compromises must, therefore, be made between the resolution of the spatial structure on the one hand and expenditure on the other.

4.5.3.4 Secondary Sampling

Secondary sampling must be carried out with the sampling density determined by primary sampling with the goal of quantitative description of the contour map in the polluted area with an acceptable error of interpolation. This **secondary sampling** can be carried out by the following procedures:

- applying rigid plans with an optimized sampling distance over the whole territory
- stratifying the whole area to find more homogeneous subparts by applying techniques of uni- and multivariate analysis of variance (see also Sections 2.3, 3.3.9, and

5.6); the secondary sampling must then be carried out with optimized sampling distances in each subpart
- applying hypothesis-directed plans in combination with optimized sampling distance, the emission hypothesis, and environmental judgment; the disadvantage of this hypothesis-directed strategy is its partially subjective character

According to FLATMAN et al. [1988], secondary sampling with the goal of map making consists of the following steps:

- Boundary fixing
The extent of the sampling grid has first to be chosen.

- Compositing and nugget
The next step is choosing the sample support [STARKS, 1986]. A compromise between the number of subsamples which can be mixed for a gross sample and the resolution of the microinhomogeneity has to be made.

- Distance between sampling locations
The required distance has to be chosen from the empirical semivariogram or (if the first sampling was done equidistantly) by autocorrelation analysis also (see example for soil sampling in Sections 9.1 and 9.4). Clearly, the required distance depends on the relationship between nugget effect and sill. The length determination is described in detail by YFANTIS et al. [1987].

- Grid orientation and shape
If the correlation in space is anisotropic, an optimum sampling strategy should reflect this. The sides of a rectangular grid have to be in the same ratio as the ranges of correlation for the corresponding semivariograms [DAVID, 1977; BARNS, 1980]. In practice, the number of samples is often too low for detailed quantitative description of this anisotropy. Regularly spaced sampling grids are often used to search for toxic materials in soils, landfills, and dumps. PARKHURST [1984] explains that triangular grids are preferable to square grids because triangles give a better coverage. A disadvantage is that a triangular grid might be slightly more difficult to lay out than a square one. Taking samples in a triangular grid is recommended if the nugget effect is relatively large [FLATMAN et al., 1988].

4.5.4 Sampling Programs

Sampling programs may have exploratory or monitoring aims. An **exploratory procedure** determines the degree of pollution and the length, area, or volume of the environmental compartment under investigation. **Monitoring procedures** help the observation of temporal changes in environmental loading and ranges of pollution. An exploratory study is usually the preliminary step to further monitoring.

The sampling program must include all the activities of the entire sampling procedure, planned and documented in detail in the sampling protocol [KRATOCHVIL and TAYLOR, 1981]. The traceability of the sampling procedure has to be guaranteed. Detailed information on the requirements of sampling checklists is given in the literature [KEITH, 1988; FORTUNATI et al., 1994]. The following **steps of the sampling program** have to be specified.

Formulation of the Purpose of Sampling
What knowledge has to be obtained? What are the objectives of the study? Some important aims of environmental investigations are [NOTHBAUM et al., 1994]:

– Determination of medium values
  This task is of interest if, for example, an average value of a nutrient element inside an agricultural area has to be determined or the average value of a water component across a river section has to be characterized. Often, the answer obtained serves as a baseline study. By means of composite samples with an optimum distribution of the sampling points, an objective assessment is possible with relatively little expenditure.

– Determination of maximum values
  To answer this question higher expenditure is necessary. A grid plan has to be selected which guarantees the detection of existing maximum values in order to obtain information on the distribution which enables the reliable assessment of such extreme values. Subareas with higher loads must be assessed with a sufficient accuracy also.

– Determination of value distribution
  If the loading in an area results from the pollution of an emitter or discharger, the linear or spatial distribution of the parameters of interest must be determined. A relatively narrow transect or grid must, therefore, be selected. The extent of contamination can be assessed.

– Verification of an input hypothesis
  If an assessment of the degree of pollution resulting from emission from an anthropogenic source is required, the sampling program has to be adapted according to the input and distribution hypothesis. A general suggestion cannot be given.

Expenditure on sampling and the subsequent analytical determination of the components increases from the first to the last step of the above list.

The following questions should be answered before going into greater detail on the sampling process [NOTHBAUM et al., 1994]:

– Is it possible to obtain the information of interest by the planned investigation?
  The samples which have to be taken must reflect the properties of the investigated subject without distortion; in other words, the samples must be **representative** according to the purpose of the investigation. Specific examples of the statistical examination of sampling representativeness in different environmental media are given in Sections 7.1, 8.1, and 9.1. The amount of data has to be **sufficient**. A principal method

for the *a priori* estimation of the number of samples required is explained in Section 4.3.1. All parameters characteristic of the problem have to be determined. The parameters which have to be analyzed must be **indicative** for the specific case.

– Are the results of the investigation **reliable** and **valid**?
The errors of the whole analytical process must be quantified. The parameters characterizing the pollution and relevant interacting parameters (for example soil parameters or meteorological influences) must both be determined.

– Is it possible to obtain the required knowledge with less **expenditure**?
The sampling procedure should be optimized. In general, a sequential procedure according to Fig. 4-9 may reduce the expenditure.

Development of Steps for the Sampling and Analysis Plan
– Record search (preliminary data collection). After information has been gathered concerning the identification of contaminants or potential contaminants at a site, and the existing levels of these contaminants and their distribution, it is necessary to combine this *a priori* knowledge with the pedogenic, geological, and/or hydrological characterization of the area in question to determine the potential dispersion of the contaminants over the investigated site [BUSCHE and BURDEN, 1991].
The most important information is visual inspection of the area to be investigated, study of results from previous investigations (archive material), interrogation of workers in emitting plants or of inhabitants of polluted sites, and information on geohydrological and climatic conditions.
A detailed description of information required before commencing soil pollution studies is given in the literature [NETHERLANDS NORMALISATIE INSTITUUT, 1991; ARBEITSGRUPPE BODENKUNDE, 1982; BLUME et al., 1989].
– Primary sampling
– Secondary sampling

Quality Assurance Program
– Quality aims (often fixed in standards)
– Quality control and assurance of the sampling process
– Selection of the corresponding blanks
– Selection of corresponding control sites

Instruction and Training of the Personnel dealing with Sampling
An excellent example of a whole sampling program for the assessment of polluted soils is given by FORTUNATI et al. [1994]. The authors describe in detail the development of a problem-adapted sampling strategy for assessment of dioxin pollution in the surroundings of Seveso (Italy). The design of a decision-support system and a case study on the sampling of aquatic sediments in lakes is presented by WEHRENS et al. [1993]. The different steps of the sampling program are described in detail taking into account constraints such as the desired precision and the maximum sampling costs.

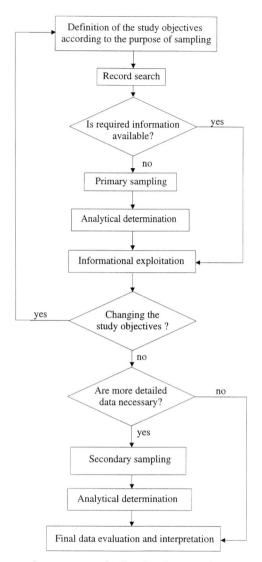

Fig. 4-9. Sampling strategy for assessment of polluted environmental compartments

The principal design of a sampling plan for the assessment of polluted environmental compartments is given in Fig. 4-9. In principle, the steps of the sampling strategy are transferable to all problems in space-time reality.

The definition of the study objectives according to the purpose of sampling should specify *a priori* the extent of the contamination (it may be changed as a result of the sampling process), the parameters which have to be analyzed, and the accuracy required. The record search as the second step of sampling strategy means the study of previous investigations, references from the emitting or discharging plant, and interrogation of workers or neighbors. The following primary sampling has to be carried out in a "raw" quadratic grid with empirical fixing of the sampling step width by analogy with similar investigations and in relation to possible expenditure.

On the basis of the analytical results from the primary sampling, the following investigations should be carried out in order to obtain a maximum of information:

– homogeneity testing
– stratification
– definition of more homogeneous subareas for further investigation
– working out the problem-adapted sampling plan with a sampling density according to the loading state of the subarea

If secondary sampling is required to provide a more detailed assessment of the problem under investigation, the analytical results from this second sampling step may be finally evaluated and interpreted together with the results from primary sampling:

– Analysis of the representativeness of the samples taken by applying correlation analysis (see Sections 6.6 and 9.1.3.3) or the kriging estimation (see Sections 4.4.3 and 9.4.3.2): If the samples are not representative, secondary sampling must be repeated. Samples at additional sampling locations must be taken to obtain a closer grid.
– Estimation of loaded areas and contamination map-making: Application of geostatistical methods (see Sections 4.4 and 9.4) for the assessment of the area and the degree of pollution.
– Identification of the pollutant pattern: Application of multivariate statistical methods (see, for example, Section 9.2) for the detection of emitters or origins.

At this point, it is very important to direct attention to the fact that each sampling program has to be performed sequentially. The amount of sampling and expenditure on the investigation can often be reduced considerably if some pollution hypotheses can be excluded in an early stage of the investigation. At each stage of the sampling program one should attempt to change the old questions or to formulate new ones.

References

AbfKlärV. Klärschlammverordnung – AbfKlärV vom 15.4.1992. Bundesgesetzblatt, Teil I (**1992**) 912
Akin, H., Siemes, H.: Praktische Geostatistik, Springer, Berlin, Heidelberg, New York, London, Paris, Tokyo, **1988**
Arbeitsgruppe Bodenkunde der geologischen Landesämter und der Bundesanstalt für Geowissenschaften in der Bundesrepublik Deutschland: Bodenkundliche Kartieranleitung, 3. Aufl., Stuttgart (in Komm.): E. Schweizbart'sche Verlagsbuchhandlung, **1982**
Assmus, R.A., Fixen, P.E., Evenson, P.D.: J. Fertilizer Issues 2 (**1985**) 136
Barcelona, M.J. in: Keith, L.H. (Ed.): Principles of Environmental Sampling, American Chemical Society, Washington, D.C., **1988**, pp. 3
Barcelona, M.J., Lettenmaier, D.P., Schlock, M.R.: Environ. Monitor. Assess. 12 (**1989**) 149

Barnard, T.E.: in: Einax, J. (Ed.): Chemometrics in Environmental Chemistry. Statistical Methods, Vol. 2, Part G, in the series: Hutzinger, O. (Ed.): Handbook of Environmental Chemistry, Springer, Berlin, **1995**, pp. 1

Barns, M.G.: The Use of Kriging for Estimating the Spatial Distribution of Radionuclides and other Spatial Phenomena; TRAN STAT (Statistics for Environmental Studies) No. 13, Battelle Memorial Institute, Richland, **1980**, pp. 1

Benedetti-Pichler, A. in: Berl, W.M. (Ed.): Physical Methods of Chemical Analysis, Vol. 3, Academic Press, New York, **1956**, p. 183

Berrow, M.L.: Anal. Proc. 25 (**1988**) 116

Blume, H.P., Burghardt, W., Cordsen, E., Finnern, H., Fried, G., Grenzius, R.: Empfehlungen des Arbeitskreises Stadtböden der deutschen bodenkundlichen Gesellschaft für die bodenkundliche Kartieranleitung urban, gewerblich und industriell überformter Flächen, UBA-Texte, Berlin, **1989**, pp. 18

Borgman, L.E., Quimby, W.F. in: Keith, L.H. (Ed.): Principles of Environmental Sampling, American Chemical Society, Washington, D.C., **1988**, pp. 25

Brands, G.: Fresenius' Z. Anal. Chem. 314 (**1983**) 6

Bryan, R.J. in: Stern, A.C. (Ed.): Air Pollution, Vol. III, 3rd Ed., Academic Press, New York, San Francisco, London, **1976**, pp. 343

Busche, F.D., Burden, D.S.: Hazard. Mater. Control 4 (**1991**) 35

Committee on Environmental Improvement of the American Chemical Society: Anal. Chem. 52 (**1980**) 2242

Danzer, K., Singer, R.: Mikrochim. Acta (**1985**) I, 219

David, M.: Geostatistical Ore Reserve Estimation, Elsevier, Amsterdam, Oxford, New York, **1977**

de Wijs, H.J.: Geologie en Mijnbouw 13 (**1951**) 365 cited from [AKIN and SIEMES, 1988]

de Wijs, H.J.: Geologie en Mijnbouw 15 (**1953**) 12 cited from [AKIN and SIEMES, 1988]

DIN 38414, Teil 7, 1/**1983** (Normenausschuß Wasserwesen im DIN Deutsches Institut für Normung e.V.): Deutsche Einheitsverfahren zur Wasser-, Abwasser- und Schlammuntersuchung, Schlamm und Sedimente (Gruppe S), Aufschluß mit Königswasser zur nachfolgenden Bestimmung des säurelöslichen Anteils von Metallen

DIN 4021, 10/**1990** (Normenausschuß Bauwesen im DIN Deutsches Institut für Normung e.V.): Aufschluß durch Schürfe und Bohrungen sowie Entnahme von Proben

Doerffel, K.: Statistik in der analytischen Chemie, 5. Aufl., VCH, Weinheim, **1990**, pp. 135

Doerffel, K., Eckschlager, K.: Optimale Strategien in der Analytik, Deutscher Verlag für Grundstoffindustrie, Leipzig, **1981**

Einax, J., Soldt, U.: Fresenius' J. Anal. Chem. 351 (**1995**) 48

Exner, J.H., Keffer, W.D., Gilbert, R.O., Kinnison, R.R.: Hazard. Waste Hazard. Mater. 2 (**1985**) 503

Finnern, H.: in: Wolf, K., van den Brink, W.-J., Colon, F.-J. (Eds.): Contaminated Soil '88, Vol. 1, Kluwer, Dortrecht, **1988**

Flatman, G.T., Englund, E.J., Yfantis, A.A. in: Keith, L.H. (Ed.): Principles of Environmental Sampling, American Chemical Society, Washington, D.C., **1988**, pp. 73

Fortunati, G.U., Banfi, C., Pasturenzi, M.: Fresenius' J. Anal. Chem. 348 (**1994**) 86

Fränzle, O. in: Markert, B. (Ed.): Environmental Sampling for Trace Analysis, VCH, Weinheim, New York, Basel, Cambridge, Tokyo, **1994**, pp. 305

Gardner, M.J.: Anal. Proc. 32 (**1995**) 115

Garfield, F.M.: J. Assoc. Off. Anal. Chem. 72 (**1989**) 405

Gilbert, R.O., Simpson, J.C.: Environ. Monit. Assess. 5 (**1985**) 113

Green, R.H.: Sampling Design and Statistical Methods for Environmental Biologists, Wiley, New York, **1979**

Gudernatsch, H.: Forum Städte-Hygiene 33 (**1982**) 129

Gy, P.M.: Sampling of Particulate Materials, Theory and Practice, 2nd Ed., Elsevier, Amsterdam, **1982**

Gy, P.M.: Anal. Chim. Acta 190 (**1986**) 13

Gy, P.M.: Process Control Qual. 1 (**1990**) 15

Gy, P.M.: Heterogeneity–Sampling–Homogenization, Elsevier, Amsterdam, **1991 a**

Gy, P.M.: Mikrochim. Acta (**1991 b**) II, 457

Gy, P.M. (Ed.): Sampling of Heterogeneous and Dynamic Material System. Theories of Heterogeneity, Sampling and Homogenizing. Data Handling in Science and Technology, Elsevier, Amsterdam, **1992**

Håkanson, L.: Water Resources Res. 20 (**1984**) 41

Hannapel, S.: in: Markert, B. (Ed.): Environmental Sampling for Trace Analysis, VCH, Weinheim, New York, Basel, Cambridge, Tokyo, **1994**, pp. 493

Henrion, R., Henrion, G., Szukalski, K., Fabian, I., Thiesies, A., Heininger, P.: Fresenius' J. Anal. Chem. 340 (**1991**) 1

Hildebrandt, G.: Dt. Lebensm.-Rundsch. 83 (**1987**) 205

Hoffmann, G., Schweiger, P.: Staub-Reinhalt. Luft 41 (**1981**) 443

Hoffmann, P.: Nachr. Chem. Tech. Lab. 40 (**1992**) M2

Holstein, F. in: Probenahme – Theorie und Praxis, Heft 36 der Schriftenreihe der Gesellschaft Deutscher Metallhütten- und Bergleute, Verlag Chemie, Weinheim, Deerfield Beach/Florida, Basel, **1980**, pp. 93

Horwitz, W.: Pure Appl. Chem. 62 (**1990**) 1193

Ingamells, C.O.: Geochim. Cosmochim. Acta 38 (**1974a**) 1225

Ingamells, C.O.: Talanta 21 (**1974b**) 141

Ingamells, C.O.: Talanta 23 (**1976**) 263

Ingamells, C.O., Switzer, P.: Talanta 20 (**1973**) 547

ISO 1213-2: Solid Mineral Fuels – Vocabulary – Part 2: Terms Relating to Sampling, Testing and Analysis, **1992**

ISO/DIS 5667-1–13: Water Quality – Sampling, **1991**

ISO 9359: Air Quality – Stratified Sampling Method for Assessment of Ambient Air Quality, **1989**

ISO/DIS 10381-1–6: Soil Quality – Sampling, **1995**
  Part 1: Guidance on the Design of Sampling Programs
  Part 2: Guidance on Sampling Techniques
  Part 3: Guidance on Safety
  Part 4: Guidance on the Procedure for the Investigation of Natural, Near-natural, and Cultivated Sites
  Part 5: Guidance on the Procedure for the Investigation of Soil Contamination of Urban and Industrial Sites
  Part 6: Guidance on the Collection, Handling and Storage of Soil for the Assessment of Aerobic Microbial Processes in the Laboratory

ISO 11074-2: Soil Quality – Vocabulary – Part 2: Terms and Definitions Relating to Sampling, Committee Draft, in preparation

ISO/DIS 11464: Soil Quality – Pretreatment of Samples for Physico-Chemical Analyses, **1995**

ISO/DIS 14507: Soil Quality – Sample Pretreatment for Determination of Organic Contaminants in Soil, **1995**

Journel, A.G.: Math. Geol. 15 (**1983**) 445

Journel, A.G. in: Keith, L.H. (Ed.): Principles of Environmental Sampling, American Chemical Society, Washington, D.C., **1988**, pp. 45

Journel, A.G., Huigbregts, Ch.: Mining Geostatistics, Academic Press, London, New York, San Francisco, **1978**

Kateman, G. in: Topics in Current Chemistry: Chemometrics and Species Identification, Vol. 141, Akademie-Verlag, Berlin, **1987**, pp. 43

Kateman, G., Pijpers, F.W.: Quality Control in Analytical Chemistry, Wiley, New York, Chichester, Brisbane, Toronto, **1981**, pp. 15

Kateman, G., Pijpers, F.W.: Chemom. Int. Lab. Syst. 4 (**1988**) 187

Keith, L.H. (Ed.): Principles of Environmental Sampling, American Chemical Society, Washington, D.C., **1988**

Kraft, G.: Erzmetall 31 (**1978**) 53

Kraft, G. in: Kienitz, H., Bock, R., Fresenius, W., Huber, W., Tölg, G. (Eds.): Analytiker-Taschenbuch, Bd. 1, Akademie-Verlag, Berlin, **1980**, pp. 3

Kraft, G.: in: Markert, B. (Ed.): Environmental Sampling for Trace Analysis, VCH, Weinheim, New York, Basel, Cambridge, Tokyo, **1994**, pp. 3

Kratochvil, B., Taylor, J.K.: Anal. Chem. 53 (**1981**) 924A
Kratochvil, B., Wallace, D., Taylor, J.K.: Anal. Chem. 56 (**1984**) 113R
Kratochvil, B., Goewie, C.E., Taylor, J.K.: Trends Anal. Chem. 5 (**1986**) 253
Krieg, M.: Dissertation, Friedrich-Schiller-Universität, Jena, **1995**
Krieg, M., Einax, J.: Fresenius' J. Anal. Chem. 348 (**1994**) 490
Krige, D.G.: M.Sc. Thesis, University of Witwatersrand, **1951a** cited from [AKIN and SIEMES, 1988]
Krige, D.G.: J. Chem. Metall. Min. Soc. S. Africa 52 (**1951b**) 119 cited from [AKIN and SIEMES, 1988]
Krishnaiah, P.R., Rao, C.R. (Eds.): Handbook of Statistics, Vol. 6, Sampling, Elsevier, Amsterdam, New York, Oxford, **1988**
Landesanstalt für Ökologie, Landschaftsentwicklung und Forstplanung, Nordrhein-Westfalen: Mindestuntersuchungsprogramm Kulturboden – zur Gefährdungsabschätzung von Altablagerungen und Altstandorten im Hinblick auf eine landwirtschaftliche oder gärtnerische Nutzung, Recklinghausen, **1988**
Lemmer, I.C.: Math. Geol. 18 (**1986**) 589
Liebich, V., Ehrlich, G., Stahlberg, U., Kluge, W.: Fresenius' Z. Anal. Chem. 335 (**1989**) 945
Liebich, V., Ehrlich, G., Herrmann, U., Siegert, L., Kluge, W.: Fresenius' J. Anal. Chem. 343 (**1992**) 251
Markert, B. (Ed.): Environmental Sampling for Trace Analysis, VCH, Weinheim, New York, Basel, Cambridge, Tokyo, **1994**
Massart, D.L., Vandeginste, B.G.M., Deming, S.N., Michotte, Y., Kaufman, L.: Chemometrics: A Textbook, Elsevier, Amsterdam, Oxford, New York, Tokyo, **1988**
Matheron, G.: Ann. Mines 9 (**1957**) 566 cited from [AKIN and SIEMES, 1988]
Matheron, G.: Econ. Geol. 58 (**1963**) 1246 cited from [AKIN and SIEMES, 1988]
Matheron, G.: Les variables regionalisèes et leur estimation, Masson, Paris, **1965** cited from [AKIN and SIEMES, 1988]
McBratney, A.B., Webster, R., Burgess, T.M.: Comput. Geosci. 7 (**1981**) 331
Minkkinen, P.: Anal. Chim. Acta 196 (**1987**) 237
Myers, D.E.: Chemom. Int. Lab. Syst. 11 (**1991**) 209
Netherlands Normalisatie Instituut: Dutch Draft Standard – Soil: Investigation Strategy for Exploratory Survey, 1st Ed., UDC 628.516, 9/**1991**
Nothbaum, N., Scholz, R.W., May, T.W.: Probenplanung und Datenanalyse bei kontaminierten Böden, Erich Schmidt Verlag, Berlin, **1994**
Paetz, A., Crößmann, G.: in: Markert, B. (Ed.): Environmental Sampling for Trace Analysis, VCH, Weinheim, New York, Basel, Cambridge, Tokyo, **1994**, pp. 321
Parkhurst, D.F.: Environ. Sci. Technol. 18 (**1984**) 521
Petersen, R.G., Calvin, L.D.: in: Klute, A. (Ed.): Methods of Soil Analysis, Part 1, 2nd Ed., American Society of Agronomy Inc., Soil Science Society of America, Madison, **1986**, pp. 33
Pijpers, F.W.: Anal. Chim. Acta 190 (**1986**) 79
Pitard, F.F.: Pierre Gy's Sampling Theory–Sampling Practice, Vols. 1 and 2, CRC Press, Boca Raton, Florida, **1989**
Reed: School of Mines Quarterly 3 (1881/82) 253 and 6 (1884/85) 351 cited from KRAFT, G. in: Ges. Deutscher Metallhütten- und Bergleute e.V. (Ed.): Probenahme Theorie und Praxis, Verlag Chemie, Weinheim, Deerfeld, Beach/Florida, Basel, **1980**, pp. 1
Rousseeuw, P.J., Leroy, A.M.: Robust Regression and Outlier Detection, Wiley, New York, Chichester, Brisbane, Toronto, Singapore, **1987**
Rump, H.H., Herklotz, K. in: Weber, H.H. (Ed.): Altlasten, Springer, Berlin, Heidelberg, New York, London, Paris, Tokyo, Hong Kong, Barcelona, **1990**, pp. 108
Sanders, T.G.: Water SA 8 (**1982**) 169
Scholz, W.S., Nothbaum, N., May, T.W.: in: Markert, B. (Ed.): Environmental Sampling in Trace Analysis, VCH, Weinheim, New York, Basel, Cambridge, Tokyo, **1994**, pp. 335
Schulin, R., Webster, R., Meuli, R.: Technical Note on Objectives, Sampling Design, and Procedures in Assessing Regional Soil Pollution and the Application of Geostatistical Analysis in such Surveys, ETH Zürich, Institute for Terrestrial Ecology, **1993**

Sharaf, M.A., Illman, D.L., Kowalski, B.R.: Chemometrics, Wiley, New York, Chichester, Brisbane, Toronto, Singapore, **1986**

Singh, A.K., Ananda, M.M.A., Sparks, A.R.: Anal. Chim. Acta 277 (**1993**) 255

Springer, J.A., McClure, F.D.: J. Assoc. Off. Anal. Chem. 71 (**1988**) 246

Starks, T.H.: Math. Geol. 18 (**1986**) 529

Stoeppler, M. (Ed.): Probennahme und Aufschluß, Springer, Berlin, Heidelberg, New York, London, Paris, Tokyo, Hong Kong, Barcelona, Budapest, **1994**

Taggart, A.: Handbook of Mineral Dressing, Ores and Industrial Minerals, Wiley, New York, **1948**, pp. 161

Thompson, M., Maguire, M.: Analyst 118 (**1993**) 1107

Thompson, M., Ramsey, M.H.: Analyst 120 (**1995**) 261

Truckenbrodt, D., Einax, J.: Fresenius' J. Anal. Chem. 352 (**1995**) 437

USEPA, United States Environmental Protection Agency, Fed. Regist. 36, No. 158 (**1971a**)

USEPA, United States Environmental Protection Agency, "Guidelines: Air Quality Surveillance Networks", No. AP-98, Research Triangle Park, North Carolina, **1971b**

Visman, J.: Mater. Res. Stand. (**1969**) 9, 51, 62

Wegscheider, W. in: Günzler, H. (Ed.): Akkreditierung und Qualitätssicherung in der Analytischen Chemie, Springer, Berlin, **1994**, pp. 61

Wehrens, R., van Hoof, P., Buydens, L., Kateman, G., Vossen, M., Mulder, W.H., Bakker, T.: Anal. Chim. Acta 271 (**1993**) 11

Yfantis, A.A., Flatman, G.T., Behar, J.V.: Math. Geol. 19 (**1987**) 183

5 Multivariate Data Analysis

5.1 General Remarks

In the simplest case **multivariate** data are obtained by acquiring measurements of several features. Because of this the data are often considered as **multidimensional**. Both notions are used without considering the number of objects characterized by the measurements. Typically, in multivariate data analysis, one has to cope with both problems: many objects and features.

In the following subsections, after a short introduction to graphical methods which enable the visualization of multivariate data, we will focus first on basic methods which analyze data without the need for additional information (or **without** *a priori* information) about the measurements or numerical data. Roughly speaking these methods handle ungrouped data. With regard to objects *cluster analysis* and with regard to features *principal components* and *factor analysis* may be useful for data analysis. An optimum combination of both aspects is possible by applying *correspondence factor analysis*, which we will merely mention.

If we do have, and **include,** *a priori* knowledge in addition to the measurements or numerical data we should use methods from the second family of basic data analysis methods. Here the data are considered to be grouped in respect of the objects, or maybe in respect of the features. Within this family we may further distinguish between non-causally and causally determined data, or by analogy with correlation and regression, we may distinguish between multivariate **relationships** and **dependencies**.

Relationships between groups of objects may be investigated by *multivariate analysis of variance and discriminant analysis*, relationships between groups of features (variables) may be explored by *canonical correlation analysis*. Discriminant analysis appears as a borderline case: if we use it after a clustering method to confirm the groups found by cluster analysis the *a priori* knowledge about the nature of the group structure of the data is weak. If discriminant analysis is applied to evaluate a group structure caused by defined or even controlled factors, for example after use of the principles of experimental design, discriminant analysis may be seen as belonging to techniques investigating causal dependencies.

Dependencies are mainly investigated by *multiple or multivariate regression* (one dependent and several independent variables), by multidimensional multivariate regression or *partial least squares regression* (several dependent and several independent variables), and by the *method of simultaneous equations* (explicitly allowing for correlated independent variables). The last mentioned method may also be interpreted as combination of correlation and regression analysis.

Sometimes *canonical correlation* or *canonical analysis* is referred to as a central technique with factor and correspondence analysis considered in one branch (having no causal concepts) and multivariate regression and discriminant analysis in the other branch (based on causal concepts).

5.2 Graphical Methods of Data Presentation

5.2.1 Introduction

Graphical methods are methods of data handling which enable large amounts of data, or sometimes even vague data, to be visualized in a clear manner. Data sets with more than two variables or data sets influenced by time or location are often complicated. The experimenter faced with such large amounts of data often feels overwhelmed; methods for searching and demonstrating the character and the interactions within the data set are, therefore, required. Because the multivariate statistical methods, described in the following sections, require mathematical knowledge of the user (employer) who has to perform the calculations, and also the acceptance of the customer who has to use the results, simple graphical representations frequently have more power of persuasion than highly sophisticated statistical methods with exact statistical parameters.

The advantage of the following methods of graphical representation is the clear and simple presentation of the essential facts. Simple charts, like bar charts, x-y scatter diagrams or pie diagrams, which are also available in 3D-form are also suited to visual representation of data. They are not described because this section is devoted to treatment of multivariate data. Graphs for control charts, particulary for quality assurance and control, can be found in [FUNK et al., 1992; AQS, 1991].

5.2.2 Transformation

Multivariate data means that more than two variables are involved. Therefore, one should compare many variables with different dimensions. The measurements have different units and different ranges of quantity. As an example, in natural water, concentrations of 400 to 600 µg L^{-1} manganese, 2 to 4 mg L^{-1} phosphate and 50 to 90 ng L^{-1} polycyclic aromatic hydrocarbons have to be demonstrated in a three-dimensional graph. Commercial graphical computer programs standardize the ranges of the single features. The axis for manganese, for example, stretches between 400 and 600 µg L^{-1}. It can be seen that without suitable scaling of the axis, a demonstration of the essential interactions is not possible. For the same reason transformation of the features is necessary for multivariate data sets. If the original feature is x the following transformations are possible.

1. Minimum/Maximum Transformation

This is a range transformation. If the x-axis has the minimum value x_{min} and the maximum value x_{max} the transformed values are obtained by

$$z_i = \frac{x_i - x_{min}}{x_{max} - x_{min}} \tag{5-1}$$

z_i – transformed value of x_i

The transformed variable z then has a minimum value of zero and a maximum value of unity.

2. Mean Centering

Single values were computed for representation as follows:

$$x_{i,\,cent} = x_i - \bar{x} \tag{5-2}$$

$x_{i,cent}$ – centered value of x_i
$\bar{x}$ – mean value of variable x

The mean of the variable x then is the zero point on the new x_{cent}-axis. Mean centering is only possible, if the variances of the different features are similar. Otherwise, both mean centering and variance scaling are necessary.

3. Mean Centering and Variance Scaling (Autoscaling)

For a normal distribution of the x values, the mean and standard deviation, as the parameters of the distribution, can be used for **standardization.**
Single values were computed as:

$$z_i = \frac{x_i - \bar{x}}{s_x} \tag{5-3}$$

z_i – autoscaled value of x_i
s_x – standard deviation of variable x

The mean of the feature x is the zero point of the z-axis; the standard deviation of the new standardized feature z is unity.

These methods of standardization are suitable if interactions between the features have to be interpreted. One hundred and more cases or objects are shown as points in two- or three-dimensional diagrams. Similarities between objects can be demonstrated as clusters in a two- or three-feature space.

In the case of fewer objects with a large number of features another kind of illustration is necessary. If the single objects have to be compared, a diagram with two or three objects as the axis and the features as points is useful. For this application, a standardization of the objects is necessary. This kind of standardization employs the same procedures as above.

Standardization is always a mathematical procedure and cannot assess the relevance of the scattering. Random errors in the measurements of a constant feature have the same weight as large natural fluctuations. There is a risk involved in graphical methods without statistical parameters.

As an example, eight features (1 to 8) were created for the description of three geometrical figures: a circle, a square, and a triangle. These eight features have a simulated minimum random error for each of the geometrical figures. Without standardization they have the original shape (Fig. 5-1).

The effect of standardization of the features by use of the minimum/maximum method is demonstrated in Fig. 5-2.

The manner in which the objects are modified should be noted. The square becomes a point because all the single features of the square are the smallest in comparison with those of the other figures. The circle and the triangle were modified by scaling with the maximum values. If the circle has the maximum value of the single features 2, 3, 7,

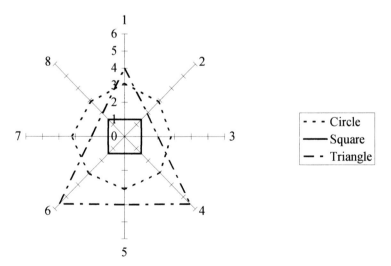

Fig. 5-1. Geometrical figures without standardization

5.2 Graphical Methods of Data Presentation

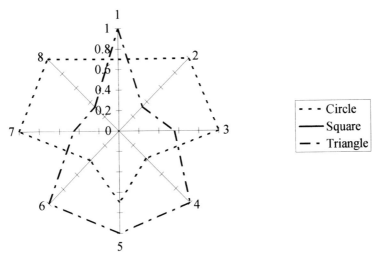

Fig. 5-2. Geometrical figures with minimum/maximum feature standardization

and 8, these values determine the range of values of the triangle. For features 1, 4, 5, and 6, the triangle determines the shape of the circle.

The standardization of objects by use of minimum/maximum method is outlined in Fig. 5-3. By this procedure values of all the features of one object were standardized between zero and one. Minimum values of each object were set to zero and maximum values to one. This particularly affects the shape of the circle because it has nearly constant values in each variable apart from the scattering. This scattering is now the character of this figure. The square has its maximum values in the diagonals, features 2, 4,

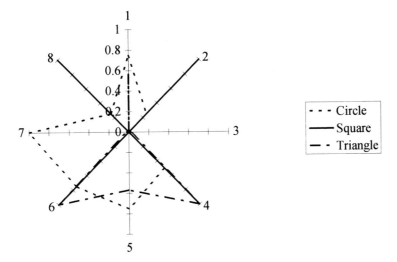

Fig. 5-3. Geometrical figures with minimum/maximum object standardization

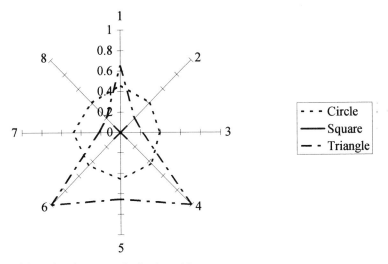

Fig. 5-4. Minimum/maximum standardization of features and objects

6, and 8. These values were set to one. The features 1, 3, 5, and 7 are the minimum values of the square and were set to zero. Therefore a cross was formed from a square. The new shape of the triangle can be explained analogously.

The simultaneous standardization of features and objects is the optimum for this example (Fig. 5-4). The minimum value of the whole data matrix is set to zero and the maximum value of the whole matrix is set to one. The original structures are more or less maintained.

This application of different methods of standardization demonstrates the possibilities and the pitfalls of these methods.

Statistical methods of data analysis generate statistical parameters which show the significance of any effects. Although standardization does not affect the statistical parameters, visual evaluation may be confusing, if methods of standardization are used incorrectly.

5.2.3 Visualization of Similar Features – Correlations

In general, the procedure used for visualization of the similarity between features, or for correlation of the features, is the same as that used for visualization of the similarity between objects. If one transposes the matrix of objects and features, the same methods may be used. It is, however, easier to imagine correlating features which describe the similarity of the functional shapes of variables and the similarity of objects, e.g. the similarity of different samples, than the other way around.

For this reason we deal with the visualization of relationships between the features. For three features only, there are many possibilities for graphical representation:

- three-dimensional scatter plot
- isolines plot
- contour plot
- three-dimensional bars

But for most cases of interest in the field of environmental research there are more than three features.

Matrix plot techniques are, therefore, an alternative. Matrix plots are normal scatter plots of two variables, each variable being plotted against each other. These scatter plots are arranged in a matrix. Fig. 5-5 is a matrix scatter plot of seven polycyclic aromatic hydrocarbons (PAH) in sediments at 45 sampling points in different rivers in Thuringia (Germany). This clearly demonstrates the relationships between the different PAH. Benzo(b)fluoranthene, benzo(k)fluoranthene, and benzo(a)pyrene have the strongest linear relationships. There is only slight correlation between naphthalene and the other features and between indeno(1,2,3-ghi)pyrene and the other features; these correlations are probably a result of the measurement error for these compounds.

In general, the matrix scatter plot is the direct graphical representation of the correlation matrix (see Section 5.4). Fig. 5-5 also demonstrates the histogram of the single features in the main diagonal as additional information.

5.2.4 Similar Objects or Groups of Objects

5.2.4.1 Nesting Techniques

For visualization of multivariate determined objects it is necessary to find possible ways of presenting a multivariate situation in a two-dimensional manner. KRZANOWSKI [1988] produced four-dimensional presentations by nesting two-dimensional objects into a two-dimensional diagram. A sketch of the principal course of the nesting technique is given in Fig. 5-6. One object is one single diagram in the total diagram. Objects are positioned in the total two-dimensional diagram according to features 1 and 2. Single two-dimensional diagrams describe features 3 and 4 of one object.

The main disadvantage of this technique is the different visual evaluation of the four equal features in the objects. The two main axes have a greater emphases than the features represented in the individual axis of the objects. The application of this method requires four-dimensional data sets; more variables cannot be treated.

Other techniques which enable representation of more than four features for different objects, e.g. stars in a two-dimensional diagram or glyphs in a two-dimensional diagram, share the same disadvantage as all nesting methods: the different visual evaluation of equal features.

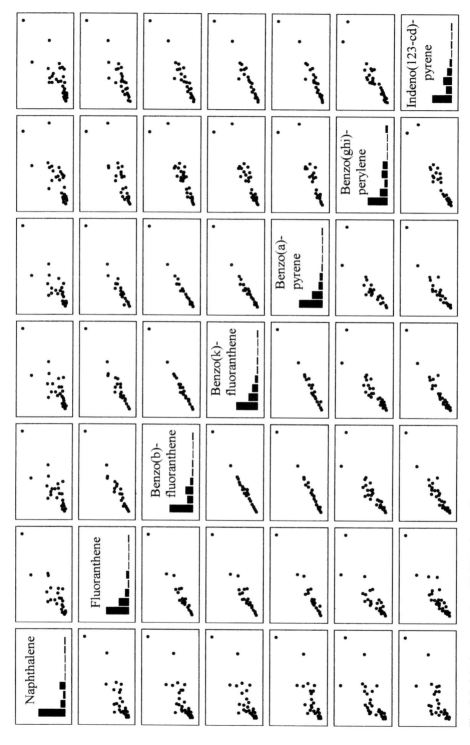

Fig. 5-5. Matrix scatter plot of PAH in river sediments

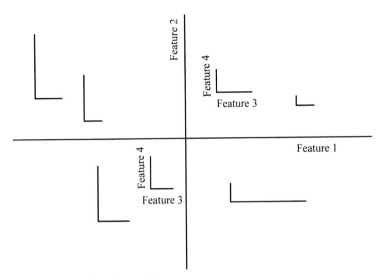

Fig. 5-6. Principal sketch of nesting technique

5.2.4.2 Star Plots

This technique allows the simultaneous treatment of about $m = 3$ to about $m = 10$ features for the visual assessment of similarities or dissimilarities of n objects. The m features are arranged as vectors with one starting point in a star. Values of the features (star axis) can be connected and a star is formed. Each object is represented as one star. Similarities between the objects are easily seen and human imagination is only necessary for pattern cognition.

Most applications require standardization and involve the potential hazard of distortion of the objects. The recognized similarities must, therefore, be proved with statistical methods. It is very dangerous to cut single features or to reduce the data set. This leads to a change of the data basis on which the standardization will be performed. A comparison with previously computed stars thus becomes impossible. Fig. 5-7 demonstrates a star plot with feature standardization for the example of the PAH in river sediments. Highly loaded sampling points are characterized by big stars and vice versa and similar patterns by similar stars. Similarities in the PAH patterns are particularly apparent for the last five objects.

Profile plotting is a similar technique to the construction of star plots. Features are arranged as bars in single diagrams for each object. Bars may also be connected by lines.

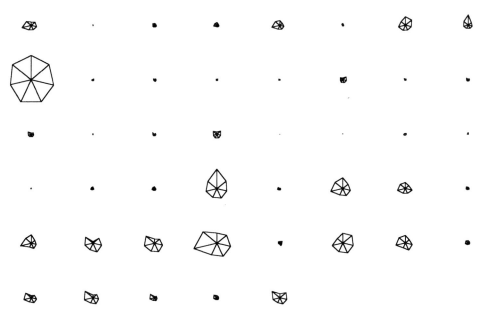

Fig. 5-7. Star plot of PAH in 45 selected river sediments from Thuringia, feature standardization. Clockwise: naphthalene (12.00), fluoranthene, benzo(b)fluoranthene, benzo(k)fluoranthene, benzo(a)pyrene, benzo(ghi)perylene, indeno(1,2,3-cd)pyrene

5.2.4.3 Pictoral Representation

CHERNOFF [1973] created an unusual graphical representation of multivariate data. He made the assumption that the "human pattern recognition ability" is best trained with human faces. Faces can be described with parameters like face width, ear level, half-face height, eccentric upper face, eccentric lower face, nose length, mouth centering, etc.

In the CHERNOFF faces representation, each parameter corresponds to one feature. Each face is one object. The CHERNOFF faces of the seven PAH in river sediments after feature standardization are demonstrated in Fig. 5-8. The information is the same as that represented in Fig. 5-7 by star plots but conclusions to be drawn are not so clear.

Similar techniques based on pictures with different parameters are described by HARTUNG and ELPELT [1992]. Castles and trees are also useful icons.

5.2.4.4 Functional Representation

Functional representation uses the parameter of a functional curve for the detection of similarities. In the 1970s, the programming of graphical methods required the expenditure of much effort, whereas functional representations were easy to program. AN-

Fig. 5-8. CHERNOFF faces of PAH in 45 selected river sediments from Thuringia, feature standardization. face width = naphthalene, ear level = fluoranthene, half-face height = benzo(b)fluoranthene, eccentric upper face = benzo(k)fluoranthene, eccentric lower face = benzo(a)pyrene, nose length = benzo(ghi)perylene, mouth centering = indeno(1,2,3-cd)pyrene

DREWS [1972], for example, uses a FOURIER curve $f(t)$ to demonstrate n-dimensional observations:

$$f(t) = \frac{x_1}{\sqrt{2}} + x_2 \sin(t) + x_3 \cos(t) + x_4 \sin(2t) + x_5 \cos(2t) + \cdots \qquad (5\text{-}4)$$

The argument t varies in the interval $-\pi$ to $+\pi$, the parameters $x_1, x_2, ..., x_5$ are the values of the different features of one object, so that one object vector is represented by $f(t)$. Curves close to each other represent similar objects.

The resulting function also possesses other interesting properties. The function preserves the total mean of the features from one object because the values of the features change the amplitudes, whereas only the number of features changes the frequencies. Also variances for one object are maintained. Distances between two functions reflect dissimilarities between the objects. The example of the circle, square, and triangle in Section 5.2.2 demonstrates the quality of this method. Eight features are included in each FOURIER curve. The circle and the square resemble each other more than the triangle (Fig. 5-9).

150   5 *Multivariate Data Analysis*

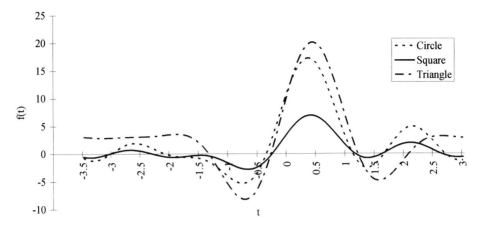

Fig. 5-9. ANDREWS plot of geometrical figures

A problem with the ANDREWS plot is that if variables of one object are permuted, a quite different picture arises because the amplitudes of the different harmonic oscillations change. It is, therefore, better if the functions are associated with principal components (see Section 5.4) of the set of features. Principal components are linear combinations of the features and explain the total variance of the data in descending order. In this manner the sequence of features, i.e. the sequence of the single principal components, should not be permuted in the ANDREWS plot.

5.2.5 Representation of Groups

5.2.5.1 Box-Whisker Plots

The box-whisker plot is not a plotting technique for multivariate representation of data, but is, rather, a very useful technique for comparing different groups of objects in relation to one feature. Every group is presented as a box in a diagram. The centre line of the box may be the group mean. The edges of the box may relate to the variance of the group. Horizontal lines represent the minimum and maximum values of the group.

The box-whisker plot is a tool for representing the distribution parameters of variables. This tool can be used to provide a visual idea in the sense of F- and t-test. Group means may be compared in relation to the group variances. A box-whisker plot may also be formed from robust statistical parameters: the median as the centre line of the box, percentiles as the edges of the box, and minimum and maximum as horizontal lines outside the box. Outliers are points outside the structure. An example of measurements of trichloroethene in river water is given in Fig. 5-10. The five groups refer to

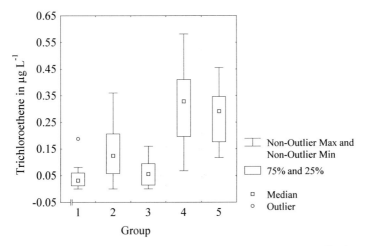

Fig. 5-10. Box-whisker plot of trichloroethene in river water at five different sampling locations

five sampling locations. Each group contains 80 single values. It is apparent that the first three sampling locations have lower trichloroethene loadings than locations 4 and 5.

5.2.5.2 Multiple Box-Whisker Plots

Multiple box-whisker plots [NAGEL et al., 1992] are used for comparing different groups of objects in relation to more than one feature. Groups are groups of single box plots. Each variable is demonstrated by a single box which reflects the parameters of the distribution of this variable. k groups are visualized by k groups of box plots. With the help of modern software, this is a rapid method of finding group differences.

To demonstrate the power of this method an example of halogenated volatile hydrocarbons in river water is described. Eleven halogenated hydrocarbons were measured using static headspace gas chromatography. The values are highly scattered. The problem is to distinguish between river sections according to the halogenated hydrocarbons. The concentrations of these compounds at five sampling points were compared in a multiple box-whisker plot (Fig. 5-11) on a robust statistical basis.

The variables tribromomethane, dibromochloromethane, bromodichloromethane, trichloromethane, and tetrachloromethane do not differ between the sampling locations, the groups. Others, trichloroethene and tetrachloroethene have maximum concentrations in groups 4 and 5 of the river section. But this visual assessment is not very rigorous.

Relationships between variables, which can be expressed as correlations, or relationships between the groups cannot be detected. Multivariate statistical methods must therefore be applied.

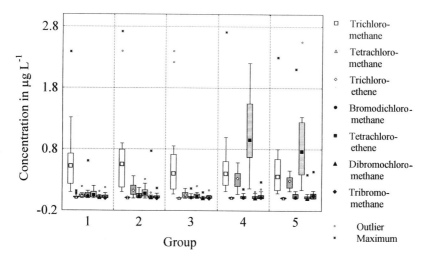

Fig. 5-11. Multiple box-whisker plot of halogenated volatile hydrocarbons in river water at five different sampling locations

The visual estimation of differences between groups of data has to be proved using multivariate statistical methods, as for example with multivariate analysis of variance and discriminant analysis (see Section 5.6).

5.2.6 Limitations

Graphical methods without statistical improvements are very suitable techniques for preliminary testing for interactions or discovery of patterns. They can give additional visual explanation to statistical tests which are an exact measure of the interesting relationships. Although graphical methods have the advantage of depicting data in a manner which is easy to understand, they have some limitations:

– As a result of false or unsuitable standardization, wrong interpretations can lead to false conclusions. The object pattern may be strongly determined by outliers.
– The simultaneous treatment of more than ten features or objects leads to a badly ordered arrangement of the data representation.

Graphical methods in connection with pattern recognition algorithms, i.e. geometrical or statistical methods, e.g. minimum spanning tree or cluster analysis, are more powerful methods for explorative data analysis than graphical methods alone.

5.3 Cluster Analysis

5.3.1 Objectives of Cluster Analysis

The notion **cluster analysis (CA)** encompasses a family of methods which are primarily useful for finding and making visible structures within observed and given data. In this respect the view of the data is optimized rather than the data being manipulated. In this sense cluster analysis can be seen as a **pattern cognition** method. Synonyms in use are numerical taxonomy (because of its biological roots and first applications) and automatic classification. Because during the analysis we neither need nor include *a priori* knowledge, even if it is available, cluster analysis is specified as unsupervised pattern cognition.

If the aim of the data analysis or the question asked about the data is clearly fixed and the result is obtained the next question will be: what is the cause of the structure found? This in turn may lead to new hypotheses about the data. Hence, methods of confirmatory data analysis ('usual statistics') may subsequently be necessary in the next step. Sometimes in recognition processes the problem of the adequate selection of objects will arise. Then an additional problem may consist in the proper choice of features to be measured in order to characterize the objects correctly.

Let us assume that a reasonable number of objects, n, and their m features have been selected and are arranged in a measuring data matrix X like:

$$X = \begin{pmatrix} x_{11} & x_{12} & \cdots & x_{1m} \\ x_{21} & x_{22} & \cdots & x_{2m} \\ \cdots & \cdots & x_{ij} & \cdots \\ x_{n1} & x_{n2} & \cdots & x_{nm} \end{pmatrix} \tag{5-5}$$

and let us proceed with steps leading from these raw data to information.

5.3.2 Similarity Measures and Data Preprocessing

In order to find structures in a data set or to reveal similarities of samples, organisms, ... which in the following are called **objects**, first of all one needs a **similarity measure**. The simplest similarity measure can be derived from geometry. Without proof one intuitively accepts that similarity and **distance** are complementary in nature and remember the law of PYTHAGORAS about the distance d of two points O_1 and O_2 in a rectangular system of two axes y and x:

$$d(O_1, O_2) = \sqrt{(y_1 - y_2)^2 + (x_1 - x_2)^2} \tag{5-6}$$

This situation is shown in Fig. 5-12. The extension of this law to more than two dimensions, to 'spatial PYTHAGORAS' leads to the EUCLIDean distance of any two objects O_i and O_k which in the following we will simply write as $d(i, k)$:

$$d(i, k) = \sqrt{\sum_{j=1}^{m} (x_{ij} - x_{kj})^2} \qquad (5\text{-}7)$$

m – number of features

Clearly for more than $m = 3$ features we cannot visualize the distance.

Eq. 5-7 appears as a special case of the so-called MINKOWSKI metrics where m still denotes the dimension of the space spanned by the m features and C is a special parameter:

$$d(i, k) = \sqrt[c]{\sum_{j=1}^{m} |x_{ij} - x_{kj}|^C} \qquad (5\text{-}8)$$

Distances with $C = 1$ are especially useful in the classification of local data as simple as in Fig. 5-12, where simply $d(1, 2) = a + b$. They are also known as Manhattan, city block, or taxi driver metrics. These distances describe an absolute distance and may be easily understood. With $C = 2$ the distance of Eq. 5-7, the EUCLIDean distance, is obtained. If one approaches infinity, $C = \infty$, in the 'maximum metric' the measurement pairs with the greatest difference will have the greatest weight. This metric is, therefore, suitable in outlier recognition.

Numerous other distance measures are known from the literature and may be used in special applications of cluster analysis [MUCHA, 1992].

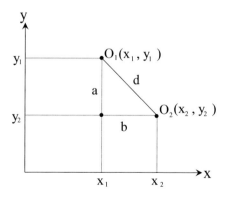

Fig. 5-12. Two objects in 2D-space (PYTHAGORAS' law)

Data Preprocessing

Obviously in Eq. 5-8 one cannot handle measurements of features with different units. It is, for example, not possible to subtract pH values from Fe content measured in %. Therefore one important aspect of **data preprocessing** is to ensure the comparability of the features. Even if no EUCLIDean measure is used one should keep this aspect in mind.

The most common possibility is **autoscaling** (see also Section 5.2.2, Eq. 5-3). 'Auto' means that the data are related to measures of their own distribution, namely the mean $\bar{x}$ and the standard deviation s of an assumed normal distribution. This is achieved by subtracting the mean $\bar{x}_j$ from each individual value x_{ij} ($i = 1, ..., n$) and dividing by s_j of the respective feature x_j ($j = 1, ..., m$):

$$z_{ij} = \frac{x_{ij} - \bar{x}_j}{s_j} \quad (5\text{-}9)$$

The values z_{ij} are then unitless and considered to be values of a new random variable z_j with a mean value of zero and a standard deviation of unity.

Certainly, several other transformations exist creating unitless numbers, remember for example the transformation used in experimental design (see Chapter 3). A similar transformation uses for each feature:

$$z_{i,j} = \frac{x_{i,j} - x_{\min,j}}{x_{\max,j} - x_{\min,j}} \quad (5\text{-}10)$$

Generally care should be taken to avoid transformations which have not been proved to leave the structure of the data unaltered.

Finally the EUCLIDean distances are calculated from the z values rather than from the raw x values. At the end of the process the mutual distances of all n objects are arranged in a squared array called a **distance matrix D**. This (n, n)-matrix is normally symmetrical with zero values on the main diagonal:

$$\mathbf{D} = \begin{pmatrix} 0 & d_{12} & d_{13} & ... & d_{1n} \\ d_{21} & 0 & d_{23} & ... & d_{2n} \\ ... & ... & ... & ... & ... \\ d_{n1} & d_{n2} & d_{n3} & ... & 0 \end{pmatrix} \quad (5\text{-}11)$$

Clustering of Features

In principle, the features describing the objects can also be subjected to cluster analysis. In this case one may think immediately of the correlation coefficient, r, or the coefficient of determination, COD, as a measure of the similarity of each pair of features. Accordingly, $1 - r$ or $1 - COD$ is useful as a measure of distance.

5.3.3 Clustering Algorithms

After selecting a measure one has to decide which **clustering algorithm** (strategy) may be appropriate. Sometimes it is necessary for the algorithm to fit the similarity measure. In most cases one wishes to use the algorithm which yields the most interpretable or plausible data structure.

Main Groups of Cluster Algorithms

Two main groups of cluster algorithms can be distinguished: **hierarchical or nonhierarchical (partitioning) techniques**.

In hierarchical techniques one can proceed in an **agglomerative** or a **divisive** way. In an agglomerative process the principal aim is to join similar objects into clusters and to add objects to clusters already found or to join similar clusters. In divisive strategies one starts with one cluster comprising all objects from which, step by step, the most 'inhomogeneous' objects are stripped, forming themselves into 'more homogeneous' clusters at lower linkage levels.

The typical output of hierarchical cluster methods is a so-called **dendrogram**, a tree-like diagram which is very useful for discussing several possible results of the clustering process. For an illustration see Fig. 5-13; the underlying example will be explained in Section 5.3.4.

As indicated, agglomerative methods start with single objects or pairs of objects; step by step clusters are formed which are finally united in one cluster. Divisive methods, on the other hand, start from the one cluster of all objects and divide it step by step. One drawback of the commonly used agglomerative methods is that clusters formed may not be broken up in a subsequent step. With certain algorithms this sometimes leads to so-called inversions in the dendrogram, i.e. crossing lines in the diagram.

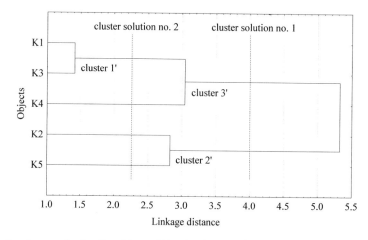

Fig. 5-13. Dendrogram for the five objects of the simple example

In contrast **nonhierarchical (partitioning) methods** [MASSART et al., 1983] allow objects to be re-arranged. This may therefore enable optimization of certain objective functions, e.g. variance criteria. One drawback with nonhierarchical partitioning techniques is that they do not produce graphical representations like dendrograms. Another drawback may be that partitioning techniques need a starting assumption of the number of clusters to be obtained. Some techniques therefore need a starting cluster structure, gained, for example, from the outcome of a hierarchical method. The centroids of each cluster could then form the initial gravity centers, e.g. in the established k-means algorithm of MAC QUEEN [STEINHAUSEN and LANGER, 1977]. Then the objects are attributed to these gravity centers in the sense of an EUCLIDean distance. Afterwards the gravity center is calculated again and some objects are possibly rearranged, and so on, until a stable solution is obtained.

Potential methods, like the CLUPOT algorithm [COOMANS et al., 1981], offer another way of nonhierarchical clustering. These methods surround the objects with 'potential spheres' which overlay to 'potential hills' if objects are similar, i.e. close together. Then a smoothing parameter enables visualization of the peaks of the bigger hills only, i.e. only a few clusters with many objects, or, looking more deeply, more clusters with fewer objects each. Therefore, similar to the hierarchical algorithms it is possible to create a number of cluster solutions differing in the number of clusters encountered. So, mainly with the last mentioned procedures one has the problem of searching for the best (robust, plausible, ...) cluster solution, i.e. for the number of clusters to interpret.

Selection and Comparison of Cluster Solutions

One has to keep in mind that groups of objects found by any clustering procedure are not statistical samples from a certain distribution of data. Nevertheless the groups or clusters are sometimes analyzed for their distinctness using statistical methods, e.g. by multivariate analysis of variance and discriminant analysis, see Section 5.6. As a result one could then discuss only those clusters which are statistically different from others.

As mentioned, hierarchical cluster analysis usually offers a series of possible cluster solutions which differ in the number of clusters. A measure of the total within-groups variance can then be utilized to decide the probable number of clusters. The procedure is very similar to that described in Section 5.4 under the name scree plot. If one plots the variance sum for each cluster solution against the number of clusters in the respective solution a decay pattern (curve) will result, hopefully tailing in a plateau level; this indicates that further increasing the number of clusters in a solution will have no effect.

RAND [1971] published an algorithm which enables comparison of cluster solutions by their similarity. So, the results of even conceptually different clustering procedures may be contrasted.

A Monte Carlo study demonstrated the problem of estimating the number of clusters [DUBES, 1987]. One principal reason for this problem is that "clustering algorithms tend to generate clusters even when applied to random data" [DUBES and JAIN, 1979]. JAIN and MOREAU [1987] therefore used the bootstrap technique [EFRON and GONG, 1983] for cluster validation.

Modes of Agglomerative Hierarchical Methods

Because of the widespread application of agglomerative hierarchical methods in the following text some details are reported to assist the selection of certain clustering strategies. Clustering algorithms themselves should fit data or problems to be solved. The different algorithms can influence the results of clustering.

For hierarchical methods there is a formula of LANCE and WILLIAMS [STEINHAUSEN and LANGER, 1977] which helps the adjustment of certain parameters for some control of the result.

If the most similar objects i_1 and i_2 are considered as one cluster i_{12} the distances of this new cluster from all the remaining objects (or clusters) k is described by the formula:

$$d(i_{12}, k) = \alpha_1 d_{i1,k} + \alpha_2 d_{i2,k} + \beta d_{i1,i2} + \gamma |d_{i1,k} - d_{i2,k}| \tag{5-12}$$

α_1 – weights the distance of the first joint object to any other object or cluster
α_2 – weights the distance of the second joint object to any other object or cluster
β – weights the distance of both objects
γ – weights the difference between the distance of both objects to any other object or cluster

Sometimes the formula is extended by an additional term (parameter) not associated with distances. Hence, the above mentioned influence on the clustering result is possible by selecting certain combinations of the parameters α, β, and γ.

Typical parameter combinations [STEINHAUSEN and LANGER, 1977] are intrinsically used if the data analyst selects a certain linkage strategy. (Software will sometimes even use other synonyms for the same parameter combination.)

Here, we want to concentrate on the most commonly used linkage strategies (Tab. 5-1).

Tab. 5-1. Modes of agglomerative hierarchical methods

Strategy	α_1	α_2	β	γ	Distance	Result
Average linkage	0.5	0.5	0	0	$\frac{1}{2}(d_{i1,k} + d_{i2,k})$	Moderate
Single linkage	0.5	0.5	0	-0.5	$\min(d_{i1,k} + d_{i2,k})$	Few large clusters
Complete linkage	0.5	0.5	0	0.5	$\max(d_{i1,k} + d_{i2,k})$	Many small clusters
WARD linkage	$f_1(n^*)$	$f_2(n^*)$	$f_3(n^*)$	0	weighted by n^*	Usually the best structured

n^* stands for an expression which incorporates the numbers of objects in the clusters joined, the sum of these numbers, and the number of the objects in the cluster to which the distance is to be computed. Because cluster analysis is a pattern cognition tool the data analyst will try as many variants as he needs to obtain the result which can be interpreted most easily.

In most applications easily interpretable results are obtained by the method of WARD where the parameters are defined by relationships between the numbers of objects (see also Section 7.2.1.2.2). If squared EUCLIDean distances are used, this strategy can also be recommended from the point of view of optimizing the variance. In contrast, some other methods, e.g. the median method and the centroid method, cannot be generally recommended, because they often form inversions.

Further details of agglomerative, and several other clustering strategies may be found in the book by MASSART and KAUFMAN [1983] or, along with remarks on the treatment of situations with missing values, in the monograph by MUCHA [1992]. Finally, it may be of interest that OZAWA [1983] even proposed a hierarchical cluster algorithm based on an asymmetric distance matrix.

5.3.4 CA Calculations Demonstrated with a Simple Example

To illustrate the typical steps of a hierarchical clustering procedure and to enable the reader to follow the calculations with a pocket calculator let us assume a very simple example with five objects (named *K1* to *K5*) and two features (x_1 and x_2) each of equal dimension.

The question is: Which objects may be regarded as similar with respect to both features?

The raw data are the following:

Object	x_1	x_2
K1	4.0	1.5
K2	4.0	6.0
K3	5.0	2.5
K4	1.5	2.0
K5	2.0	8.0

The raw (n, m) data matrix is:

$$X = \begin{pmatrix} 4.0 & 1.5 \\ 4.0 & 6.0 \\ 5.0 & 2.5 \\ 1.5 & 2.0 \\ 2.0 & 8.0 \end{pmatrix}$$

Because of the assumed equal dimensions the features will not be standardized.

The distance matrix contains the EUCLIDean distances d_{ik} of all object pairs i and k. It is a symmetrical (5, 5)-matrix with $d_{ki} = d_{ik}$, from which we write the upper triangle only:

$$D = \begin{pmatrix} 0 & 4.5 & 1.41 & 2.55 & 6.80 \\ & 0 & 3.64 & 4.72 & 2.83 \\ & & 0 & 3.54 & 6.26 \\ & & & 0 & 6.02 \\ & & & & 0 \end{pmatrix}$$

The clustering algorithm starts by joining objects $K1$ and $K3$ because their distance is the least: $d_{13} = 1.41$.

Using $\beta = \gamma = 0$ and $\alpha_1 = \alpha_2 = 0.5$ for the **average linkage strategy** in the above given LANCE and WILLIAMS expression we compute the distance of cluster $1'$ formed by $K1$ and $K3$ (1+3) to the remaining objects.

In the **new distance matrix** lines 1 and 3 and columns 1 and 3 are now deleted because they relate to clustered objects. Instead we have:

$$D = \begin{pmatrix} 0 & d_{1'2} & d_{1'4} & d_{1'5} \\ & 0 & 4.72 & 2.83 \\ & & 0 & 6.02 \\ & & & 0 \end{pmatrix}$$

The new distances are:

$$d_{1'2} = 0.5(d_{12} + d_{32}) = 0.5(4.5 + 3.64) = 4.07$$
$$d_{1'4} = 0.5(d_{14} + d_{34}) = 0.5(2.55 + 3.54) = 3.04$$
$$d_{1'5} = 0.5(d_{15} + d_{35}) = 0.5(6.80 + 6.26) = 6.53$$

In the new distance matrix $d_{25} = 2.83$ is the lowest value. Accordingly, the objects $K2$ and $K5$ are most similar and will be joined on level 2.83 to form cluster $2'$.

The **remaining distance matrix** is:

$$D = \begin{pmatrix} 0 & d_{1'2'} & 3.04 \\ & 0 & d_{2'4} \\ & & 0 \end{pmatrix}$$

Now only two new distances are to be calculated:

$$d_{2'1'} = d_{1'2'} = 0.5(d_{1'2} + d_{1'5}) = 0.5(4.07 + 6.53) = 5.30$$
$$d_{2'4} = \phantom{d_{1'2'} =} 0.5(d_{24} + d_{54}) = 0.5(4.72 + 6.02) = 5.37$$

The lowest element of this distance matrix is $d_{1'4} = 3.04$. This is the level on which object $K4$ is added to the cluster $1'$ to form cluster $3'$.

The **final distance matrix** only contains one element, $d_{3'2'}$, which is the distance from cluster $3'$ to cluster $2'$:

$$d_{3'2'} = 0.5(d_{1'2'} + d_{42'}) = 0.5(5.30 + 5.37) = 5.34$$
$$D = (5.34)$$

At this level all objects fuse to a single cluster.

(Note: If you check the results by computation numerical deviations may occur owing to rough rounding in the above example.) Knowing the levels of fusions one can now draw the **dendrogram** in Fig. 5-13.

The interpretation of the results is the last step of cluster analysis. Two possible cluster solutions may to be discussed. The **first solution** consists of two clusters obtained if the dendrogram is cut between linkage distances 3.04 and 5.34. Then we have to interpret the nature of the two sets of objects $2' = \{K2, K5\}$ and $3' = \{K1, K3, K4\}$. The **second possible solution** is obtained by cutting the dendrogram betweeen linkage distances 1.41 and 2.83 which leaves none of the former clusters unaltered. In other words the previously found clusters now appear not to be homogeneous. Because of the greater distance between 3.04 and 5.34 than between 1.41 and 2.83, however, one should rather try to interpret the nature of solution number 1.

If we look at the raw data we can easily identify the common behavior of objects $\{K1, K3\}$ on the one hand and of objects $\{K2, K5\}$ on the other: for $K1$ and $K3$, $x_{i1} > x_{i2}$ whereas for $K2$ und $K5$ the opposite situation occurs, $x_{i1} < x_{i2}$. From this inspection $K4$ should more correctly belong to cluster $2'$, which was not found by the algorithm used.

Let us not stress this simple example further but note that in cases with two variables only, a simple scattergram (of standardized values) will probably suffice.

5.3.5 Typical CA Results Illustrated with an Extended Example

Example 5-1

For demonstration we use the data of a cooperative test [DOERFFEL and ZWANZIGER, 1987]. In this interlaboratory comparison five laboratories were involved in the analysis of slag samples three times for seven chemical elements. So the (15, 7)-data matrix consists of 5 times 3 rows and 7 columns. The raw data in % are given in Tab. 5-2. The data have been preprocessed by standardization (autoscaling).

The result of the clustering of the objects (samples) is shown in the dendrogram of Fig. 5-14. Clearly one can identify several groups of laboratories. From the analytical results laboratory B is separated as outstanding (cut the dendrogram at a linkage distance of, say, 60). So, as a **first solution** we have the laboratory B values in one cluster and those from the remaining laboratories in the other cluster.

Beside this solution we can discuss a **second solution** obtained by cutting at a linkage distance of about 40. Here, the inhomogenous second cluster of the former solution splits into two clusters, namely one containing the values from laboratory E and another comprising those from the rest of the laboratories. From the aspect of precision we recognize

Tab. 5-2. Values from an interlaboratory test of five laboratories (three objects each; seven features measured) (data in %)

Lab	Sample	Code	Si	Al	Fe	Ti	Na	Mg	Ca
A	1	A1	53.3	12.4	10.3	1.2	0.3	2.8	13.9
A	2	A2	52.8	12.3	10.2	1.2	0.2	2.7	13.8
A	3	A3	52.9	12.3	10.2	1.2	0.2	2.7	13.9
B	1	B1	69.0	11.34	9.01	1.0	0.2	2.8	14.1
B	2	B2	57.0	10.35	8.37	1.0	0.2	2.5	13.7
B	3	B3	61.0	10.39	8.44	1.0	0.2	2.6	14.0
C	1	C1	53.3	12.25	10.63	1.2	0.2	2.5	13.6
C	2	C2	53.4	12.47	10.69	1.3	0.4	2.5	13.7
C	3	C3	53.2	12.18	9.85	1.2	0.2	2.3	13.5
D	1	D1	55.3	12.80	10.0	1.17	0.13	3.00	14.22
D	2	D2	54.7	12.40	9.9	1.17	0.13	2.70	13.92
D	3	D3	54.8	12.50	10.0	1.17	0.13	2.81	13.95
E	1	E1	53.9	12.60	9.6	1.4	0.18	3.4	13.00
E	2	E2	54.1	12.80	9.7	1.4	0.19	3.5	13.30
E	3	E3	53.8	12.30	9.5	1.3	0.17	3.3	12.90

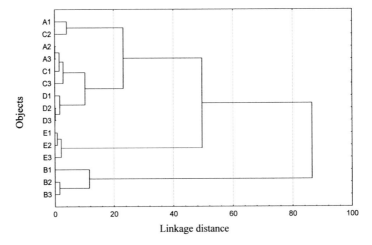

Fig. 5-14. Clustering of the objects (samples) from a cooperative test

that laboratory E is even better than laboratory B, because the samples of lab E are united at lower distances (on average).

It is now up to the analyst to interpret the dendrogram with respect to the possible causes of the structures found. For example in the case at hand a discussion with the laboratories revealed that laboratory E optimized its determination of some elements by atomic absorption spectroscopy. If we inspect the raw data in Tab. 5-2. the special location of the samples from laboratories *B* and *E* samples may be explained by the extreme Si values (laboratory *B*) on one side and the high Mg values provided by laboratory *E*.

For illustration only we show a result of feature clustering in Fig. 5-15. As a distance measure 1 − *r* was used, with *r* the bivariate correlation coefficients of each feature pair. Again we see a well structured dendrogram showing two groups of chemical elements, the set {Si, Na, Ca} and the set {Al, Fe, Ti, Mg}. Because this arrangement cannot be explained easily by inspection of the raw data table, for the moment we skip the discussion of that clustering result. (Some readers will have computed the correlation matrix and may have tried to see explanations from this matrix rather from the raw data matrix. They will probably note that the second feature contains elements for which most of the correlation coefficients are positive and significant ... We will, instead, wait for the next section and what it has to offer in respect of the discussion of feature structures.)

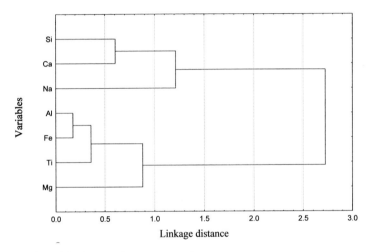

Fig. 5-15. Clustering of the features from a cooperative test

Example 5-2

In Section 5.2.5.2 we used as an example halogenated volatile hydrocarbons measured in 80 samples taken along a river in Thuringia at five different locations. The reader may believe that the clustering of data obtained over a relatively long period of time was not very instructive. But one interesting point is found by clustering the features. In Fig. 5-16 we can obviously distinguish between the chlorine-containing and the bromine-containing hydrocarbons. One explanation could be that the bromine hydrocarbons stem from an unique source (possibly from industrial and municipal water treatment) whereas the chlorine hydrocarbons may be attributed to general pollution.

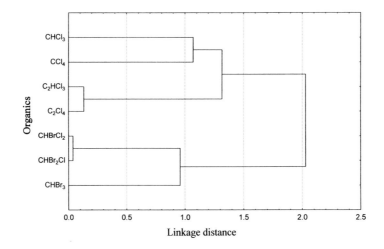

Fig. 5-16. Clustering of seven features from the example of halogenated volatile hydrocarbons in river water at different sampling locations

5.4 Principal Components Analysis and Factor Analysis

Principal components analysis (PCA) and **factor analysis (FA)** are aimed at finding and interpreting hidden complex, and possibly causally determined, relationships between features in a data set. Correlating features are converted to the so-called **factors** which are themselves noncorrelated.

The central point of such analysis is to reduce the original (m, n)-data matrix X with m features and n objects to the following component parts:

- factor loadings A
- factor scores F

$$\begin{pmatrix} x_{11} & x_{12} & \cdots & x_{1n} \\ x_{21} & x_{22} & \cdots & x_{2n} \\ \cdots & \cdots & \cdots & \cdots \\ x_{m1} & x_{m2} & \cdots & x_{mn} \end{pmatrix} = \begin{pmatrix} a_{11} & \cdots & a_{1s} \\ a_{21} & \cdots & a_{2s} \\ \cdots & \cdots & \cdots \\ a_{m1} & \cdots & a_{ms} \end{pmatrix} \cdot \begin{pmatrix} f_{11} & f_{12} & \cdots & f_{1n} \\ \cdots & \cdots & \cdots & \cdots \\ \cdots & \cdots & \cdots & \cdots \\ f_{s1} & f_{s2} & \cdots & f_{sn} \end{pmatrix} \quad (5\text{-}13)$$

$$X = A \cdot F \quad (5\text{-}14)$$

X – data matrix
A – factor loadings
F – factor scores
s – number of factors ($s = m$)

A linear combination of different factors in the matrix A with factor scores in the matrix F can reproduce the data matrix X. These factors are new synthetic variables and represent a certain quantity of features of the data set. They explain the total variance of all features in a descending order and are themselves noncorrelated. It is, therefore, possible to reduce the dimension m of the data set with a minimum loss of information expressed by the matrix of residuals E.

$$\begin{pmatrix} x_{11} & x_{12} & \cdots & x_{1n} \\ x_{21} & x_{22} & \cdots & x_{2n} \\ \cdots & \cdots & \cdots & \cdots \\ x_{m1} & x_{m2} & \cdots & x_{mn} \end{pmatrix} = \begin{pmatrix} a_{11} & \cdots & a_{1s} \\ a_{21} & \cdots & a_{2s} \\ \cdots & \cdots & \cdots \\ a_{m1} & \cdots & a_{ms} \end{pmatrix} \cdot \begin{pmatrix} f_{11} & f_{12} & \cdots & f_{1n} \\ \cdots & \cdots & \cdots & \cdots \\ f_{s1} & f_{s2} & \cdots & f_{sn} \end{pmatrix}$$

$$+ \begin{pmatrix} e_{11} & e_{12} & \cdots & e_{1n} \\ e_{21} & e_{22} & \cdots & e_{2n} \\ \cdots & \cdots & \cdots & \cdots \\ e_{m1} & e_{m2} & \cdots & e_{mn} \end{pmatrix}$$

(5-15)

$$X = A \cdot F + E \tag{5-16}$$

X – data matrix
A – factor loadings
F – factor scores
E – residuals as a result of reduction of dimensionality
s – number of factors ($s < m$)

In this manner, both PCA and FA provide a projection of the objects from the high-dimensional feature space on to a space defined by a few factors; they can also be used as a method for graphical representation of multidimensional data.

5.4.1 Description of Principal Components Analysis

The decomposition according to Eq. 5-16 is performed by a **principal axes transformation** of the correlation matrix R. The correlation matrix of the raw data is therefore the starting point of the calculations.

The mathematical algorithm is the solution of the eigenvalue problem:

$$\begin{aligned} R \cdot e_1 &= \lambda_1 \cdot e_1 \\ R \cdot e_2 &= \lambda_2 \cdot e_2 \\ &\cdots \end{aligned} \tag{5-17}$$

R – correlation matrix
e – eigenvectors
λ – eigenvalues

The nontrivial solution of this problem leads to the determinant:

$$|\mathbf{R} - \lambda \mathbf{I}| = 0 \qquad (5\text{-}18)$$

The evolution of this determinant first yields the eigenvalues. The solution of the whole eigenvalue problem provides pairs of eigenvalues and eigenvectors. The mathematical algorithm is described in detail in [MALINOWSKI, 1991]. A simple example, discussed in Section 5.4.2, will demonstrate the calculation. The following properties of these abstract mathematical measures are essential:

- eigenvalues are a measure of the extracted variance from the total feature variance s_{total}^2, expressed by the correlation matrix $\mathbf{R}$
- eigenvalues are arranged in descending order:

$$\lambda_1 > \lambda_2 > \lambda_3 > \cdots > \lambda_m \qquad (5\text{-}19)$$

- the sum of all the eigenvalues is equal to the number of features m
- eigenvalues and eigenvectors are coupled pairs
- eigenvectors are orthogonal to each other
- eigenvectors contain the nonnormalized coefficients in the matrix of factor loadings $\mathbf{A}$

The eigenvalues were normalized by dividing by the square root of λ_j; they then have variance equal to unity and one therefore gets the matrix of factor loadings $\mathbf{A}$. These factor loadings are the weights of original features in the new variables, the factors.

Therefore the new synthetic factors are noncorrelated with each other, they have themselves the variance of one and they contain a certain part of the total variance of the data set expressed by their eigenvalues. Because:

$$\sum_{j=1}^{m} \lambda_j = m \qquad (5\text{-}20)$$

the variance part of the factor j is:

$$\frac{s_j^2}{s_{total}^2} = \frac{\lambda_j}{m} \qquad (5\text{-}21)$$

Because of these rules, it is possible to reduce the number of factors. The total number of factors ($s = m$) describes the variance of the full data set. If one accepts minimal losses of variance explained by the model one can reduce the number of factors. The remaining factors are ignored.

In this section we will not go further into the discussion of how many principal components to include. Because this question is important in the interpretation of the numerical pattern (the loadings) of more than two or three principal components we will come back to this topic in the section on factor analysis.

Interpretation of the Results

The matrix of factor loadings A describes the weights of each feature in each factor. Features with low loadings have only slight influences on the factor; features with high positive or negative loadings essentially determine the factor. It is not possible to make a rule about the minimum amount of loadings which can be interpreted. One interpretation regards loadings as correlation coefficients which can be tested for significance in dependence on the degrees of freedom. Another interpretation says that all factors with an eigenvalue $\lambda > 1$ are significant and squared loadings detect the part of the variance of one feature in one factor. This variance part can be fixed to, e.g. 25% of the variance of the factor. This means $a_{ij}^2 > 0.25$ can be interpreted.

From the matrix of factor loadings one can conclude which features have common influences on the objects. Common influences are present if more than one feature have high loadings in one factor. The graphical representation of features in the space of the factors is not very useful for practical conclusions because negative correlated features seem to be uncorrelated.

If one wishes to represent the objects in the space of the factors, one has to calculate the matrix of factor scores F. The procedure is a multiple regression between the original values and the factors and is also called estimation according to BARTLETT [JAHN and VAHLE, 1970]. The graphical representation of the objects may be used to detect groups of related objects or to identify objects which are strongly related to one or more of the factors.

A simultaneous graphical representation of objects and features in the space of factors is called a biplot. Computer programs give this display as an option. If objects have similar coordinates as features this would lead to the conclusion that high values of this feature characterize this object. The authors discourage this approach because a biplot can lead to wrong conclusions:

- When objects and features are near the zero coordinates of the factors this means that objects have medium scores in the factors but features are not explained by the factors. Therefore the features have no relationship with the objects although they are close together in the coordinate system.
- Another case occurs if one factor contains two negatively correlated features. In a biplot the features will appear at the minimum and the maximum of the axis of this factor. Objects which have high factor scores in this factor only appear near the maximum of the axis.

An **optimum projection** of both data matrix aspects, namely lines and columns is provided by **correspondence factor analysis** which also offers a number of sophisticated and useful variance decompositions. For details see [GREENACRE, 1984; MELLINGER, 1987]. One of the first applications to environmental data has been reported by FEINBERG [1986].

5.4.2 PCA Calculations Demonstrated with a Simple Example

We take the same example as used in Section 5.3.4 for cluster analysis:

$$
\begin{array}{llcc}
object & & x_1 & x_2 \\
no.\ 1: & K1 & 4.0 & 1.5 \\
no.\ 2: & K2 & 4.0 & 6.0 \\
no.\ 3: & K3 & 5.0 & 2.5 \\
no.\ 4: & K4 & 1.5 & 2.0 \\
no.\ 5: & K5 & 2.0 & 8.0 \\
\end{array}
$$

From the **(m, n)-data matrix X**:

$$X = \begin{pmatrix} 4.0 & 4.0 & 5.0 & 1.5 & 2.0 \\ 1.5 & 6.0 & 2.5 & 2.0 & 8.0 \end{pmatrix}$$

first the correlation matrix **R** is calculated.

The elements of **R** are the (bivariate) correlation coefficients r_{i1i2} of each feature pair (x_{i1}, x_{i2}) introduced in Section 2.4.2 as the covariance s^2_{i1i2} of two features standardized by both standard deviations s_{i1i2}: $r_{i1i2} = s^2_{i1i2} / (s_{i1} \cdot s_{i2})$.

With the mean of x_1, $\bar{x}_1 = 3.3$, and the mean $\bar{x}_2 = 4.0$ we easily obtain the covariance $s^2_{12} = -1.125$, the standard deviations $s_1 = 1.4832$ and $s_2 = 2.8504$ and the correlation coefficient $r_{12} = -0.2661$.

Hence the **correlation matrix** is:

$$R = \begin{pmatrix} 1.0 & -0.2661 \\ -0.2661 & 1.0 \end{pmatrix}$$

The **eigenvalues**, λ, of this correlation matrix are obtained by solving the characteristic determinant (Eq. 5-18 is repeated for easier understanding), i.e.:

$$|R - \lambda I| = 0$$

where **I** is the identity matrix of order $m = 2$, containing zeroes except for all diagonal elements, which are set equal to 1.
From

$$\begin{vmatrix} 1 - \lambda & -0.2661 \\ -0.2661 & 1 - \lambda \end{vmatrix} = (1 - \lambda)^2 - 0.2661^2 = 0$$

one gets $\lambda_{1,2} = 1 \pm [1 - (1 - 0.2661)^2]^{1/2}$, i.e. $\lambda_1 = 1.2661$ and $\lambda_2 = 0.7339$ ($\lambda_1 = 1 + r_{12}$, $\lambda_2 = 1 - r_{12}$).
(Note that $\lambda_1 + \lambda_2 = m$; this is generally true for correlation matrices.)

5.4 Principal Components Analysis and Factor Analysis

The **eigenvectors** e connected with the two eigenvalues are computed from the equations:

$$R e_1 = \lambda_1 e_1$$
$$R e_2 = \lambda_2 e_2$$

So,

$$\begin{pmatrix} 1 & -0.2661 \\ -0.2661 & 1 \end{pmatrix} (e_{11} e_{12}) = 1.2661(e_{11} e_{12})$$

leads to

$$e_{11} - 0.2661\, e_{12} = 1.2661\, e_{11}$$
$$-0.2661\, e_{11} + e_{12} = 1.2661\, e_{12}$$

and hence

$$-0.2661\, e_{11} - 0.2661\, e_{12} = 0$$
$$-0.2661\, e_{11} - 0.2661\, e_{12} = 0$$

which yields the components of the first eigenvector as, e.g., $+1$ and -1.

The vector e_1 may be written:

$$e_1 = \begin{pmatrix} e_{11} \\ e_{12} \end{pmatrix} = \begin{pmatrix} 1 \\ -1 \end{pmatrix}$$

If normalized to length 1 a new vector a_1 results, for which $a_1^T a_1 = 1$:

$$a_1 = \begin{pmatrix} \dfrac{1}{\sqrt{2}} \\ -\dfrac{1}{\sqrt{2}} \end{pmatrix}$$

Writing the equation system for e_2 with λ_2 one obtains

$$e_2 = \begin{pmatrix} 1 \\ 1 \end{pmatrix} \quad \text{and} \quad a_2 = \begin{pmatrix} \dfrac{1}{\sqrt{2}} \\ \dfrac{1}{\sqrt{2}} \end{pmatrix}$$

and observes that both vectors are orthogonal: $e_1^T e_2 = 0 = a_1^T a_2$.

Now with both eigenvalues λ and factor loadings $\boldsymbol{a}$ one decomposes (or reconstructs) matrix $\boldsymbol{R}$ according to

$$\boldsymbol{R} = \lambda_1\,\boldsymbol{a}_1\,\boldsymbol{a}_1^T + \lambda_2\,\boldsymbol{a}_2\,\boldsymbol{a}_2^T$$

from which one understands the meaning of λ as the weight of the importance of each factor.

$$\boldsymbol{R} = 1.2661 \begin{pmatrix} 0.5 & -0.5 \\ -0.5 & 0.5 \end{pmatrix} + 0.7339 \begin{pmatrix} 0.5 & 0.5 \\ 0.5 & 0.5 \end{pmatrix}$$

$$= \begin{pmatrix} 0.63305 & -0.63305 \\ -0.63305 & 0.63305 \end{pmatrix} + \begin{pmatrix} 0.36695 & 0.36695 \\ 0.36695 & 0.36695 \end{pmatrix} = \begin{pmatrix} 1 & -0.2661 \\ -0.2661 & 1 \end{pmatrix}$$

From this sum one can see that the first matrix fraction alone is a poor approximation of the total correlation matrix. The fraction of the variance accounted for is only about 63%:

$$\frac{\lambda_1}{\lambda_1 + \lambda_2} = 0.633$$

The **factor loadings table** is now:

	factor 1 ($\boldsymbol{a}_1$)	factor 2 ($\boldsymbol{a}_2$)
x_1	0.7071	0.7071
x_2	−0.7071	−0.7071

One can see the opposite character of the features in factor 1.

Let us now come to the **graphical aspects**. The graphical display of the objects as factor scores is now possible after multiplying the data matrix $\boldsymbol{X}$ with each of the vectors $\boldsymbol{a}$ (factor loadings).

Because we derived two principal components from the two features we obtain from

$$\boldsymbol{f}_1 = \boldsymbol{a}_1^T \boldsymbol{X} \quad \text{and} \quad \boldsymbol{f}_2 = \boldsymbol{a}_2^T \boldsymbol{X}$$

the necessary coordinates with respect to both component axes:

$$\boldsymbol{f}_1 = (1/\sqrt{2}\ -1/\sqrt{2})\,\boldsymbol{X} = (1.77\quad -1.41\quad 1.77\quad -0.35\quad -4.24)$$
$$\boldsymbol{f}_2 = (3.89\quad 7.07\quad 5.30\quad 2.47\quad 7.07)$$

The display of the five objects in the space of the two principal component axes is shown in Fig. 5-17.

Comparison with the information from the dendrogram of the cluster analysis shows that we have obtained exactly the same result: K1 and K3 may probably be circled as one group (cluster), K2 and K5 may then form the second group of objects, and K4 seems to belong to the first group rather than to the second.

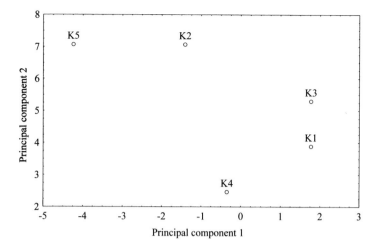

Fig. 5-17. Five objects displayed in the plane of two principal components

5.4.3 Description of Factor Analysis

The reader has seen in Section 5.4.1 that the total number of eigenvalues and eigenvectors, which is the same as the number of features, reproduces the correlation matrix and therefore describes the total variance of the data. The general model of both PCA and FA described in Section 5.4.1 is therefore called **complete factor solution**.

When analyzing real data sets one has to **find common factor structures** which explain the main part of the variance of the data. Therefore in **factor analysis** the total variance of the data is divided by the **reduced factor solution** into the three parts:

- common feature variance
- specific feature variance
- residuals or error

The common feature variance originates from correlating features. **Specific feature variance and residuals or error are now expressed by the matrix E:**

$$\begin{pmatrix} x_{11} & x_{12} & \cdots & x_{1n} \\ x_{21} & x_{22} & \cdots & x_{2n} \\ \cdots & \cdots & \cdots & \cdots \\ x_{m1} & x_{m2} & \cdots & x_{mn} \end{pmatrix} = \begin{pmatrix} a_{11} & \cdots & a_{1s} \\ a_{21} & \cdots & a_{2s} \\ \cdots & \cdots & \cdots \\ a_{m1} & \cdots & a_{ms} \end{pmatrix} \cdot \begin{pmatrix} f_{11} & f_{12} & \cdots & f_{1n} \\ \cdots & \cdots & \cdots & \cdots \\ f_{s1} & f_{s2} & \cdots & f_{sn} \end{pmatrix}$$

$$+ \begin{pmatrix} e_{11} & e_{12} & \cdots & e_{1n} \\ e_{21} & e_{22} & \cdots & e_{2n} \\ \cdots & \cdots & \cdots & \cdots \\ e_{m1} & e_{m2} & \cdots & e_{mn} \end{pmatrix}$$

(5-22)

$$X = A \cdot F + E \tag{5-23}$$

X – data matrix
A – factor loadings
F – factor scores
E – residuals

The communality is introduced as a mathematical measure of this common feature variance. The **communality** is the part of the variance of one feature which is described by the common factor solution in the factor analysis. High communalities, h_i^2, mean that this feature variance is highly explained by the factor solution. Low communalities for one feature detect either a specific feature variance or high random error.

Such a definition of the aim of factor analysis affects also the eigenvalue solution as the way of extracting the factors. Therefore the diagonal elements of the original correlation matrix R, which are all identical to unity were substituted into the communalities of the features:

$$R = \begin{pmatrix} 1 & \cdots & \cdots & \cdots \\ \cdots & 1 & \cdots & \cdots \\ \cdots & \cdots & \cdots & \cdots \\ \cdots & \cdots & \cdots & 1 \end{pmatrix} \rightarrow \begin{pmatrix} h_1^2 & \cdots & \cdots & \cdots \\ \cdots & h_2^2 & \cdots & \cdots \\ \cdots & \cdots & \cdots & \cdots \\ \cdots & \cdots & \cdots & h_m^2 \end{pmatrix} \tag{5-24}$$

with $h_i^2 < 1$.

Because the correlation matrix is the starting point of the analysis and communalities are unknown, communalities have first to be estimated and then iterated by solving a few eigenvalue solutions. Computer programs perform estimations and iterations with different algorithms but solutions differ only minimally.

Then, the table of the factor loadings has the following structure:

Feature	Factor 1	Factor 2	Factor s	Communality
i	a_{i1}	a_{i2}	a_{is}	h_i^2
eigenvalue	λ_1	λ_2	λ_s	$\Sigma \lambda_j$

Some regularities of the common factor solution are the following:

(1) The sum of the squared factor loadings of one feature in every factor is equal to the communality of this feature.

$$\sum_{j=1}^{s} a_{ij}^2 = h_i^2 \tag{5-25}$$

(2) The sum of the communalities of all features is equal to sum of the eigenvalues of factors which contribute to the common factor solution.

$$\sum_{i=1}^{m} h_i^2 = \sum_{j=1}^{s} \lambda_j \tag{5-26}$$

(3) The sum of squared factor loadings of one factor is equal to its own eigenvalue.

$$\sum_{i=1}^{m} a_{ij}^2 = \lambda_j \tag{5-27}$$

The Number of Factors to Extract
In data analysis it is rather interesting to separate noisy contributions of data which should be explained by the last factors with the smallest eigenvalues. For display purposes one naturally has to use $s = 2$ or $s = 3$. For detailed discussion of the numerical results instead one normally has to interpret $s > 2$ factors. But the question is how many factors should one try to interpret?

An empirical measure of how sufficient a decomposition with s factors will be is available with the *IND* function defined by MALINOWSKI [1977]. The minimum value of *IND* indicates the probable number of relevant factors.

$$IND = \frac{RE}{(m-s)^2} \quad \text{with} \quad RE^2 = \frac{1}{n(m-s)} \sum_{j=s+1}^{m} \lambda_j \tag{5-28}$$

m – number of features
n – number of objects
s – number of extracted factors
RE – 'real error'

Another possibility is to include a number of factors such that variance extraction reaches a predefined level, say 90%. If factor analysis starts from the correlation matrix a simple decision is to include all factors associated with eigenvalues larger than the average value of 1.

The decrease of the eigenvalues can also be followed graphically using the so-called **scree test**. If one uses a scattergram where the eigenvalues are plotted as ordinate values versus their number it is possible in most cases to see the heavy decay of eigenvalues to a low plateau level where eigenvalues do not change very much (see next section). The beginning of this level may indicate the number of factors to be included.

A more complex method is described by WOLD [1978], who used **cross-validation** to estimate the number of factors in FA and PCA. WOLD applied the NIPALS (non linear iterative partial least squares) algorithm and also mentioned its usefulness in cases of incomplete data.

Rotation of the Factor Solution

The rotation of the factor solution is a tool to assist the ability to interpret factor loadings. The factor structures computed as explained above often have a shape which assigns nearly all features to the first factor. Interpretation of the contribution of features to this factor is, therefore, usually not possible. Although rotation of the coordinate system of factors will not affect the position of the objects relative to each other, it can be used to simplify the factor structure. Now one has to decide between the possibilities for rotation – they are infinite. THURSTONE (1965) [WEBER, 1974] developed some criteria for the so-called **simple structure**:

(1) Each row in the factor loadings matrix should have at least one value near zero, i.e. each feature should have at least one zero loading in the extracted factors.
(2) In the case of s extracted factors, each column of the factor loadings matrix should have s zero loadings, i.e. each factor should contain s zero loadings.
(3) Each pair of extracted factors should contain different features which have zero loadings.
(4) Only a minimum number of features should have nonzero loadings in more than one factor.

The short expression of these ideal structure criteria is the minimization of medium factor loadings to the benefit of near zero and high factor loadings. After rotation the descending order of the eigenvalues of factors may be revoked.

These criteria lead to different numeric transformation algorithms. The main distinction between them is **orthogonal and oblique rotation**. Orthogonal rotations save the structure of independent factors. Typical examples are the **varimax, quartimax, and equimax** methods. Oblique rotations can lead to more useful information than orthogonal rotations but the interpretation of the results is not so straightforward. The rules about the factor loadings matrix explained above are not observed. Examples are **oblimax** and **oblimin** methods.

Target Factor Rotation

Target factor analysis is a suitable method for receptor modeling. The essence of this method is the **correlation of one special influence, the target, with one factor from the factor solution**. The factor solution is constrained into a shape such that the target correlates with one factor but not with the other factors. It is, therefore, possible to estimate the part of the variance from this special influence and to find other correlating features.

Imagine an industrial stack which contaminates the environment. The pollution of the air is measured as metal concentrations in suspended dust. Using the metal emission profile of the dust from the stack as a target it is possible to estimate the part of the pollution variance which results from this industrial stack.

The mathematical way of such analysis is the following: Single factor loadings a_{ij} are the same as correlation coefficients between the original feature i and the factor j. It is, therefore, possible to define the interaction between factors from the factor solution

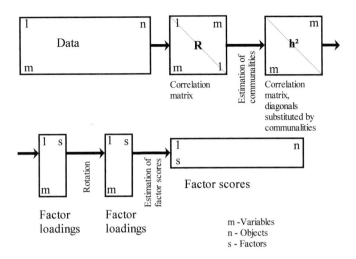

Fig. 5-18. Schematic diagram of the principal course of factor analysis

and the target as high factor loadings by the typical features of the target. The factor rotation is carried out with the criteria of maximum loadings of target features in one factor and minimum loadings of these elements in the other factors. The influence of the target on different objects can be estimated by factor scores as explained above.

The principal course of factor analysis is summarized schematically in Fig. 5-18.

5.4.4 Typical FA Results Illustrated with an Extended Example

Again we use the extended example first analyzed by cluster analysis in Section 5.3.5.

The **correlation matrix** is given in Tab. 5-3.

Remember that one can test the significance of all the correlation coefficients by comparison with one critical value (see Section 2.4.2, example). The significant correlations at a risk of an error of the first kind of 5% are printed in bold in Tab. 5-3.

After the **diagonalization** of this matrix the **eigenvalues** shown in Tab. 5-4 are obtained. From the values we see that most of the variance is carried by the first three factors. Almost 50% of variance is extracted by the first factor. Every single eigenvalue stands for more than 10% contribution. The cumulative percentage indicates that more than three factors may be necessary to account for more than 90% of the variance. In contrast the MALINOWSKI index indicates only one factor as being worth discussing. Fig. 5-19 illustrates the decay of the eigenvalues by means of a scree plot. The corresponding **eigenvectors** are shown in Tab. 5-5.

Multiplying by the square roots of the respective eigenvalues gives the **loadings** listed in Tab. 5-6. If we concentrate on features which contribute to the factors by more than

Tab. 5-3. Correlation matrix for seven features (lower triangle)

	Si	Al	Fe	Ti	Na	Mg	Ca
Si	1.000						
Al	**−0.604**	1.000					
Fe	**−0.638**	**+0.822**	1.000				
Ti	**−0.666**	**+0.803**	**+0.567**	1.000			
Na	−0.103	−0.022	+0.311	+0.144	1.000		
Mg	−0.063	+0.429	−0.065	**+0.609**	−0.302	1.000	
Ca	+0.395	−0.237	0.000	**−0.694**	−0.020	**−0.517**	1.000

Tab. 5-4. Eigenvalues, λ_i, of the correlation matrix given in Tab. 5-3

i	λ_i	% variance	Cumul. %	IND
1	3.463	49.47	49.47	0.53
2	1.680	24.00	73.47	0.61
3	0.984	14.05	87.52	0.73
4	0.620	8.86	96.38	0.81
5	0.193	2.75	99.14	1.09
6	0.043	0.61	99.74	3.35
7	0.018	0.26	100.00	

50%, i.e. where the squared coefficients exceed approximately 0.7, we may interpret the first factor as determined mainly by the elements Si, Al, Fe, and Ti. We note that most of the elements found in the second cluster (see also Section 5.3.5) load this factor. Na and Mg dominate the following two factors. The importance of the element Ca is spread over the first three factors.

It is to be seen that four elements contribute to the first factor. One also can find many medium factor loadings in the matrix. Rotation of the factor solution is necessary for better interpretation.

Let us now compare the loadings obtained after utilizing a standard rotational procedure (Varimax). In Tab. 5-7 we find the first factor loaded only by Al and Fe. (Using 'insider knowledge', should we relate this to the fact that these are the only elements in the interlaboratory comparison determined by a gravimetric method ...?) All other factors express the importance of single elements. The sole exception is Ti! If we keep our limit of 0.7 for a coefficient Ti does not contribute to any of the factors significantly.

In Fig. 5-20 the 15 samples of the interlaboratory test are displayed using the rotated factor solution. We can clearly identify the special role of laboratory *B*. The first factor describes the variance part between the single measurements *B2* and *B3* and the

5.4 Principal Components Analysis and Factor Analysis

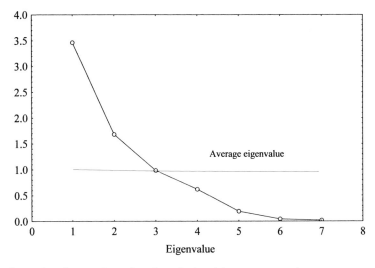

Fig. 5-19. Scree plot of seven eigenvalues from the interlaboratory comparison

Tab. 5-5. Eigenvectors, *EV*, associated with the eigenvalues

	EV 1	EV 2	EV 3	EV 4	EV 5	EV 6	EV 7
Si	−0.418	−0.202	0.003	**0.646**	**0.587**	−0.086	0.122
Al	**0.478**	0.096	0.372	0.247	0.186	−0.334	**−0.646**
Fe	**0.388**	**0.475**	0.232	0.125	0.319	**0.588**	0.327
Ti	**0.518**	−0.118	−0.144	0.102	0.008	**−0.558**	**0.613**
Na	0.062	**0.466**	**−0.711**	**0.430**	−0.224	0.003	−0.196
Mg	0.271	**−0.580**	0.069	**0.459**	**−0.448**	**0.417**	−0.015
Ca	−0.316	**0.396**	**0.526**	0.316	**−0.518**	−0.226	0.217

Tab. 5-6. Unrotated loadings of five factors, *F*, and communalities h_i^2

	F 1	F 2	F 3	F 4	F 5	h_i^2
Si	**−0.778**	0.262	−0.003	0.509	0.258	0.999
Al	**0.890**	−0.124	−0.369	0.195	0.082	0.988
Fe	**0.722**	−0.616	−0.230	0.098	0.140	0.983
Ti	**0.964**	0.153	0.143	0.080	0.004	0.980
Na	0.115	−0.604	**0.705**	0.339	−0.099	0.999
Mg	0.504	**0.752**	−0.068	0.362	−0.197	0.993
Ca	−0.589	−0.514	−0.522	0.249	−0.227	0.997

Tab. 5-7. Varimax rotated loadings of five factors, F, and communalities h_i^2

	F 1	F 2	F 3	F 4	F 5	h_i^2
Si	−0.447	0.256	0.029	**−0.857**	0.022	0.999
Al	**0.904**	−0.110	0.070	0.211	0.318	0.988
Fe	**0.926**	0.037	−0.210	0.264	−0.105	0.983
Ti	0.579	−0.535	−0.131	0.320	0.460	0.980
Na	0.091	−0.033	**−0.985**	0.026	−0.138	0.999
Mg	0.103	−0.296	0.194	−0.037	**0.929**	0.993
Ca	−0.014	**0.948**	0.020	−0.177	−0.263	0.997

rest of the measurements. One can, therefore, conclude that *B2* and *B3* have minimal values in factor 1, i.e. in Al and Fe values. Factor 2 describes the variance part between laboratories *D* and *E*. Laboratory *D*, therefore, has high Ca values and *E* has low Ca values.

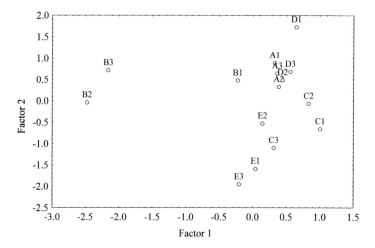

Fig. 5-20. Graphical representation of fifteen objects (samples) from the interlaboratory comparison in the plane of the first two factors

5.5 Canonical Correlation Analysis

5.5.1 Description of Canonical Correlation Analysis

Canonical correlation analysis (CCA) is a method for searching for interactions between two data sets, the matrices X and Y. These data sets may have different numbers of features but the same number of objects. Using canonical analysis one creates a set of canonical variables f for the data set X and a set of canonical variables g for data set Y similar to the factors in factor analysis. The canonical variables f and g should have the following properties:

(1) f and g are standardized variables
(2) when $i = j$ a maximum correlation exists between f_i and g_j
(3) when $i \neq j$ a minimum correlation exists between f_i and g_j
(4) f and g are noncorrelated themselves, i.e. they are orthogonal

The starting point of analysis is the data matrix composed from X and Y:

$$(XY) = \begin{pmatrix} x_{11} & x_{12} & \cdots & x_{1m_x} & y_{11} & y_{12} & \cdots & y_{1m_y} \\ x_{21} & x_{22} & \cdots & x_{2m_x} & y_{21} & y_{22} & \cdots & y_{2m_y} \\ \cdots & \cdots & \cdots & \cdots & \cdots & \cdots & \cdots & \cdots \\ x_{n1} & x_{n2} & \cdots & x_{nm_x} & y_{n1} & y_{n2} & \cdots & y_{nm_y} \end{pmatrix} \quad (5\text{-}29)$$

The correlation matrix of this composed matrix:

$$R_{(XY)} = \begin{pmatrix} R_{XX} & R_{XY} \\ R_{YX} & R_{YY} \end{pmatrix} \quad (5\text{-}30)$$

is used to find canonical variables by solution of an eigenvalue problem:

$$(R_{XY} R_{YY}^{-1} R_{YX} R_{XX}^{-1} - \lambda I) f_e = 0 \quad (5\text{-}31)$$

- Similar to PCA, the eigenvalue solution leads to eigenvalues λ and eigenvectors f_e which describe the total variance of the composed correlation matrix.
- The number of eigenvectors f_e is equal to the number of features in X.
- The normalized eigenvectors f are the canonical variables for the matrix X similar to the factor loadings in PCA.
- g are the canonical variables for the matrix Y. The single canonical variable f_k is connected with the single canonical variable g_k for the matrix Y. Canonical variables g_k for Y can be calculated from:

$$g_k = \frac{R_{YY}^{-1} R_{YX} f_k}{\sqrt{\lambda_k}} \quad (5\text{-}32)$$

- Square roots of the eigenvalues λ are the canonical correlations between the canonical variables f and g.
- The number of canonical variables g_k for the matrix Y is equal to the number of features in Y.

Interpretation of Results

One has now extracted synthetic variables which explain most of the interactions between the data sets X and Y. But in most cases one needs an interpretation of these canonical variables in relation to the original features:

- The correlations between f and the original features in X, or between g and the original features in Y, are called **intra-set loadings**. They express the part of the variance which is explained by the canonical variables of its own feature set, and are also called the **extraction measures**.
- The correlations between the original features of one set and the canonical variables of the second set are called **inter-set loadings**. They are also **redundancy measures** and demonstrate the possibility of describing the first data set with the features of the second set. The inter-set loadings characterize the overlapping of both sets or, in other words, the variance of the first set explained by the second set of features.

5.5.2 Typical CCA Results Illustrated with an Extended Example

Example 5-1 (continued)

We select two sets of features from our interlaboratory experiment. The first set is X = {Ti, Fe} and the second is Y = {Al, Ca, Mg}.

The numerical results of CCA are the following: There are two *canonical eigenvalues* (roots), λ_1 = 0.921 and λ_2 = 0.851. The *overall canonical correlation* of X and Y (computed from the scores associated with the first eigenvalues of both sets) is 0.9596 and is highly significant. The second canonical correlation coefficient, (which may also be computed from the square root of λ_2) yields 0.922 and is also significant.

The *factor structure* for two canonical variables (the loadings) is given in Tab. 5-8.

These contributions represent the overall correlations with each canonical feature. Again we find factor patterns which are not very pronounced. The extraction and the redundancy measures are reported in Tab. 5-9. From the total values of the variance explained we see that both sets are well represented by their canonical variables. On the other hand the redundancy measure (90% or 72%) indicates that both feature sets may be of equal practical weight.

Tab. 5-8. Factor structure for two canonical variables, CV

	X Set		Y Set		
	Fe	Ti	Al	Mg	Ca
CV 1	−0.752	−0.970	−0.925	−0.488	0.579
CV 2	−0.659	0.245	−0.255	0.686	−0.687

Tab. 5-9. Variance of each feature set explained by the respective canonical variable, CV, and redundancy in the respective variable (variance explained by the other feature set)

	X Set Variance	Redundancy	Y Set Variance	Redundancy
CV 1	0.753	0.693	0.476	0.439
CV 2	0.247	0.211	0.336	0.285
total	1.000	0.904	0.812	0.724

By analogy with factor analysis we can now display the objects of the data set with those canonical features which extract the main portion of data variation. Fig. 5-21 shows the samples from the interlab test in the plane of the two first canonical variables. In addition to the already known special role of laboratories *B* and *E* we note some indication of the separation of the other laboratories also.

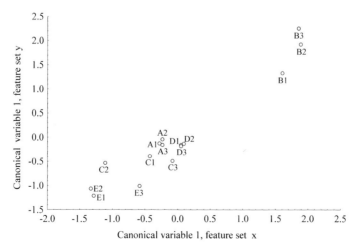

Fig. 5-21. Graphical representation of fifteen objects (samples) from the interlaboratory comparison in the plane of the two first canonical variables (each derived from two feature sets)

Example 5-2 (continued)

The example first mentioned in Section 5.3.5 will now be examined. There, by cluster analysis we found clear distinction was possible between the two feature sets X = {4 chlorine hydrocarbons} and Y = {3 bromine hydrocarbons} in 80 samples from a river in Thuringia. The overall canonical correlation of 0.2796 is not significant. Hence, in principle we can skip discussion of further details.

The only point to note is how the pattern is contrary to the above example. In the example with halogenated volatile hydrocarbons we find a total redundancy of 3.0% in the X set (variance explained by the Y set) and 2.7% in the Y set (variance explained by the X set). The extracted variance proportions are 73.3% for the X set and 100% for the Y set. Whereas the last-mentioned proportions are 'normal', we confirm from the redundancies that we really have independent sets of different halogenated hydrocarbons.

For practical purposes, in contrast with the above example this means that we have to discuss and use both feature sets.

5.6 Multivariate Analysis of Variance and Discriminant Analysis

5.6.1 General Description

Because multivariate mean comparison by variance analysis is strongly related to the task of discriminating means, or the respective classes of objects, in applications both aspects are seen at the same time. Therefore the name **multivariate variance and discriminant analysis (MVDA)** is commonly used.

1. Multivariate Analysis of Variance

Analysis of variance in general serves as a statistical test of the influence of random or systematic factors on measured data (test for random or fixed effects). One wants to test if the feature mean values of two or more classes are different. Classes of objects or clusters of data may be given *a priori* (supervised learning) or found in the course of a learning process (unsupervised learning; see Section 5.3, cluster analysis). In the first case variance analysis is used for class pattern confirmation.

As we know from the section on ANOVA (analysis of variance; see Section 2.3) in **univariate cases**, where only one feature is investigated, the sum of the squares of deviations of all n measuring values from the total mean is split into a part determined by random influences (experimental uncertainty; within-class scattering) and into a part influenced systematically by defined or attributable factors (between-class scattering). The ratio F_{exp} of between-class and within-class scattering, expressed by the respective variances (sum of squares of deviations divided by the associated degrees of freedom) is then compared against a theoretical (tabulated) value F_{crit}. If $F_{exp} \geq F_{crit}$ it is justified to assume statistically significant differences of the data classes.

In principle, with this F-test we test deviations of class mean values from a total mean value.

The simultaneous comparison of the mean values of a set of features, i.e. two vectors of means in the simplest case of *two classes*, we will call **multivariate analysis of variance** (MANOVA).

In the general case, for *several classes*, the sums of squares representing between-class and within-class scattering are then extended to matrices $\boldsymbol{B}$ and $\boldsymbol{W}$:

$$\boldsymbol{B} = \sum_{k=1}^{k_{total}} n_k (\bar{\boldsymbol{x}}_k - \bar{\boldsymbol{x}})(\bar{\boldsymbol{x}}_k - \bar{\boldsymbol{x}})^T \tag{5-33}$$

$$\boldsymbol{W} = \sum_{k=1}^{k_{total}} \sum_{n=1}^{n_k} (\boldsymbol{x}_{kn} - \bar{\boldsymbol{x}}_k)(\boldsymbol{x}_{kn} - \bar{\boldsymbol{x}}_k)^T$$

k_{total} – number of classes
n_k – number of objects in class k
$\bar{\boldsymbol{x}}$ – total mean, vector of m mean values computed from all objects
$\bar{\boldsymbol{x}}_k$ – mean of class k, vector of m mean values computed for each class

Matrix $\boldsymbol{B}$ expresses the variance between the means of the classes, matrix $\boldsymbol{W}$ expresses the pooled within-classes variance of all classes. The two matrices $\boldsymbol{B}$ and $\boldsymbol{W}$ are the starting point both for multivariate analysis of variance and for discriminant analysis.

The multivariate test criterion uses the ratio between the two matrices $\boldsymbol{B}$ and $\boldsymbol{W}$. A multivariate F_{exp} can be computed in a manner similar to the univariate F_{exp}:

$$F_{exp} = C_1 \cdot sp\,(\boldsymbol{B} \cdot \boldsymbol{W}^{-1}) \tag{5-34}$$

C_1 – coefficient which contains several degrees of freedom in respect of classes and features
sp – spur of the matrix, i.e. the sum of the diagonal elements of the quadratic matrix $\boldsymbol{B} \cdot \boldsymbol{W}^{-1}$

Remark on the multivariate degrees of freedom used in this section: Because we are only interested in presenting the main points of MVDA, we shall not report the degrees of freedom in detail but rather indicate by a subscript number the different expressions of such degrees of freedom. The exact value of C_1, or the involved degrees of freedom, depends on the special combination of the feature number m and the class number solely for the example under study.

If F_{exp} is high the variance between the groups is higher than the averaged variance within the groups. This occurs when groups are nearly separated. F_{exp} will be low if the variance between groups is nearly the same as the variance within the groups. Then groups cannot be distinguished.

Several statistics for multivariate tests are known from the literature [AHRENS and LÄUTER, 1981; FAHRMEIR and HAMERLE, 1984]; the user of statistical packages may find several of them implemented and will rely on their performing correctly. Other, different, tests for separation of groups are used to determine the most discriminating results in discriminant analysis with feature reduction.

2. Discriminant Analysis

In **discriminant analysis**, in a manner similar to factor analysis, new synthetic features have to be created as linear combinations of the original features which should **best indicate the differences between the classes, in contrast with the variances within the classes**. These new features are called **discriminant functions**. Discriminant analysis is based on the same matrices B and W as above. The above tested groups or classes of data are 'modeled' with the aim of reclassifying the given objects with a low error risk and of classifying ('**discriminating**') another objects using the model functions.

The mathematical procedure for finding such discriminant functions is to solve the eigenvalue problem of the quotient $B \cdot W^{-1}$, i.e. to find the characteristic roots and eigenvectors as known from principal components analysis:

$$\begin{aligned} (B \cdot W^{-1}) \cdot e_1 &= \lambda_1 \cdot e_1 \\ (B \cdot W^{-1}) \cdot e_2 &= \lambda_2 \cdot e_2 \\ &\dots \end{aligned} \quad (5\text{-}35)$$

The nontrivial solution leads to the evolution of the determinant:

$$|B \cdot W^{-1} - \lambda I| = 0 \quad (5\text{-}36)$$

The solution provides pairs of eigenvalues and eigenvectors as known from principal components analysis. Eigenvectors, e, are themselves noncorrelated, i.e. they are orthogonal. Eigenvalues λ provided from this equation express the part of extracted variance of the matrix $B \cdot W^{-1}$. The first eigenvalue λ_1 is equal to the part of the variance extracted by the first eigenvector, the second eigenvalue λ_2 represents the variance extracted by the second eigenvector, and so on. The sum of all the eigenvalues represents the total variance in $B \cdot W^{-1}$.

The coefficients for the discriminant functions are, therefore, the coefficients of e. These **unstandardized coefficients** are the multipliers of the original variables which yield the discriminant functions. An interpretation of the magnitude of the unstandardized coefficients is not useful and could lead to wrong conclusions because of the different magnitudes of the features.

Therefore **standardized coefficients, a,** are useful information for assessment of the nature of the class differences:

$$a = \frac{e}{\sqrt{e^T e}} \qquad (5\text{-}37)$$

Another possibility of interpreting the discriminant functions is to calculate the correlation between these functions and the original features.

3. Discriminant Scores of the Objects

The coordinates of the objects in the new **space of discriminant functions** can be calculated, by analogy with principal components analysis, from the original feature values of the objects.

With the n_{df} discriminant functions the n_{df} **discriminant scores** of each object may be calculated. As an example the first discriminant score of an object i is given by:

$$df_{1i} = e_{11} x_{1i} + e_{12} x_{2i} + \cdots + e_{1j} x_{ji} + e_{1m} x_{mi} \qquad (5\text{-}38)$$

df_{1i} – discriminant function 1 of object i
e_{1j} – coefficients for all original features ($j = 1, ..., m$) of discriminant function 1
x_{ji} – values of original features ($j = 1, ..., m$) of object i

4. Classification of 'Unknown' Objects

In the case of two groups one describes the groups by the discriminant scores of one discriminant function. The coordinates of the two class means with their variances are known and the new object may easily be discriminated, in other words classified to one of the groups. This is achieved simply by attributing the 'unknown' object to that class into which confidence region it falls. This may be tested using:

$$F_k = C_2 \sum_{t=1}^{n_{df}} (df_{ti} - df_{tk})^2 \qquad (5\text{-}39)$$

C_2 – coefficient combining several degrees of freedom
df_{ti} – discriminant feature score of object i; discriminant function t
df_{tk} – discriminant feature score of the mean of class k
  (This value is calculated by inserting the original feature mean values into Eq. 5-38.)
F_k – experimental F-value of class k
n_{df} – number of discriminant features

$$C_2 = \frac{n - k_{total} - n_{df} + 1}{n_{df}(n - k_{total})} \frac{n_k}{n_k + 1} \qquad (5\text{-}40)$$

k_{total} – total number of classes
n – total number of objects
n_{df} – number of discriminant features
n_k – number of objects in class k

With Eq. 5-40 every class has attributed to it, an n_{df}-dimensional scatter region. If F_k of the object under question is less than or equal to $F(n_{df}; n - k_{total} - n_{df} + 1; q = 1 - \alpha)$ then the object belongs to class k. In the case of overlapping regions, the class with the lowest F-value is usually preferred.

5. Classification Error

To judge the performance of the discriminant functions and the classification procedure *in respect of future samples* one can calculate **misclassification probabilities** or **error rates**. But these probabilities cannot be calculated in general because they depend on the unknown density functions of the classes. Instead we can usually utilize a measure called **apparent error rate**. The value of this quantity is easily calculated from the classification or confusion matrix *based on the samples of the training set*. For example with two classes we can have the following matrix:

Actual membership of class		Predicted membership of class		
		1		2
1:	$n_1 =$	$n_{1,right}$	+	$n_{2,wrong}$
2:	$n_2 =$	$n_{1,wrong}$	+	$n_{2,right}$

With the number n_{wrong} of misclassified samples from the training set the apparent error rate, *AER*, is:

$$AER = \frac{n_{1,wrong} + n_{2,wrong}}{n_1 + n_2} \qquad (5\text{-}41)$$

Here one has the same problem as with the calculation of calibration errors only using the members of the set of calibration measurements. One gets an optimistic estimate of the future error.

For a more realistic estimate of the future error one splits the total data set into a training and a prediction part. With the training set the discriminant functions are calculated and with the objects of the prediction or validation set, the error rate is then calculated. If one has insufficient samples for this splitting, other methods of **cross-validation** are useful, especially the 'holdout method' of LACHENBRUCH [1975] which is also called 'jackknifing' or **'leaving one out'**. The last name explains the procedure: For every class of objects the discriminant function is developed using all the class members except one. The left out object is then reclassified. The procedure is then repeated until each object of the respective group has been reclassified. Finally the estimate of the actual error rate is calculated by use of Eq. 5-41.

6. Feature Reduction

As in factor analysis, the discriminant feature space may have a lower dimension n_{df} than the original feature space. With respect to the classification into a certain number of classes the following number of discriminant functions is necessary:

$$n_{df} = \min(k_{total} - 1, m) \qquad (5\text{-}42)$$

k_{total} – number of classes modeled
m – number of original features
n_{df} – total number of discriminant functions (discriminant features)

(The mathematical reason is that rank $(B) \leq k_{total} - 1$ and rank $(W) \leq m$.)

At this stage, however, discriminant analysis as well as factor analysis, do not provide a real reduction in dimensions from the practical (experimental) point of view because in the linear combinations used in both methods we still need **all the original features**.

A **real reduction in dimension** is possible in discriminant analysis on a statistical basis because we can delete features bearing redundant information, i.e. which are highly correlated to others, in an eliminating process which finds an optimum feature set with a statistically sufficient discriminating power and a risk of error which is still acceptable. Hence, only with DA can we offer real economical advantages.

If one actually wants to reduce an original feature set, for example in order to reduce experimental effort, one can follow two main strategies.

In the first strategy, a **forward strategy**, one could start with the feature with the highest F-value and then successively add further features. In the second strategy, a **backward strategy**, one could start with all features. Then the feature with the lowest discriminating power is deleted first from the entire set, and so on.

The only question is when to stop such a strategy. From the statistical point of view one may keep those features in an optimum feature set for which the F-value (see Eq. 5-34) is still significant, e.g. at the significance level of 5%. Another way is to stop the selection procedure when the reclassification error reaches a minimum or a constant value.

The following statistical measures are those most commonly found in software packages. First we mention HOTELLING's T^2 for the 2-class case which is based on a generalized distance measure, the MAHALANOBIS distance D^2, and from which a F-test can be derived:

$$D^2 = (n - k_{total})(\bar{x}_1 - \bar{x}_2)^T W^{-1}(\bar{x}_1 - \bar{x}_2) \qquad (5\text{-}43)$$

D^2 – MAHALANOBIS distance
$\bar{x}$ – vectors containing the mean values of classes 1 and 2
W – pooled within-class scatter matrix, see Eq. 5-33, under the assumption that the covariance matrices of the classes are equal (in the statistical sense):

$$T^2 = \frac{1}{n_1 + n_2 - 2} \frac{n_1 n_2}{n_1 + n_2} D^2 \tag{5-44}$$

n_1, n_2 – number of objects in classes 1 and 2
T^2 – HOTELLING's T^2

The formula holds for the *comparison of two classes*, and the F-value calculated according to Eq. 5-45 is tested against $F_{crit}(m, n_1 + n_2 - m - 1; q = 1 - \alpha)$.

$$F = C_3 T^2 \quad \text{with} \quad C_3 = \frac{n_1 + n_2 - m - 1}{m} \tag{5-45}$$

This test now offers the possibility of stopping the feature selection process to keep sufficient power of discrimination, which is assured as long as F exceeds the critical value.

Other measures use the spur of the matrix product (BW^{-1}) as described in Eq. 5-34 or are based on the statistics of WILKS:

$$\Lambda_{WILKS} = \frac{|W|}{|B + W|} = \prod_{t=1}^{n_{df}} \frac{1}{1 + \lambda_t} \tag{5-46}$$

Because of the use of determinants it is also called the determinant criterion, for which test the user also needs special table values or subprograms for the calculation of the respective F-approximation.

It is, however, very common to use the so-called **partial lambda** which enables evaluation of the process of selection of original variables for calculation of an optimum set of features. This partial lambda is the ratio of two WILKS lambda values, Λ_{set1} and Λ_{set2}, where Λ_{set1} holds for the smaller set of features and Λ_{set2} is computed after adding a feature to the former set:

$$\Lambda_{partial} = \frac{\Lambda_{set2}}{\Lambda_{set1}} \tag{5-47}$$

For this measure a respective F-value may be calculated and tested:

$$F = C_4 \frac{1 - \Lambda_{partial}}{\Lambda_{partial}} \quad \text{with} \quad C_4 = \frac{n - k_{total} - m_1}{k_{total} - 1} \tag{5-48}$$

m_1 – number of features in the smaller set of features

The principal course of MVDA is summarized in Fig. 5-22.

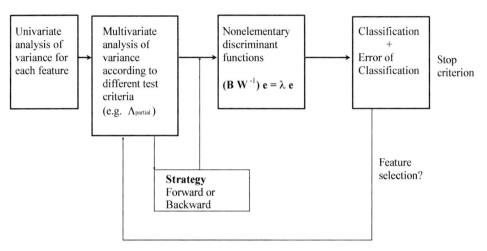

Fig. 5-22. Schematic diagram of the course of multivariate analysis of variance and discriminant analysis

5.6.2 DA Calculations Demonstrated with a Simple Example

This section is dedicated to the interested reader who wants to follow the major steps of discriminant analysis and also to inform him of further mathematical details. (For the calculations we use a pocket calculator and sometimes rounded values; small differences may, therefore, occur if other precision or programmed algorithms are used.)

Again we start from the **raw data** we first used in cluster analysis, Tab. 5-10.

We have five objects each with two features. The objects are classified according to the result of cluster solution 1; class 1 contains objects $K1$, $K3$ and $K4$ ($n_1 = 3$), class 2 has the two elements $K2$ and $K5$ ($n_2 = 2$). With the objects written as

$$x_{K1} = (4.0 \quad 1.5), \quad x_{K2} = (4.0 \quad 6.0), \ldots$$

and the class means and the total mean written as

$$\bar{x}_1 = (3.5 \quad 2.0) \quad \bar{x}_2 = (3.0 \quad 7.0) \quad \bar{x} = (3.3 \quad 4.0)$$

we obtain the **between-class scatter matrix** B and the **within-class scatter matrix** W as

$$B = \begin{pmatrix} 0.3 & -3.0 \\ -3.0 & 30 \end{pmatrix}, \quad W = \begin{pmatrix} 8.5 & -1.5 \\ -1.5 & 2.5 \end{pmatrix}$$

(Note the linear dependence of lines or rows in B which indicates that B is not of full rank.)

Tab. 5-10. Example taken in this section; the units of features x_1 and x_2 are arbitrary but identical

Object	Class	x_1	x_2
K1	1	4.0	1.5
K2	2	4.0	6.0
K3	1	5.0	2.5
K4	1	1.5	2.0
K5	2	2.0	8.0

Multivariate Analysis of Variance

As we mentioned in the preceding section, multivariate analysis of variance, like discriminant analysis, uses the scatter matrices B and W.

Unfortunately, test statistics are not easy to understand and critical values are normally not available from standard F-tables. However, in our two-class case we have the advantage that F_{exp} of Eq. 5-34 is F-distributed with

$$C_1 = \frac{n_1 + n_2 - m - 1}{m} \tag{5-49}$$

With sp(BW^{-1}) = 12.9868, F_{exp} = 12.9868 (C_1 = 1) is to be compared with F_{crit} (2; 2; 0.95) = 19. Therefore, at a significance level of α = 5%, we have to conclude that the variance encountered in B and W occurs only by chance, in other words, from the statistical point of view the classes are not separable.

Although the above 'overall' F-test signaled no significant class differences we compute the specific effects of the features, see Tab. 5-11. In our example we can take the mean squared effects from the diagonal elements of B and the mean squared errors from the diagonal elements of W, divided by $n - m = 3$. The respective ratios then yield

Tab. 5-11. Results from analysis of variance; specific effects of the features

	Mean squared effect	Mean squared error	F_{exp}	α_{calc}
Feature 1	0.3	2.83	0.106	76.6%
Feature 2	30	0.83	36.0	0.9%

the F_{exp}-values which can be compared with $F_{crit}(1; 3; 0.95) = 10.1$. From this test or from the calculated values of α_{calc} we find no significance for feature 1 but a significant influence of feature 2.

Therefore, only in respect of feature 2 is the variation due to class structure significantly higher than the variation due to experimental error. In other words, if we want a better classification model we should omit feature 1 and keep feature 2.

Discriminant Functions

From Eqs. 5-35 and 5-36 we obtain the eigenvalues and the associated eigenvectors e.

For numerical reasons, instead of Eq. 5-36 we use the equivalent form

$$|B - \lambda W| = 0$$

We formulate the characteristic equation:

$$\left| \begin{pmatrix} 0.3 & -3 \\ -3 & 30 \end{pmatrix} - \lambda \begin{pmatrix} 8.5 & -1.5 \\ -1.5 & 2.5 \end{pmatrix} \right| = 0$$

and find:

$$(0.3 - 8.5\,\lambda)(30 - 2.5\,\lambda) - (-3 + 1.5\,\lambda)^2 = -246.75\,\lambda + 19\,\lambda^2 = 0$$
$$\lambda_1 = 12.9868 \quad \lambda_2 = 0$$

(According to the rank of B we find only one eigenvalue differing from zero.)

As in Section 5.4.2 (PCA) we compute the eigenvector and its components associated with λ_1 and finally find the normalized coefficients for the discriminant function

$$e = \begin{pmatrix} 1 \\ -6.68 \end{pmatrix} \quad a = \begin{pmatrix} 0.148 \\ -0.989 \end{pmatrix}$$

With these coefficients the coordinates of the objects are computed; they are given in Tab. 5-12.

Tab. 5-12. Raw, R, and standardized, S, object coordinates (discriminant feature scores)

Object Class	K1 1	K2 2	K3 1	K4 1	K5 2	$\bar{x}_1$	$\bar{x}_2$	$\bar{x}$
R, manual	−14.0	−44.0	−21.6	−14.8	−55.4	−16.8	−49.7	−30.0
S, manual	−2.1	−6.5	−3.2	−2.2	−8.2	−2.5	−7.4	−4.44
R, software	2.77	−2.44	1.44	2.62	−4.40	2.28	−3.42	0.0
S, software	−2.75	−7.50	−4.10	−2.55	−9.03	−3.13	−8.27	−5.19

The raw and the standardized coordinates are calculated both manually and using software [STATISTICA, 1995]. As in most cases where the reader of publications wants to reproduce the results, surprisingly we get a different result. In our case this is because most software (SPSS, STATISTICA, ...) calculates a constant along with the raw coefficients. At the same time this demonstrates that there are several ways of finding discriminant functions. So, in some instances it may be convenient to use so-called elementary discriminant functions [AHRENS and LÄUTER, 1981] or to try quadratic discrimination (see [FAHRMEIR and HAMERLE, 1984]).

What counts is the best discrimination of the classes modeled. In our example with all modes of score calculation we can see the same *relative position* of all objects and means. Because of simplicity of the example the reader may quickly create a draft graph of the objects.

At this stage we should confess that we are stressing statistical models very much because our example has so few objects compared to the number of features. Especially for the nonelementary discrimination functions df discussed above there are recommendations for ensuring a ratio between the number of objects n and the number of original features m of $n/m > 3$ or even $n_k/m > 3$. Therefore attempts are made to develop classifiers which work well in the case of $n/m < 3$. One example is the EUCLIDean distance classifier of MARCO et al. [1987], the additional advantages of which are: no covariance matrix is necessary, inversions are skipped, and correlated training data cause no problems.

5.6.3 Typical DA Results Illustrated with an Extended Example

Example 5-1 (continued)

Case 1: Given Classes

In this example let us stress again the interlaboratory experiment results first introduced in Section 5.3.5. In some variables we have added a digit in the 5th decimal place for a few objects only in order to avoid numerical problems as a result of identical values.

If we calculate the four nonelementary discriminant functions df we find the following fractions of data variance explained: 77.3% (by one function), 98.8% (by two functions), and 99.9% (by three discriminant functions). Hence we do not expect severe biased projections of our data on to the plane. In Fig. 5-23 we find some overlapping laboratories, however. In the 3D-plot of Fig. 5-24 a good, separated display of all laboratories' data is indicated. So far, the data projection is satisfactory.

Another aspect is how good the discriminant functions are for classification. With all seven original features we do not have any misclassification. Now a laboratory expert may ask if it is really necessary to use all the measurement information, in other words, if it is possible to reach equally good reclassification with fewer than seven features.

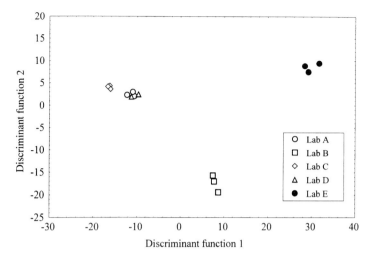

Fig. 5-23. Graphical representation of fifteen objects (samples) from the interlaboratory comparison in the plane of the first two discriminant functions (seven original variables)

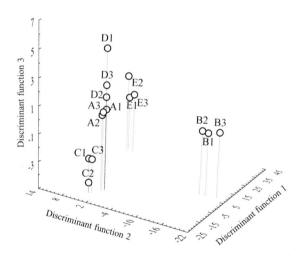

Fig. 5-24. Graphical representation of fifteen objects (samples) from the interlaboratory comparison in the space of the first three discriminant functions (seven original variables)

In Tab. 5-13 we report the results of both mentioned strategies for the selection process. In both procedures the WILKS lambda varies monotonously and each set has a significant meaning. We may, therefore, stop the selection process following the misclassification rate. In the forward strategy the first zero error rate appears with the feature set {Ti, Mg, Ca} in step 3 (Fig. 5-25) whereas in the backward strategy the zero error rate is obtained with the remaining elements {Si, Ca, Al, Mg} in step 3. Now it is up to the expert to decide which feature set to retain in the future.

Tab. 5-13. Results of reclassification as a function of different feature sets; comparison of the forward and the backward selection strategy

Forward Strategy

Step	Element added	Misclassifications for laboratory				
		A	B	C	D	E
1	Ti	0	0	2	0	1
2	Mg	0	0	0	1	0
3	**Ca**	**0**	**0**	**0**	**0**	**0**
4	Fe	0	0	0	0	0
5	Na	0	0	0	0	0
6	Si	0	0	0	0	0
7	Al	0	0	0	0	0

Backward Strategy

Step	Element removed	Misclassifications for laboratory				
		A	B	C	D	E
1	Na	0	0	0	0	0
2	Ti	0	0	0	0	0
3	**Fe**	**0**	**0**	**0**	**0**	**0**
4	Si	0	0	0	2	0
5	Ca	1	0	0	1	0
6	Al	1	2	0	1	0
7	Mg	–	–	–	–	–

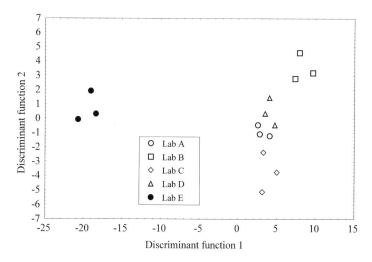

Fig. 5-25. Graphical representation of fifteen objects (samples) from the interlaboratory comparison in the plane of the first two discriminant functions (three optimum variables)

We should, however, remark that in reality the data analyst will look for more samples, will probably try several cross-validation procedures, will test the classification functions using independent test sets, and so on.

As already mentioned, in this respect DA is a *pattern recognition and learning tool*. Finally the analyst will rely on a result only if it is obtained by several different methods.

Case 2: Classes Found by CA

In the previous example the data situation was adequately investigated by multivariate analysis of variance and discriminant analysis.

Let us now imagine another situation with the same data set. In Section 5.3.5, Fig. 5-14, we discussed the clustering of some laboratories' data. We distinguished three clusters {B1, B2, B3}, {E1, E2, E3}, and {A1-3, C1-3, D1-3}. Another application of discriminant analysis would, therefore, be to confirm this cluster solution or structure by multivariate analysis of variance or to find good separation functions for this solution by discriminant analysis. Again two feature selection strategies may be used to find the best discriminating features. In contrast to the above result in Example 5-1 we find two nonoverlapping optimum feature sets: {Ti, Fe} and {Al, Ca, Mg}. Then the question which may be interesting to answer is how much the two sets correlate. This question is properly answered by canonical correlation analysis.

5.7 Multivariate Modeling of Causal Dependencies

In the following we will meet causal models of increasing 'integration' or complexity. In **multiple regression** we try to model the dependence of one variable, y, on several influencing variables, x. There are mathematical conditions for reliable estimation of the weights of the independent variables (estimation of regression coefficients):

- that the variables x are real independent variables
- that the variables x have to be uncorrelated
- that variables x are measured with negligible error

If the variables are correlated, the occurring **problem of multicollinearity** may be circumvented by performing a principal components calculation with the variables x. This will create independent ('orthogonal') variables and one can continue the regression analysis using the scores (see Section 5.4) instead of the original x values. This method is known as **principal components regression**.

If, on the other hand, the condition of error-free variables x is violated, one can use orthogonal estimation of the regression coefficients.

If we extend the set of the dependent y variables from one to a number, n_y, variables depending on the number, m_x, of variables x, the procedure for estimating the coefficients in the system of equations is more complicated. Such an analysis can be carried out with the PLS-algorithm, see Section 5.7.2. If the conditions for x as above are to be met for y variables, in principle we have uncoupled equations for the single y variables and may apply multiple regression analysis in a sequential manner for all n_y variables y.

The highest 'integration' of all variables and a certain combination of both sets of variables are possible by application of the **method of simultaneous equations** which is also called the **structural equations method** or **path analysis**. In this method the correlation of independent variables is explicitly accounted for. It is, furthermore, allowed that some independent variables are also considered as dependent variables within the same system of equations. As a result of both advantages the system equations are coupled at least by common error variables.

So far, the mentioned methods have one basic assumption in common: all start from linear models.

For some years a family of nonlinear methods, called (artificial) **neural networks**, has gained some importance in chemistry in general [ZUPAN and GASTEIGER, 1993] and in analytical chemistry in particular [JANSON, 1991; KATEMAN, 1993]. These networks are black boxes trained in a learning phase to give the best fit output to given responses.

The principal advantage of modeling nonlinear dependencies is sometimes compensated by the lack of interpretation of the weights.

5.7.1 Multiple Regression

Theoretical Basis
From Section 2.4 we know that univariate regression models describe the relationship of one centered variable, y, on another variable x_1, e.g. by the model:

$$y = b_1 x_1 \qquad (5\text{-}50)$$

In the current section we extend this type of a causal dependence to include a number, m, of independent variables, x, and use model equations like:

$$y = b_1 x_1 + b_2 x_2 + \cdots + b_j x_j + \cdots + b_m x_m \qquad (5\text{-}51)$$

Coefficients b can be estimated if at least m observations of y and all variables x are available. Let us assume that we have $n \geq m$ objects characterized by measurements of y and of x_1 to x_m. Then the first measurement (object) would obey the equation:

$$y_1 = x_{11} b_1 + x_{12} b_2 + \cdots + x_{1m} b_m \qquad (5\text{-}52)$$

and instead of all *n* measurements in the form:

$$
\begin{aligned}
y_1 &= x_{11} b_1 + x_{12} b_2 + \cdots + x_{1m} b_m \\
y_2 &= x_{21} b_1 + x_{22} b_2 + \cdots + x_{2m} b_m \\
&\cdots \\
y_n &= x_{n1} b_1 + x_{n2} b_2 + \cdots + x_{nm} b_m
\end{aligned}
\quad (5\text{-}53)
$$

we write in matrix notation:

$$\boldsymbol{y} = \boldsymbol{X}\boldsymbol{b} \quad (5\text{-}54)$$

which also enables the estimated coefficients to be written in the highly abbreviated form:

$$\hat{\boldsymbol{b}} = (\boldsymbol{X}^T\boldsymbol{X})^{-1}\boldsymbol{X}^T\boldsymbol{y} \quad (5\text{-}55)$$

As we realise, the matrix $(\boldsymbol{X}^T\boldsymbol{X})$ must not be singular, otherwise the inversion process is not executable. If the matrix is nearly singular we will get more or less biased results for the coefficient vector $\boldsymbol{b}$.

The analyst may, in principle, use two ways of avoiding such ill-conditioned matrices:

- Firstly, one can design the experiments statistically, so that the factors x are independent. In the practice of environmental modeling it will even be difficult to adjust sampling conditions according to experimental design.
- The second approach is therefore to apply principal components decomposition to produce orthogonal variables x' which will enable the proper estimation of $\boldsymbol{b}$.

The regression analysis of multicollinear data is described in several papers e.g. [MANDEL, 1985; HWANG and WINEFORDNER, 1988]. HWANG and WINEFORDNER [1988] also discuss the principle of ridge regression, which is, essentially, the addition of a small contribution to the diagonal of correlation matrix. The method of partial least squares (PLS) described in Section 5.7.2 is one approach to solving this problem.

Feature Reduction

Sometimes the question arises whether it is possible to find an optimum regression model by a feature selection procedure. The usual way is to select the model which gives the minimum predictive residual error sum of squares, *PRESS* (see Section 5.7.2) from a series of calibration sets. Commonly these series are created by so-called **cross-validation** procedures applied to one and the same set of calibration experiments. In the same way *PRESS* may be calculated for a different sets of features, which enables one to find the 'optimum set'.

Example 5-3

In Section 3.3.6 we have already performed an example of multivariate regression, another application will be discussed in Section 10.2. In the current example let us again stress the small data set with five objects as last mentioned in Section 5.6.2.

Model Without Intercept

Regression analysis with the dependent class variable y where $y = 1$ or $y = 2$ should yield results similar to those from discriminant analysis [LACHENBRUCH, 1975]. Therefore let us first try to predict class memberships by the two variables, x_1 and x_2, without using an intercept in the regression model.

According to Eq. 5-53 we write

$$1 = 4.0\,b_1 + 1.5\,b_2$$
$$2 = 4.0\,b_1 + 6.0\,b_2$$
$$1 = 5.0\,b_1 + 2.5\,b_2$$
$$1 = 1.5\,b_1 + 2.0\,b_2$$
$$2 = 2.0\,b_1 + 8.0\,b_2$$

and calculate the generalized inverse as

$$(\boldsymbol{X}^T\boldsymbol{X})^{-1}\boldsymbol{X}^T = \begin{pmatrix} 0.107 & 0.024 & 0.123 & 0.014 & -0.080 \\ -0.045 & 0.040 & -0.045 & 0.010 & 0.115 \end{pmatrix}$$

From this we easily obtain

$$\hat{\boldsymbol{b}} = \begin{pmatrix} 0.1321 \\ 0.2300 \end{pmatrix}$$

With these coefficients according to Eq. 5-54 we obtain the following predictions of class memberships

$$\hat{\boldsymbol{y}} = \boldsymbol{X}\hat{\boldsymbol{b}} = \begin{pmatrix} 0.87 \\ 1.91 \\ 1.23 \\ 0.66 \\ 2.10 \end{pmatrix}$$

which yields approximately a 'standard error of calibration' of $s(e) = 0.26$ 'class units'.

Model With Intercept

In a second attempt let us try a regression model with an intercept. Using software [STATISTICA, 1995] we quickly obtain the model as

$$class = y = 0.5527 + 0.0282\, x_1 + 0.1885\, x_2$$

with an error of $s(e) = 0.21$ 'class units'. We note that the inclusion of an intercept reduced the prediction error in the 'calibration step'. If we take care of the parameter statistics we see that only the coefficient of x_2 has a significant meaning, which is in accordance with previous findings. Here, we remember Section 2.4.6 where we spoke of a discrepancy between statistical significance and numerical correctness. If we now calculate the predicted class membership with only the coefficient of x_2 we get residuals resulting in $s(e) = 0.75$ 'class units'.

In addition we confirm that the raw discriminant coefficients from STATISTICA [1995] $e^T = (-0.1732 - 1.157)$ and the regression coefficients b_1 and b_2 are, indeed, proportionally related.

5.7.2 Partial Least Squares Method

Description of the PLS Method

Partial least squares regression (PLS) [WOLD et al., 1984] is a **generalized method of least squares regression**. This method uses latent variables u_1, u_2, ..., i.e. matrix U, for separately modeling the objects in the matrix of dependent data Y, and t_1, t_2, ..., i.e. matrix T, for separately modeling the objects in the matrix of independent data X. These latent variables U and T are the basis of the regression model. The starting points are the centered matrices X and Y:

$$\begin{pmatrix} y_{11} & \cdots & y_{1m_y} \\ y_{21} & \cdots & y_{2m_y} \\ \cdots & \cdots & \cdots \\ \cdots & \cdots & \cdots \\ y_{n1} & \cdots & y_{nm_y} \end{pmatrix} = Y \qquad \begin{pmatrix} x_{11} & x_{12} & \cdots & x_{1m_x} \\ x_{21} & x_{22} & \cdots & x_{2m_x} \\ \cdots & \cdots & \cdots & \cdots \\ \cdots & \cdots & \cdots & \cdots \\ x_{n1} & x_{n2} & \cdots & x_{nm_x} \end{pmatrix} = X \qquad (5\text{-}56)$$

$$\updownarrow \qquad\qquad\qquad\qquad \updownarrow$$
$$U \qquad\qquad\qquad\qquad\; T$$

$$U = A \cdot T + E \qquad (5\text{-}57)$$

The latent variables, calculated by means of an iterative process, have the following properties:

- regression errors between U, i.e. $u_1, u_2, ..., u_s$, and T, i.e. $t_1, t_2, ..., t_s$, have a minimum
- $u_1, u_2, ..., u_s$ are orthogonal linear combinations of the features in Y and model the objects of Y, and $t_1, t_2, ..., t_s$ are orthogonal linear combinations of the features in X and model the objects of X
- in the case of $i = j$ a maximum correlation exists between u_i and t_j
- the pairs u_1 and t_1, u_2 and t_2, ... explain the covariance between X and Y in descending order

In a manner similar to the methods of PCA or FA, PLS extracts **linear combinations of essential features which model the original data X and Y**. But, in contrast with the methods above, **PLS also models the dependence of the two data sets**. This type of model is well suited for modeling and simulating environmental relationships or for multivariate calibration.

The advance of the PLS method is the nonproblematic handling of **multicollinearities**. In contrast with the other methods of multivariate data analysis the PLS algorithm is an iterative algorithm which makes it possible to treat data which have more features than objects [GELADI, 1988].

Errors of the Model

The goodness of fit of PLS models is calculated as an error of the prediction, in a manner similar to the description in ordinary least squares methods. Using the so-called **cross-validation test** one can determine the number of significant vectors in U and T and also the error of prediction.

According to the cross-validation test one uses the matrix Y and deletes a certain number of values, for example in accordance with the following scheme:

$$Y' = \begin{pmatrix} \cdot & y & y & \cdot & y & y \\ y & \cdot & y & y & \cdot & y \\ y & y & \cdot & y & y & \cdot \\ \cdot & y & y & \cdot & y & y \\ y & \cdot & y & y & \cdot & y \\ y & y & \cdot & y & y & \cdot \end{pmatrix} \rightarrow \begin{pmatrix} \hat{y} & y & y & \hat{y} & y & y \\ y & \hat{y} & y & y & \hat{y} & y \\ y & y & \hat{y} & y & y & \hat{y} \\ \hat{y} & y & y & \hat{y} & y & y \\ y & \hat{y} & y & y & \hat{y} & y \\ y & y & \hat{y} & y & y & \hat{y} \end{pmatrix} \quad (5\text{-}58)$$

$\cdot$ – values omitted
$\hat{y}$ – predicted values

The PLS model is calculated without these values. The omitted values are predicted and then compared with the original values. This procedure is repeated until all values have been omitted once. Therefore an error of prediction, in terms of its dependence on the number of latent variables, is determined. The **predicted residual error sum of squares (*PRESS*)** is also the parameter which limits the number of latent vectors u and t:

$$PRESS(t_n) = \sum_i^n \sum_j^{m_y} e_{ij}^2 \quad (5\text{-}59)$$

$$\frac{PRESS(t_k)}{PRESS(t_{k+1})} > 1 \qquad (5\text{-}60)$$

PRESS – predicted residual error sum of squares

If an additional latent vector does not improve the error of prediction it has no adequate effect on this model. Noise cannot be predicted and therefore a minimum value of *PRESS* can be reached. OSTEN [1988], however, stated *PRESS* as having 'poor statistical properties' and proposed improvements of this criterion.

Interpretation of Results

The PLS method is very well suited **for modeling** of any relationships. One may simulate changes in *X* and then observe the changes in *Y*. For given observations in *Y* one can calculate the changes in *X* in a manner similar to that achieved by use of calibration (see Sections 8.3.2.2 and 8.3.3).

It is, furthermore, possible to interpret the latent vectors *t* or *u*. The latent vectors have got scores for each object, as in factor analysis. These scores can be used to display the objects. Another possibility is to compute the correlation between original features and the latent vectors to assess the kind of interacting features for both data sets.

5.7.3 Simultaneous Equations and Path Analysis

Within the framework of this short introduction to essential methods it is not possible to present all the basic details of path analysis. The interested reader will find details in the books of KMENTA [1971], MARDIA et al. [1979], or JOHNSON and WICHERN [1982]. GELADI [1988] mentioned early applications of PLS (see Section 5.7.2) as a path modeling device. The principle and the advantages are indicated in the introduction to the Section 5.7.

Let us only give a résumé of the main steps:

(1) Set-up of the path diagram
    The path diagram marks the correlations of all primary independent variables and their assumed causal influence on other 'independent' variables and on dependent variables. In addition, all relationships between dependent variables are indicated.
(2) Set-up of the model equations
    There are some mathematical conditions to be met: the number of paths leading to each resulting variable must not exceed the number of equations; if the numbers are equal the system is exactly determined and computational (numerical!) problems should be minimal.
(3) Standardization of all variables
    This step ensures the comparability of all variables. In the following step the calculation of the correlation matrix is simplified.

(4) Computation of the correlation matrix of all variables
(5) Calculation of the path coefficients
    The path coeffcients are standardized regression coefficients and are, therefore, directly comparable.
(6) Assessment of the paths
    In this step all paths of different lengths connecting the primary variables with the final dependent variables are given a numerical value. In this way the importance of each path is weighted and may be discussed. The contributions of all path lengths characterizing the direct and the indirect connection of any two variables should add up to the correlation coefficient of these two features.
(7) Interpretation of the coefficients in the model equations

References

Ahrens, H., Läuter, J.: Mehrdimensionale Varianzanalyse - Hypothesenprüfung, Dimensionserniedrigung, Diskrimination, Akademie-Verlag, Berlin, **1981**

Andrews, D.F.: Biometrics 28 (**1972**) 125

AQS: AQS (Analytische Qualitätssicherung)-Merkblätter für die Wasser-, Abwasser- und Schlammuntersuchung, A-2, Erich Schmidt, Berlin, **1991**

Chernoff, H.: J. Am. Stat. Assoc. 70 (**1973**) 548

Coomans, D., Derde, M., Massart, D.L., Broeckaert, I.: Anal. Chim. Acta 133 (**1981**) 225

Doerffel, K., Zwanziger, H.: Fresenius' Z. Anal. Chem. 329 (**1987**) 1

Dubes, R.C.: Pattern Recognition 20 (**1987**) 645

Dubes, R.C., Jain, A.K.: Pattern Recognition 11 (**1979**) 235

Efron, B., Gong, G.: Am. Stat. 37 (**1983**) 36

Fahrmeir, L., Hamerle, A.: Multivariate statistische Verfahren, Walter de Gruyter, Berlin, **1984**

Feinberg, M.: Anal. Chim. Acta 191 (**1986**) 75

Funk, W., Dammann, V., Donnevert, G.: Qualitätssicherung in der Analytischen Chemie, VCH, Weinheim, New York, Basel, Cambridge, **1992**, pp. 56

Geladi, P.: J. Chemometrics 2 (**1988**) 231

Greenacre, M.J.: Theory and Applications of Correspondence Analysis, Academic Press, **1984**

Hartung, J., Elpelt, B.: Multivariate Statistik, 4. Aufl., R. Oldenbourg, München, Wien, **1992**, pp. 593

Hwang, J.D., Winefordner, J.D.: Prog. Anal. Spectrosc. 11 (**1988**) 209

Jahn, W., Vahle, H.: Die Faktorenanalyse und ihre Anwendung, Verlag Die Wirtschaft, Berlin, **1970**, pp. 147

Jain, A.K., Moreau, J.V.: Pattern Recognition 20 (**1987**) 547

Janson, P.E.: Anal. Chem. 63 (**1991**) 357A

Johnson, R.A., Wichern, D.W.: Applied Multivariate Statistical Analysis, Prentice-Hall, New Jersey, **1982**

Kateman, G.: Chemom. Int. Lab. Syst. 19 (**1993**) 135

Kmenta, J.: Elements of Econometrics, Macmillan Company, New York, **1971**

Krzanowski, W.J.: Principles of Multivariate Analysis: A Users Perspective, Oxford University Press, New York, **1988**

Lachenbruch, P.A.: Discriminant Analysis, Hafner Press, London, **1975**

Malinowski, E.R.: Anal. Chem. 49 (**1977**) 612

Malinowski, E.R.: Factor Analysis in Chemistry, 2nd. Ed., Wiley, New York, Chichester, Brisbane, Toronto, Singapore, **1991**

Mandel, J.: J. Res. Nat. Bur. Stand. 90 (**1985**) 465

Marco, V. R., Young, D. M., Turner, D. W.: Commun. Statist. - Simula. 16 (**1987**) 485

Mardia, K.V., Kent, J.T., Bibby, J.M.: Multivariate Analysis, Academic Press, London, **1979**, pp. 191

Massart, D.L., Kaufman, L.: The Interpretation of Analytical Data by the Use of Cluster Analysis, Wiley, New York, **1983**

Massart, D.L., Plastria, F., Kaufman, L.: Pattern Recognition 16 (**1983**) 507

Mellinger, M.: Chemom. Int. Lab. Syst. 2 (**1987**) 61

Mucha, H.J.: Clusteranalyse mit Mikrocomputern, Akademie Verlag, Berlin, **1992**

Nagel, M., Hothorn, L., Hartmann, P. in: Enke, H., Gölles, J., Haux, R., Wernecke, K.-D. (Hrsg.): Methoden und Werkzeuge für die exploratorische Datenanalyse in den Biowissenschaften, Gustav Fischer, Stuttgart, Jena, New York, **1992**, pp. 75

Osten, D.W.: J. Chemometrics 2 (**1988**) 39

Ozawa, K.: Pattern Recognition 16 (**1983**) 201

Rand, W.M.: J. Amer. Stat. Assoc. 66 (**1971**) 846

STATISTICA 5.0 for Windows, StatSoft, Tulsa OK, **1995**

Steinhausen, D., Langer, K.: Clusteranalyse, Walter de Gruyter, Berlin, New York, **1977**

Weber, E.: Einführung in die Faktorenanalyse, Gustav Fischer, Jena, **1974**, pp. 128

Wold, S.: Technometrics 20 (**1978**) 397

Wold, S., Ruhe, A., Wold, H., Dunn, W.J.: SIAM J. Sci. Stat. Comput. 5 (**1984**) 735

Zupan, J., Gasteiger, J.: Neural Networks for Chemists. An Introduction, VCH, Weinheim, **1993**

6 Basic Methods of Time Series Analysis

6.1 Introduction

Until recently mathematical methods of time series analysis in the environmental sciences have only been used quite rarely; the methods have mostly been applied in economic science. Consequently, the mathematical fundamentals of time series analysis are mainly described in textbooks and papers dealing with statistics and econometrics [FÖRSTER and RÖNZ, 1979; COX, 1981; SCHLITTGEN and STREITBERG, 1989; CHATFIELD, 1989; BROCKWELL and DAVIS, 1987; BOX and JENKINS, 1976; FOMBY et al., 1984; METZLER and NICKEL, 1986; PANDIT and WU, 1990]. This section explains the basic methods of time series analysis and their applicability in environmental analysis.

In general, time series analysis has the following main purposes:
– display of the series
– preprocessing of the data
– modeling and describing of the series
– forecasting with suitable models
– control of predicted values

In practical environmental analysis and valuation complex questions concerning environmental relationships in time often have to be answered.

Most of the above purposes can be fulfilled with modern time series models but the environmental scientist also needs a guide to clear interpretation of the results of the applied mathematical methods. The practician should be given time series analytical methods, i.e. mathematical techniques or computer programs which are relatively simple to use, as a tool for his daily work. The interpretation of the computations must be easy. More complicated time series models are, therefore, not included.

The various time series analytical methods were applied to the same environmental example – a series of nitrate concentrations in a storage reservoir. A comparison of the power of time series methods is, therefore, possible.

6.2 Example: Nitrate Loadings in a Drinking Water Reservoir – Description of the Problem

In Thuringia (Germany), there are many storage reservoirs which supply drinking water. Nitrate concentrations in storage reservoirs, particularly in agricultural areas, may be too high. The nitrate content can increase to values higher than 50 mg L^{-1}, the legally binding value for human drinking water in Germany [TRINKWASSERVERORD-NUNG, 1990; E.U.-TW, 1980]. The main source of nitrate is agricultural fertilization. The wash-out effect from agricultural areas is strong in autumn and winter, but minimal in summer because of nitrogen-consuming assimilation by plants.

The example examined in this section is a storage reservoir system (Fig. 6-1) with a main feeder stream (70% of the total feeder stream at the Läwitz water gauge) and several smaller feeder streams into a first storage reservoir (Zeulenroda). The second storage reservoir (Weida) is directly connected to the first reservoir. Because only a few small direct feeder streams flow into this second reservoir, the first reservoir has a large influence on the quality of water in the second reservoir.

This second reservoir contains crude water which is treated to produce drinking water in that region. For this reason, it is very important to observe the limit for nitrate concentrations in drinking water.

The nitrate concentration was measured monthly from August 1976 to November 1992. Sampling for nitrate measurements was conducted at the water gauge in Läwitz and

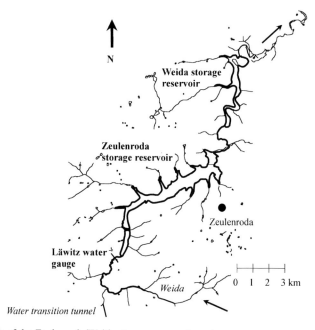

Fig. 6-1. Sketch of the Zeulenroda/Weida storage reservoir system

sampling of the two reservoirs was performed in different horizontal layers. The mean nitrate concentration in a storage reservoir was calculated after weighting the nitrate concentration in each horizontal layer with its volume. The nitrate values for the two storage reservoirs are the layer volume weighted average concentrations.

The time series plots of the nitrate concentrations of the feeder stream, the storage reservoir in Zeulenroda and the storage reservoir in Weida (Fig. 6-2) show periodic fluctuations with maxima near, or higher than, the limit of 50 mg L^{-1}. The time axes in all following figures are the x-axes with monthly scaling for every year. The notation "1.80" means January in the year 1980.

Two points of view have influenced the planning of the calculations:
− Sampling of storage reservoirs is complicated. In winter, sampling from boats is difficult if not impossible.
− Sampling to determine a mean nitrate concentration of the reservoir (e.g. sampling in different horizontal layers, many samples to analyze) is expensive.

For this reason, and because of the relevance of the second storage reservoir (Weida) for human drinking water, the following examinations concentrate only on the relationship between the feeder stream at the water gauge at Läwitz and the second storage reservoir in Weida − the drinking water storage reservoir.

The aim of the following calculations is to reduce the expense of sampling and analysis and to discover valid time series models at critical dates which make it possible to decide whether the water from the reservoir in Weida can be used as human drinking water, or if other sources have to be included.

Not all the time series analytical methods demonstrated in this section are suitable for solving this problem. But for comparison of efficiency of the different methods this example will be used throughout the section and will serve as a typical time series.

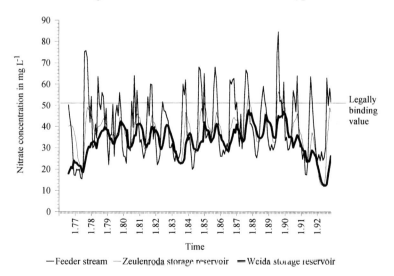

Fig. 6-2. Time series plot of the feeder stream and the storage reservoirs at Zeulenroda and at Weida

6.3 Plotting Methods

6.3.1 Time Series Plot

The first step in the analysis of time series $x(t)$ is always to draw a plot:

$$x = f(t) \tag{6-1}$$

x – variable, dependent on time
t – time
$x(t)$ – time series

Plotting gives a general idea about the shape of the time series. It may give **visual information about any periodicity, trends, fluctuation, and outliers**. It is, furthermore, possible not only to detect a dependence of the mean of $x(t)$ on time (trend), but also the dependence of the variance of $x(t)$ on time (requires a transformation). After visually examining this plot, one may decide the next steps of the analysis.

The nitrate concentrations of the feeder stream, the Zeulenroda reservoir and the drinking water reservoir at Weida are shown in the plot of the time series example (Fig. 6-2). One may observe the periodic fluctuations with the above described maxima in autumn and winter and minima in summer. In addition to the periodic fluctuations, nitrate concentrations in the feeder stream are scattered very much at random. An attenuating effect can be seen in the Zeulenroda reservoir. The greatest attenuation of scattering can be seen in the drinking water reservoir. Another feature to observe is a time delay between nitrate maxima in the feeder stream and those in the first reservoir and in the second, drinking water, reservoir. In time series analysis such a delay is called lag, and is given the symbol τ.

The following time series analytical examinations will only deal with the time series at the feeder stream gauge at Läwitz and the drinking water reservoir at Weida. This is the only point of interest for practical conclusions in relation to the supply of humans with drinking water.

6.3.2 Seasonal Sub-Series Plot

The seasonal sub-series plot displays the values of a **time series** with seasonal fluctuations **arranged according to time**, e.g. a year. The values from each **sub-period** (e.g. all January values in each year, all February values, …) were summarized by computing a mean of the sub-periods and the deviations from the mean. The horizontal line represents the average of each sub-period of a time series – here the single months of each year between 1976 and 1992. Vertical lines signify the actual annual differences of values from the mean.

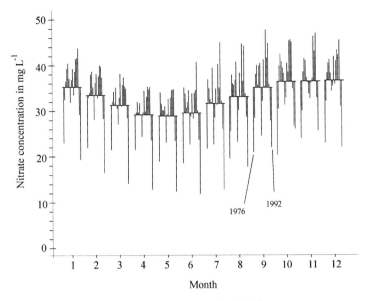

Fig. 6-3. Seasonal sub-series plot of nitrate time series in the drinking water storage reservoir

In the example of the second storage reservoir at Weida (Fig. 6-3), this method of seasonal sub-series plotting underlines the annual periodicity of the nitrate concentration in this reservoir. The minimum monthly mean of nitrate concentration occurs in May, the maximum from November to January. The last two to three years (last vertical lines at each month) show strong negative deviations from the mean. This means a decreasing nitrate level during these years. The reason for this is the dry summers during 1990, 1991, and 1992, rendering the wash-out effect weaker than in previous years.

6.4 Smoothing and Filtering

The main purpose of smoothing a series is **to strip away the random fluctuations**. The analogous procedures which are used for smoothing can also be used for filtering the periodicities. Use of such techniques achieves short-term forecasts with a memory of the last values of the series.

6.4.1 Simple Moving Average

Moving average is the simplest technique for smoothing. The operator determines a **time window within which the values are averaged**. For the next value the window is lagged one step ahead and the mean is again calculated. The time window will then be moved step by step over the total time series. The most important parameter is the length of the time window. Windows which are too small are ineffective for smoothing and windows which are too long can destroy the character of the time series. It is possible to center the time window:

$$\hat{x}(t) = \frac{1}{5}[x(t-2) + x(t-1) + x(t) + x(t+1) + x(t+2)] \tag{6-2}$$

or to use the time window only backward:

$$\hat{x}(t) = \frac{1}{5}[x(t-4) + x(t-3) + x(t-2) + x(t-1) + x(t)] \tag{6-3}$$

A more general formula for the same thing is:

$$\hat{x}(t) = \frac{1}{\tau_{max}+1} \sum_{\tau=0}^{\tau_{max}} x(t-\tau) \tag{6-4}$$

τ_{max} – width of the time window

In our example, two centered windows are applied. With the symmetrical window with 5 steps (5 months were averaged) a smoothing effect can be seen (Fig. 6-4). The most extreme points are cut off and yearly periods are indicated.

As a second possibility, a symmetrical window with 13 steps was also chosen (Fig. 6-5). This size is larger than one period in the series, because there are twelve months in the year. A good starting point for estimating a trend can therefore be achieved. In this application, the negative trend of the last three years is detectable without seasonal fluctuations.

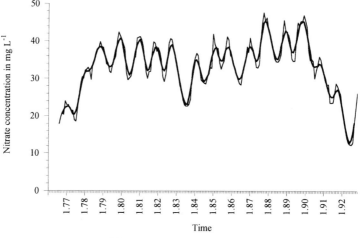

Fig. 6-4. Moving average with a 5-step window

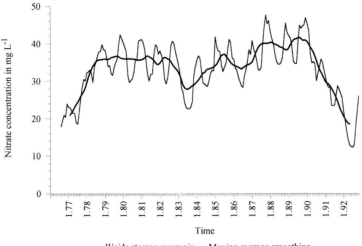

Fig. 6-5. Moving average with a 13-step window

6.4.2 Exponential Smoothing

The exponential smoothing technique is a more involved method of smoothing. Here, in contrast with simple moving average procedures, the values $x(t)$ in the lagged time window do not have the same weights in exponential smoothing techniques. **The weights of the single values, joining the time window, exponentially decrease**. One could call this a time window with a memory. This memory keeps the most recent values rather than previous values. This "memory" function is of an exponential type. The power of the memory is the weight, α.

Simple exponential smoothing of a time series means the representation of a value at a specific time by an exponential weighted sum of recent values:

$$\hat{x}(t) = \frac{1}{\sum_{\tau=0}^{n_{\max}-1}(1-\alpha)^\tau} \cdot \sum_{\tau=0}^{n_{\max}-1}(1-\alpha)^\tau \cdot [x(t-\tau-1)] \quad (6\text{-}5)$$

$$(0 \leq \alpha < 1)$$

$n_{\max}$ — number of cases which join the series
α — smoothing parameter

Low values of α (minimum value zero) introduce a **long memory** effect, **higher** α (maximum value <1) create a **short memory**. Fig. 6-6 explains the effect of the parameter α in a small simulated model time series.

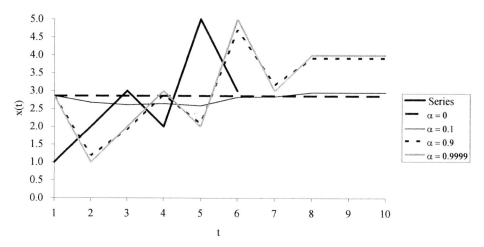

Fig. 6-6. Sketch of the effect of the smoothing parameter α

To estimate α for the best fitted time series it is necessary to calculate the sum of squared residuals. The best fit is that with a minimum sum of squared residuals.

Time series without systematic changes (trend or seasonal fluctuations), i.e. with a fixed level, are best approximated by the mean of the series, i.e. $\alpha = 0$. The mean over the full time range gives a minimum sum of squared differences between the mean and the original series (squared residues). All cases have the same weight, because α is equal to zero.

If the time series contains a trend or seasonal fluctuations, these effects may also be smoothed, e.g. the slope of different time ranges is smoothed or the amplitude of periodic fluctuations is smoothed. For model building it is necessary to declare the type of trend (linear, quadratic, or exponential) and the length of the periodicity.

The starting point for the computation of the exponential smoothing model with trend and seasonal effects is the **additive component model**:

$$\hat{x}(t) = l(t) + m(t) + sea(t) + e(t) \qquad (6\text{-}6)$$

$l(t)$ – level component with the smoothing parameter α
$m(t)$ – trend component with the smoothing parameter β
$sea(t)$ – seasonal component with the smoothing parameter γ
$e(t)$ – error

Models for **multiplicative effects** are also available. Multiplicative seasonal factors are suitable if the variance $s^2[x(t)]$ increases as the mean increases:

$$\log[\hat{x}(t)] = \log[l(t)] + \log[m(t)] + \log[sea(t)] + \log[e(t)] \qquad (6\text{-}7)$$

The constants chosen by the operator determine the length of memory. The constants (α for the level component, β for the trend, γ for the seasonal component) must lie in the interval between 0 and <1. Values close to zero imply a heavy weight of all the previous values – "a long memory". Values near one approximate the series by itself – "a short memory".

The trend component is, therefore, a weighted mean from the previous slopes of the series. The seasonal component, $sea(t)$, is the weighted mean from the seasonal deviations from the mean seasonal value of the series. This means that all monthly values are averaged in year x. This is the starting point for calculating the seasonal component, $sea(t)$, e.g. for January. It is the deviation of the January value from the total mean for one year. The seasonal components for each month were smoothed exponentially, smoothing over all of the years.

Forecasting with Exponential Smoothing

Exponential smoothing is intended for calculation of **one step ahead forecasts**. All further forecasts $\hat{x}(t+2)$, $\hat{x}(t+3)$, ... relate to the recent forecasted value $\hat{x}(t+1)$, $\hat{x}(t+2)$, ... and also, in dependence on the value of the smoothing parameter, to more recent, real values:

$$\hat{x}(t+1) = \frac{1}{\sum_{\tau=0}^{n_{max}-1}(1-\alpha)^\tau} \cdot \sum_{\tau=0}^{n_{max}-1}(1-\alpha)^\tau [x(t-\tau)]$$

$$\hat{x}(t+2) = \frac{1}{\sum_{\tau=0}^{n_{max}-1}(1-\alpha)^\tau} \cdot \sum_{\tau=0}^{n_{max}-1}(1-\alpha)^\tau [x(t+1-\tau)]$$

(6-8)

n_{max} – number of cases which join the series
α – smoothing parameter

The weight of values which are forecasted increases step by step. The predicted values do not, therefore, change any more. Fig. 6-6 demonstrates the effect: the predicted time series is a constant.

This type of model is well suited for fitting and smoothing as well as for forecasting one step or one period ahead. Long-term forecasts show only constant series, a constant trend or a constant seasonality of the series.

Application of Exponential Smoothing

Using the nitrate time series example, the effect of a trend model with additive seasonality is shown in Fig. 6-7.

The constants α, β, and γ were optimized by stepwise variation and comparison of the sum of quadratic residues. The general smoothing parameter α for this model is one, i.e. the best fitting results are obtained if all former values have the same weight. The seasonal smoothing parameter, γ, is not relevant. Its variation between one and zero does not induce changes in the errors. The trend parameter, β, gives the best results when its value is zero. The fit by smoothing is very close to the original values. The

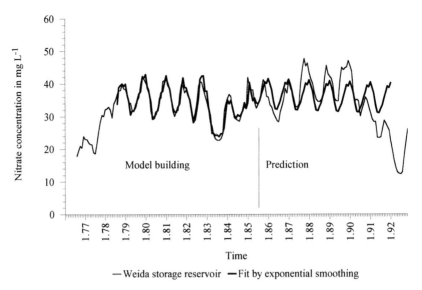

Fig. 6-7. Fit by exponential smoothing with parameters $\alpha = 1$, $\beta = 0$, $\gamma = 0$

application of this exponential smoothing model for forecasting allows a prediction period of one. Subsequent prediction periods are only a replay of the previous period.

6.4.3 Simple and Seasonal Differencing and the CUSUM Technique

Simple differencing, $x_1(t)$, Eq. 6-9 enables the operator to remove a linear trend from the time series. Twofold differencing, $x_2(t)$, Eq. 6-10 removes a quadratic trend:

$$x_1(t) = x(t) - x(t-1) \tag{6-9}$$
$$x_2(t) = x_1(t) - x_1(t-1) \tag{6-10}$$

Differencing by one is similar to the first derivation of the time series. Because of the digital data structure, it is not a differentiation but a differencing. The differencing method is, therefore, a plot of the dependence of the slope of the series on time. This example (Fig. 6-8) does not indicate any trend with this method of trend-detecting. But maximum and minimum slope values can be observed every year.

Using seasonal differences, for a difference of 12 for 12 months in the year, the **seasonal effect can be removed**:

$$x_{12}(t) = x(t) - x(t-12) \tag{6-11}$$

The remaining series from seasonal differencing is a good starting point for trend detection methods (Fig. 6-9).

6.4 Smoothing and Filtering   215

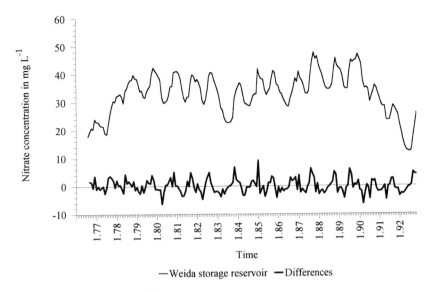

Fig. 6-8. Time series after simple differencing

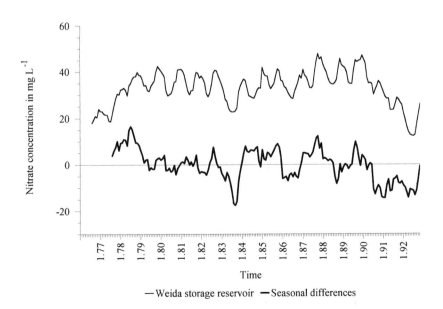

Fig. 6-9. Time series after seasonal differencing

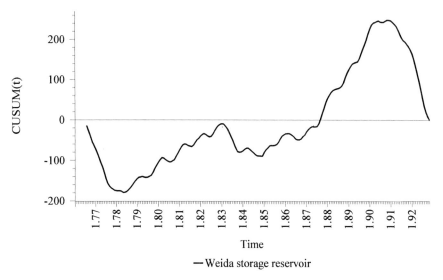

Fig. 6-10. CUSUM graph of the time series

If the procedure is reversed the so-called cumulative sum (CUSUM) integrates the deviations from the mean of the series [DOERFFEL et al., 1990; 1991]:

$$CUSUM(t) = \sum_{i=1}^{t} x(i) - t \cdot \bar{x} \qquad (6\text{-}12)$$

In a random time series, values $x(t)$ scatter around the mean of the series. A drift of the time series can be seen as a trend of the CUSUM graph.

The CUSUM technique clearly displays the decreasing trend of the series in the last three years (Fig. 6-10).

6.4.4 Seasonal Decomposition

The technique of seasonal decomposition uses the same additive and multiplicative models as in exponential smoothing, but **without the smoothing procedure**.

Additive model:

$$\hat{x}(t) = l(t) + m(t) + sea(t) + e(t) \qquad (6\text{-}13)$$

$l(t)$ – level component
$m(t)$ – trend component
$sea(t)$ – seasonal component
$e(t)$ – error

Multiplicative model:

$$\log[\hat{x}(t)] = \log[l(t)] + \log[m(t)] + \log[sea(t)] + \log[e(t)] \qquad (6\text{-}14)$$

The only aim is **to detect the differences between the seasonal mean and the total mean of the series** or to detect the factors between the seasonal mean and the total mean. Then it is possible to deseasonalize the series using these differences or factors.

In the example, the following seasonal differences are obtained for the additive model (Tab. 6-1).

Tab. 6-1. Seasonal differences in an additive model from the nitrate time series example

Month	Seasonal differences	Month	Seasonal differences
January	3.34	July	−3.62
February	1.92	August	−1.45
March	0.16	September	0.68
April	−2.02	October	2.56
May	−4.11	November	3.87
June	−4.43	December	3.07

By looking at these simple means, one can clearly see the months with maximum nitrate concentrations (November to January), and the mean difference between the monthly means and the total mean.

6.5 Regression Techniques

6.5.1 Trend Evaluation with Ordinary Least Squares Regression

Regression techniques are most frequently used for detection of trends in a series. For evaluating nonparametric trend tests see BERRYMAN et al. [1988]. This example series concerning the nitrate concentrations in the storage reservoir will be tested for any trends over the full time of observation.

Values on the x-axis in time series analysis are mostly data like day, month, or year. In order to test a trend, one needs a **linearly increasing** x-axis for that particular section of the series which has to be tested for a trend. So one has to create a series with running numbers **as the independent variable**.

The second task is **to deseasonalize the series**. Differencing by 12 can be used to achieve this. To evaluate a trend, regression analysis as basically described in Section

2.4 was performed for the time interval between August 1989 and the end of the series. The model equation was:

$$\hat{x}(t-12) = 6.2556 - 0.0618\, num \qquad (6\text{-}15)$$

regression standard error: 6.029 mg L^{-1} NO$_3^-$
standard error of the slope: 0.0084 mg L^{-1} NO$_3^-$
standard error of the intercept: 0.981 mg L^{-1} NO$_3^-$

num – running numbers
x(*t*-12) – deseasonalized series *x*(*t*)
$\hat{x}$(*t*-12) – estimated value *x*(*t*-12) by model computation

A significant slope signifies a trend. For the appropriate test see Section 2.4. After seasonal differencing the example (Fig. 6-11) shows a significant trend with a slope of –0.0618, i.e. a small negative trend or a decreasing nitrate concentration over the total time range of the observations.

Another possibility is to smooth the seasonal fluctuations using the moving average procedure, but this leads to lack of sharpness because a certain number of values in the lagged time window is included. Seasonally adjusted series achieved by seasonal decomposition are also good starting points for trend searching.

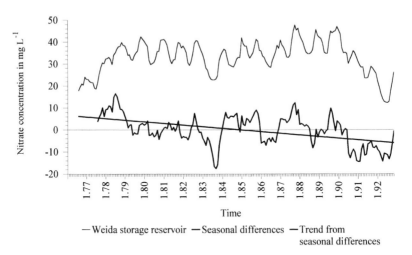

Fig. 6-11. Trend evaluation using seasonal differences

6.5.2 Least Squares Regression with an Explanatory Variable

Regression analysis in time series analysis is a very useful technique if an explanatory variable is available. **Explanatory variables may be any variables with a deterministic relationship to the time series.** VAN STRATEN and KOUWENHOVEN [1991] describe the dependence of dissolved oxygen on solar radiation, photosynthesis, and the respiration rate of a lake and make predictions about the oxygen concentration. STOCK [1981] uses the temperature, biological oxygen demand, and the ammonia concentration to describe the oxygen content in the river Rhine. A trend analysis of ozone data was demonstrated by TIAO et al. [1986].

In the time series example it is fortunate that the nitrate concentration of the feeder stream in the storage reservoir system is known. Now we have to test if this parameter can explain the nitrate concentration in the storage reservoir. Approximately 70% of the total amount of water which is flowing into the storage reservoir system was measured as originating from the feeder stream from the Läwitz gauge.

The time series plot (Fig. 6-2) displays a lag between the nitrate concentration maxima in the feeder stream and those in the drinking water reservoir. The cross-correlation function (see Section 6.6.1, Fig. 6-15) between the feeder stream and second storage reservoir therefore enables the analyst to detect the real lag. This lag (between 2 and 3 months) gives the analyst an idea of the possible length of the prediction period. High nitrate concentrations in the feeder stream lead to high nitrate concentrations in the drinking water reservoir which can be detected two or three months later. The time series of nitrate in the feeder stream was, therefore, lagged with 2 and cut at the end. The resulting regression model is:

$$\hat{x}(t) = a_0 + a_1 x(t-2) + e(t) \qquad (6\text{-}16)$$

$\hat{x}(t)$ – fit for the time series nitrate in the drinking water reservoir
$x(t)$ – time series of the feeder stream as the independent variable
$e(t)$ – error
a_0 – intercept
a_1 – slope

$$a_0 = 22.53 \text{ mg L}^{-1} \text{ NO}_3^-$$
$$a_1 = 0.265$$

standard error of regression: 6.23 mg L^{-1} NO$_3^-$
standard error of a_0: 1.32 mg L^{-1} NO$_3^-$
standard error of a_1: 0.031

Fig. 6-12 illustrates the regression model fit of the nitrate concentration in the storage reservoir. The fitted value of the time series obtained by regression cannot be used to estimate relevant changes in the nitrate concentration, especially in the cases of extreme

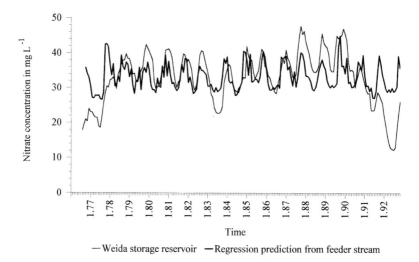

Fig. 6-12. Regression model fit of the time series

maxima of nitrate concentrations, e.g. in the winters of 1988 and 1989, or extreme minimum situations, e.g. in the summers of 1991 and 1992.

6.5.3 Least Squares Regression with Dummy Variables (Multiple Least Squares Regression)

Dummy variables are artificial variables included in multiple least squares regression models (see also Section 5.7.1) which should explain the time series [SPSS, 1990]. These dummy variables may express different kinds of event which can affect the series, e.g. changes in environmental laws, political events, changes in measuring equipment, etc.. The dummy variable is equal to zero before the special event and equal to one or a linearly increasing function from the moment the event begins. If this dummy variable in a multiple least squares model is significant, the expressed event has a significant effect on the series.

In the case study, dummy variables were used to evaluate seasonality and the trend over a period of three years with a dry summer (1989, 1990, 1991). To evaluate the seasonality, an additional series is assigned for each month; the series is equal to one or zero. This means the addition of 11 new series or dummy variables (the twelfth month variable is redundant) for a multiple regression. To evaluate the trend the twelfth dummy variable, "dry summer", is equal to one in 1989, equal to two in 1990, and equal to three in 1991. The following new dummy variables were created:

jan, feb, mar, ... , summ, and *num* (for running numbers) by the following rule:

if month = 1 (January) then *jan* = 1 otherwise *jan* = 0
if month = 2 (February) then *feb* = 1 otherwise *feb* = 0

⋮   ⋮   ⋮

if (year = 1989 then *summ* = 1, year = 1990 then *summ* = 2,
year = 1991 then *summ* = 3) otherwise *summ* = 0

Such analysis demands many cases in the series similar to multivariate model computations. Now a multiple regression analysis with the independent variables *jan, feb, mar, ... , summ* and the number variable *num*, and the nitrate time series as the dependent variable is started.

$$\hat{x}(t) = function\ of\ [jan(t),\ feb(t),\ mar(t),\ \ldots,\ summ(t),\ num(t)] \quad (6\text{-}17)$$

Significant variables found are *summ, jul, jun, may, apr, aug,* and *num*. The resulting model is:

$$\hat{x}(t) = 30.59 - 8.01 \cdot summ - 5.91 \cdot jul - 6.52 \cdot jun \\ - 6.16 \cdot may - 3.99 \cdot apr - 4.44 \cdot aug + 0.08 \cdot num \quad (6\text{-}18)$$

standard error of the estimate: 4.69 mg L^{-1} NO$_3^-$.

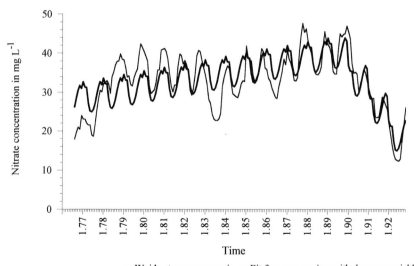

Fig. 6-13. Fit of the time series from regression with dummy variables

Summer months have high negative coefficients; the coefficients for winter are not significant. Over the full time range, a small positive trend (see the coefficient of the variable *num*) is significant; within the dry summer years, a strong negative trend (coefficient of *summ*) is detectable. Fig. 6-13 demonstrates the precision of the fit by means of this regression method.

Another means of determining the effect of the seasonality is to include a harmonic term, e.g. $\sin(\omega t)$ and $\cos(\omega t)$, into the multiple regression [TÜMPLING, 1973]. One month refers to a segment of $\omega = 30°$ and the significance of seasonal effects can be determined.

Forecasting with such multiple regression models is possible if the values of the significant independent variables are known and the type of the model remains valid.

6.6 Correlation Techniques

6.6.1 Autocorrelation, Autoregression, Partial Autocorrelation, and Cross-correlation Function

Very useful tools for analyzing time series are the autocorrelation function, the cross-correlation function, and the partial autocorrelation function [SCHLITTGEN and STREITBERG, 1989; DOERFFEL and WUNDRACK, 1986]. The interpretation of the patterns of these functions provides the experienced user with substantially more information about the time series than plotting methods.

The aim of correlation analysis is to compare one or more functions and to calculate their relationship **with respect to a change of τ** (lag) in time or distance. In this way, memory effects within the time curve or between two curves can be revealed. Model building for the time series is easy if information concerning autocorrelation is available.

In order to imagine such memory effects, it is useful to plot $x(t)$ against previous values, such as $x(t-1)$ (Fig. 6-14):

$$x(t) = f[x(t-1)] \tag{6-19}$$

It can be seen that the series is correlated with itself by a lag of one. The series has a "memory" for the first preceding value. **All cases of the series are dependent on the preceding value**. The statistical expression for a relationship between two variables is the correlation coefficient (see Section 2.4.2). The resulting correlation coefficient between the variables $x(t)$ and $x(t-1)$ is the autocorrelation coefficient [KATEMAN, 1987] for the lag $\tau = 1$ in this case. If such a relationship between the values $x(t)$ and $x(t-1)$ exists, we can formulate a linear regression expression:

$$x(t) = a_0 + a_1 x(t-1) + e \tag{6-20}$$

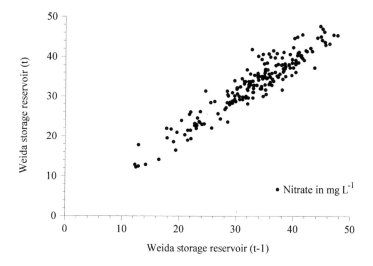

Fig. 6-14. Autocorrelation within the example time series

An estimation of actual values is also possible with the help of previous values. A more general definition of such an **autoregressive process** is:

$$x(t) = a_0 + a_1 x(t-1) + a_2 x(t-2) + \cdots + a_p x(t-p) + e \qquad (6\text{-}21)$$

The current value of a time series is a linear combination of a number of previous observations of the time series. **The number of significant coefficients, a, is called the order of the autoregressive process.**

The dependence of the autocorrelation coefficients on lag is the autocorrelation function $r_{xx}(\tau)$:

$r_{xx}(0)$ – correlation between $x(t)$ and $x(t)$, $r_{xx}(0) = 1$
$r_{xx}(1)$ – correlation between $x(t)$ and $x(t-1)$
$r_{xx}(2)$ – correlation between $x(t)$ and $x(t-2)$
...
$r_{xx}(\tau)$ – correlation between $x(t)$ and $x(t-\tau)$

One definition of the autocorrelation function, r_{xx}, using autoscaled values (mean = 0, standard deviation = 1) is:

$$r_{xx}(\tau) = \frac{1}{n} \sum_{\tau=0}^{n_{\max}} x(t) \cdot x(t+\tau) \qquad (6\text{-}22)$$

n – number of cases included (decreases as τ increases)
$n_{\max}$ – total number of cases

The autocorrelation function of autocorrelated time series (first or higher order) has an exponentially decreasing shape. An autocorrelation of the order unity means only a correlation between $x(t)$ and $x(t-1)$. Because of the same correlation between $x(t-1)$ to $x(t-2)$ it seems that a correlation between $x(t)$ and $x(t-2)$ is also detectable. For this reason, it is very difficult to find the correct order of the autocorrelation process. A useful tool in this case is the partial correlation coefficient.

The general **partial correlation coefficient** between the variables x and y **is the correlation coefficient without any influence of other variables**, e.g. z. One calculates the correlation function without the influence of correlation transfer effects. The notation is $r_{xy.z}$.

The partial autocorrelation coefficient in time series analysis is the autocorrelation between the variables $x(t)$ and $x(t-\tau)$ with the exclusion of the influences of the variable $x(t-1)$, $x(t-2)$, ..., $x(t-\tau+1)$. For the special case of autocorrelation between $x(t)$ and $x(t-2)$, the partial autocorrelation coefficient can be calculated as follows:

$$r_{x(t)x(t-2).x(t-1)} = \frac{r_{x(t)x(t-2)} - r_{x(t)x(t-1)} \cdot r_{x(t-1)x(t-2)}}{\sqrt{[1 - r^2_{x(t)x(t-1)}][1 - r^2_{x(t-1)x(t-2)}]}} \qquad (6\text{-}23)$$

The partial correlation function overcomes the correlation transfer effect as described above and shows, in contrast to the autocorrelation function, only one spike at $\tau = 1$ for first order autoregressive processes and spikes at $\tau = 1$ and $\tau = 2$ for second order autoregressive processes, and so on.

The use of **the cross-correlation function** enables not only the determination of **the relationship between two different time series**, $x(t)$ and $y(t)$, as the ordinary correlation coefficient does, but also **in relation to a time lag**:

$r_{xy}(0)$ – ordinary correlation coefficient between $x(t)$ and $y(t)$
$r_{xy}(1)$ – correlation between $x(t)$ and $y(t-1)$
$r_{xy}(2)$ – correlation between $x(t)$ and $y(t-2)$
...
$r_{xy}(\tau)$ – correlation between $x(t)$ and $y(t-\tau)$

Cross-correlation function for autoscaled values:

$$r_{xy}(\tau) = \frac{1}{n} \sum_{\tau=-(n_{max}-1)}^{n_{max}-1} x(t)y(t+\tau) \qquad (6\text{-}24)$$

n – number of joining cases (decreasing with growing τ)
n_{max} – total number of cases

The time series example used in this section requires one to determine the time lag between the storage reservoir feeder stream and the drinking water storage reservoir. The cross-correlation function was then calculated between the nitrate concentrations in the

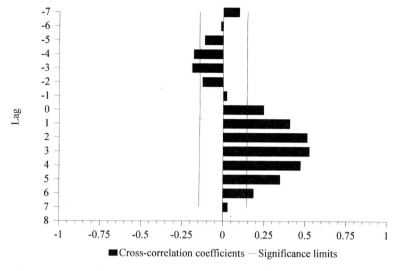

Fig. 6-15. Cross-correlation functions between nitrate data on the feeder stream and the drinking water storage reservoir

feeder stream, the gauge at Läwitz, and the average nitrate concentration in the drinking water reservoir. A broad maximum between lag two and three was obtained (Fig. 6-15). This gives an idea of the mean time (between two and three months) which is needed for water to flow between the feeder stream and the second drinking water reservoir.

Conventional testing tables for correlation coefficients, as described in Section 2.4.2, can be used to test the significance of autocorrelation or cross-correlation coefficients in terms of their dependence on the degrees of freedom. In the following figures, these critical values for a 5% risk of an error of the first kind are called significance limits.

6.6.2 Autoregression Analysis – Regression with an Explanatory Variable

Autocorrelated variables and autocorrelated errors, which occur frequently in time series analysis, **violate the general regression assumption of uncorrelated errors**. Regression analysis as described in Section 6.5.2 will fail. The autocorrelation affects the significance levels and the goodness of fit statistics. If one of the explanatory variables is itself autocorrelated, the regression will fail.

The **autoregression technique** as the alternative **was developed for** such **autocorrelated variables and errors**, which are frequently available in time series analysis.

Autoregression is, in general, a multiple regression between independent and dependent variables where accidental similarities as a result of autocorrelation (seasonal effects, storage reservoir-smoothing effect) are excluded.

Autoregressive processes are stochastic processes with a memory effect. The basic equation is known from regression analysis:

$$\hat{x}(t) = a_0 + a_1 x(t-1) + \cdots + a_p x(t-p) + e \qquad (6\text{-}25)$$

p – order of the autoregressive process
e – residuals

This means the dependence of an actual value on p preceding values.

The autoregression model for a relationship between two variables $x(t)$ and $y(t)$ and autocorrelated errors is as follows:

$$y(t) = a_0 + a_1 x(t) + e(t) \qquad (6\text{-}26)$$

$$e(t) = a_2 e(t-1) + e_{res}(t) \qquad (6\text{-}27)$$

$y(t)$ – dependent variable
$e(t)$ – autocorrelated error
e_{res} – residual error

For practical computations one has to determine the order of the autoregressive process.

First, the autocorrelation function must be computed. In the example plot a strong seasonal effect could be seen in the explanatory variable (nitrate concentration in the feeder stream) as well as in the dependent variable (nitrate concentration in the drinking water reservoir) (Fig. 6-2). The autocorrelation function (Fig. 6-16) has, therefore, the expected exponentially decreasing shape and, because of the seasonal fluctuations, increasing values at $\tau = 12, 24, \ldots$ A better tool for determining the order is the partial autocorrelation function. This function shows the partial correlation between $x(t)$ and $x(t-\tau)$ and ignores the influences of other variables, e.g. $x(t-\tau+1)$. It reveals the order one by the spike at $\tau = 1$ in Fig. 6-17.

The storage reservoir and the feeder stream both show the order one for autoregression, but a time lag of two months between the two series is detected by the cross-correlation function (Fig. 6-15). For this reason, it is necessary to modify the general model to:

$$y(t) = a_0 + a_1 x(t-2) + e(t) \qquad (6\text{-}28)$$

$$e(t) = a_2 e(t-1) + e_{res}(t) \qquad (6\text{-}29)$$

$y(t)$ – dependent variable, here fitted for the time series at the drinking water reservoir
$x(t{-}2)$ – independent variable, here lagged and cut by 2, time series at the feeder stream

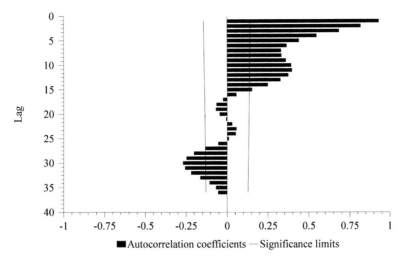

Fig. 6-16. Autocorrelation function of the nitrate time series at the drinking water reservoir

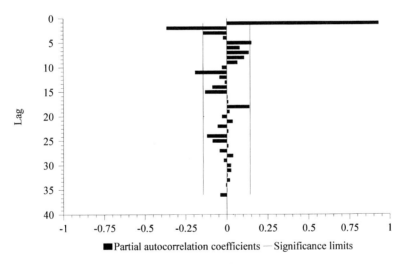

Fig. 6-17. Partial autocorrelation function of the nitrate time series at the drinking water reservoir

The period between August 1976 and August 1985 was used. Within this period, the regression parameters were computed for the following variables: nitrate in the drinking water reservoir as the dependent variable and, lagged with 2, nitrate in the feeder stream as the independent variable:

Standard error: 2.32   mg L^{-1} NO$_3^-$
a_0:   30.031   mg L^{-1} NO$_3^-$
a_1:   0.0518
a_2:   0.894

The prediction period shown in Fig. 6-18 uses this model with the actually given predictor variable x – nitrate concentration in the feeder stream. Real prediction is possible over a period of two months by using this model because the independent predictor variable is $x(t-2)$.

The fit of autoregression and the real concentration in the second reservoir is displayed in Fig. 6-18. Notice how closely the fitted values in the model building period correspond to the original series and how close the predicted values are to real values.

A special model from this type (autoregression with an explanatory variable), an autoregression model combined with a moving average model was applied by VAN STRATEN and KOUWENHOVEN [1991] to the time dependence of dissolved oxygen in lakes.

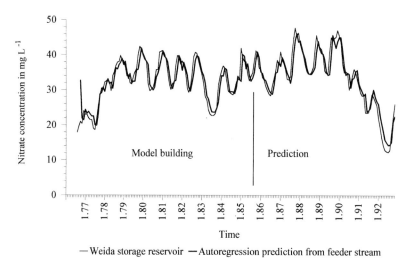

Fig. 6-18. Autoregression fit of the example time series

6.6.3 Multivariate Auto- and Cross-correlation Analysis

Frequently, concentration variations in environmental matrices not only concern themselves with one-dimensional cases, e.g. the time series of one parameter as discussed before, but also with many parameters which change simultaneously. In environmental analysis in particular, such time or local changes of environmental contaminants are very relevant [GEISS and EINAX, 1992]. Multivariate time series models are available, e.g. for the autoregressive moving average [COX, 1981; PRIESTLEY, 1981] or multi-way decomposition [STÄHLE, 1991].

A relevant question is the plotting of multivariate auto- or cross-correlation functions to **determine multivariate relationship to lag**.

In environmental analysis there are two problems which need to be solved:
- large variation of trace concentrations in environmental compartments, e.g. air, water, and soil because of natural inhomogeneities
- complex problems of interactions and the common appearance of different parameters

For this purpose the well-known univariate correlation analysis was changed to the more general multivariate case [GEISS and EINAX, 1991; 1996]. Multivariate correlation analysis enables inclusion of all interactions within the variables and the exclusion of the share of the variance resulting from the variable noise.

One advantage of multivariate correlation is the possibility of simultaneous handling of all variables in the time or local series. This enables all interactions within the variables in the series and between the series which are dependent upon lag to be taken into consideration.

Multivariate cross-correlation between $x(t)$ and $y(t)$ means that all relationships between the various variables within $x(t)$ (e.g. $x_1, x_2, \ldots, x_n$), within $y(t)$ (e.g. $y_1, y_2, \ldots, y_n$), and furthermore the relationships between $x(t)$ and $y(t)$, which are dependent upon lag are considered.

Multivariate autocorrelation of $x(t)$ contains all relationships between the various variables within $x(t)$ (e.g. $x_1, x_2, \ldots, x_n$) and also the relationships of $x(t)$ which are dependent upon lag.

The multivariate autocorrelation matrix $R_{xx}(\tau)$ (for centered values) is defined as:

$$R_{xx}(\tau) = (x(t)\, x(t+\tau)^T) \tag{6-30}$$

$$R_{xx}(\tau) = \begin{pmatrix} x_1(t) \\ x_2(t) \\ x_3(t) \end{pmatrix} \cdot (x_1(t+\tau)\ x_2(t+\tau)\ x_3(t+\tau))$$

$$= \begin{pmatrix} x_1(t)x_1(t+\tau) & x_1(t)x_2(t+\tau) & x_1(t)x_3(t+\tau) \\ x_2(t)x_1(t+\tau) & x_2(t)x_2(t+\tau) & x_2(t)x_3(t+\tau) \\ x_3(t)x_1(t+\tau) & x_3(t)x_2(t+\tau) & x_3(t)x_3(t+\tau) \end{pmatrix} \tag{6-31}$$

The matrix $R_{xx}(\tau)$ is not a symmetrical matrix. It contains cross-correlation coefficients between the variables x_1, x_2, and x_3 as off-diagonal elements. A symmetrical matrix is, therefore, produced by averaging the cross-correlation elements:

$$\text{symmetrical } R_{xx} = (R_{xx} + R_{xx}^T)/2 \tag{6-32}$$

The multivariate autocorrelation function should contain the total variance of these autocorrelation matrices in dependence on the lag τ. Principal components analysis (see Section 5.4) is one possibility of extracting the total variance from a correlation matrix. The total variance is equal to the sum of positive eigenvalues of the correlation matrices. This function of matrices is, therefore, reduced into a univariate function of multivariate relationships by the following instruction:

$$|R_{xx}(\tau) - \lambda I| = 0 \tag{6-33}$$

$$R_{xx}(\tau) = \sum_{\lambda > 0} \lambda_i \tag{6-34}$$

I – unity matrix
λ_i – eigenvalues

The multivariate cross-correlation function was treated in the same way:

$$R_{xy}(\tau) = (x(t)\, y(t+\tau)^T) \tag{6-35}$$

$$|R_{xy}(\tau) - \lambda I| = 0 \tag{6-36}$$

$$R_{xy}(\tau) = \sum_{\lambda > 0} \lambda_i \tag{6-37}$$

The auto- and cross-correlation matrices are now reduced to one-dimensional simply called multivariate auto- and cross-correlation functions.

For multivariate calculations, many data sets are necessary. According to HORST's rule [WEBER, 1974], for m variables, at least $3\,m$ values should be available for the calculation.

When testing the significance of the multivariate correlation value it is necessary to take three influences into consideration:

Multiplication of Probabilities
In a multivariate relationship, there were multiplied statistical probabilities of some realizations to a multivariate statistical probability, e.g. realizations of two variables with a single probability of 0.95 have the simultaneous probability of $0.95 \cdot 0.95 = 0.9025$.

Therefore, a probability of 0.95^m is expected for the multivariate case with m variables and a statistical significance of 0.95.

Multivariate Testing by the Spur Criterion
In the multivariate case, the significant cross-correlation or autocorrelation coefficients for each variable add up to the significant multivariate correlation value.

The Calculation of Degrees of Freedom

$$f = n \cdot m - m \cdot m = m \cdot (m - n) \tag{6-38}$$

f – degrees of freedom
m – number of variables
n – number of objects

In order to determine n variables, the same number of objects as the number of variables is necessary. The remaining objects are the degrees of freedom.

Example 6-1

The concentration of 23 elements in deposited dust according to Section 7.2.2.2.2 [EINAX et al., 1994] was analyzed.
In order to declare a multivariate probability $P = 0.977$ for testing the multivariate correlation coefficient, the following univariate probability is required:

$$P = \sqrt[23]{0.977} = 0.999 \tag{6-39}$$

This value of probability is the basis for finding the significance limit in tables of univariate correlation coefficients, e.g. in [SACHS, 1992].

The number of degrees of freedom depends on the lag τ:

$f = 23(60-23) = 851 \quad \tau = 0$
$f = 23(50-23) = 621 \quad \tau = 10$
$\vdots$

$r_{crit} = 0.115$ for $\tau = 0$ ($f = 851$) and $P = 0.999$

The following critical multivariate correlation coefficient results:

$R_{crit} = 2.645$ for $\tau = 0$ and $P = 0.977$

These critical values depend on τ, because of the resulting degrees of freedom, and on the error probability.

Example: Multivariate cross-correlation for the computation of transport rates of dissolved metals in river water

The transport rates of substances in river water are of economic importance, e.g. in potentially hazardous or pollution incidents. These transport rates have usually been measured by expensive tracer experiments with salts, colored compounds, or radioactive isotopes which can themselves be pollutants if the measurements are repeated frequently.

The cross-correlation function of the concentrations of any compound, which are observable in a stream and which scatter like heavy metals, is a possible means of measuring transport rates in flowing systems. The maximum of the cross-correlation function indicates the time which is needed for transporting these compounds. For this purpose, the time series of heavy metal concentrations (Cd, Cr, Cu, Fe, and Zn) at two sampling points in the river 4.5 km apart taken at exactly the same time were cross-correlated (Tab. 6-2) [EINAX et al., 1994].

Tab. 6-2. Scheme of cross-correlated time series of metal concentrations in a river

Time	Time series 1 Bridge 1 (x variables)					Time series 2 Bridge 2 (y variables)				
	Cd	Cr	Cu	Fe	Zn	Cd	Cr	Cu	Fe	Zn
6.00	$x_1(0)$	$x_2(0)$	$x_3(0)$	$x_4(0)$	$x_5(0)$	$y_1(0)$	$y_2(0)$	$y_3(0)$	$y_4(0)$	$y_5(0)$
6.30	$x_1(1)$	$x_2(1)$	$x_3(1)$	$x_4(1)$	$x_5(1)$	$y_1(1)$	$y_2(1)$	$y_3(1)$	$y_4(1)$	$y_5(1)$
7.00	$x_1(2)$	$x_2(2)$	$x_3(2)$	$x_4(2)$	$x_5(2)$	$y_1(2)$	$y_2(2)$	$y_3(2)$	$y_4(2)$	$y_5(2)$
⋮	⋮	⋮	⋮	⋮	⋮	⋮	⋮	⋮	⋮	⋮
13.00										

First, all time series of the five elements were univariate cross-correlated, e.g. each time series from the first sampling point in the river was cross-correlated with the time series from the second sampling point. Single time series of trace concentrations of metals in the river show a distinctly scattered pattern. There is a large fluctuation in the univariate cross-correlation functions for the five elements and, therefore, no useful information is obtained (Fig. 6-19).

It has been possible by means of multivariate cross-correlation analysis to include the interaction between the metals which arise as a result of emission and transformation. The assumption is that the metals are transported at the same rate. The multivariate cross-correlation function $R_{xy}(\tau)$ expresses a broad maximum at $\tau = 3$, i.e. 1.5 h (Fig. 6-20). This means that the mean transport rate of the metals is 3 km h^{-1}. Hydrological data gathered on the same day renders this result plausible:

maximum flow rate = 4.28 km h^{-1}
medium flow rate = 2.64 km h^{-1}

6.6 Correlation Techniques   233

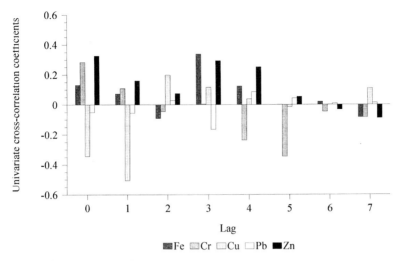

Fig. 6-19. Univariate cross-correlation functions of metal concentrations in a river

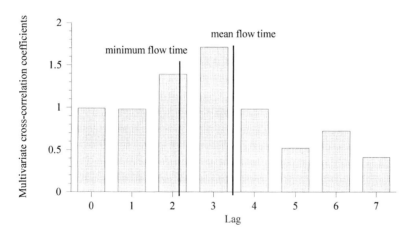

Fig. 6-20. Multivariate cross-correlation function of metal concentrations in a river

By means of the multivariate cross-correlation function, transport rates in streams can be calculated from natural compound concentrations in water without adding any tracer substances.

6.7 ARIMA Modeling

ARIMA is the acronym for **<u>a</u>utoregressive <u>i</u>ntegrated <u>m</u>oving <u>a</u>verage**. ARIMA is one of the most powerful methods in time series analysis for describing and forecasting [SCHLITTGEN and STREITBERG, 1989; BROCKWELL and DAVIS, 1987] if no explanatory variables are available.

ARIMA modeling in the analysis of water quality data is discussed in detail by ZETTERQVIST [1989]. Similar models were applied to the analysis of acid rain data [TERRY et al., 1985] and to the analysis of air quality data [JOHNSON and WIJNBERG, 1981]. In general, these models are very rarely used in environmental analysis. We want to demonstrate the power of this kind of time series model and will also supply candidates suitable for use of these models for evaluation of time series.

In order to calculate the fitted values of the drinking water in the storage reservoir by ARIMA modeling, the data set was shortened for the explanatory variable, the feeder stream. All following time series analytical procedures only use the values from the nitrate concentration series in the drinking water reservoir.

6.7.1 Mathematical Fundamentals

ARIMA connects both **autoregressive** and **moving average** models and includes **integrating** effects, e.g. trends or seasonal effects.

First an **ARMA** (**<u>a</u>uto<u>r</u>egressive <u>m</u>oving <u>a</u>verage**) model will be explained without taking into account trends and seasonal effects in order to get a better understanding of the method.

The **first element** of an ARMA process **is the autoregression** (AR process); it can be described as:

$$x(t) = a_1 x(t-1) + a_2 x(t-2) + \cdots + e_t \qquad (6\text{-}40)$$

Eq. 6-40 refers to the **time series as a linear combination of one or more previous values and an error**, which is a disturbance at time t. The number of significant coefficients a_i in Eq. 6-40 is called the order of the autoregression process (see also Section 6.6.2). It can be determined by partial autocorrelation analysis.

For more effective notation it is necessary to introduce the backshift operator B:

$$B^1 x(t) = x(t-1) \qquad (6\text{-}41)$$

$$B^{12} x(t) = x(t-12) \qquad (6\text{-}42)$$

$$(1 + b_1 B^1) \cdot e(t) = e(t) + b_1 e(t-1) \qquad (6\text{-}43)$$

B^n – backshift operator with n steps

6.7 ARIMA Modeling

One can imagine this operator as a lag which is observed in the time series. The new notation for autoregressive processes as shown in Eq. 6-40, using the back shift operator is:

$$x(t) = a_1 B^1 x(t) + a_2 B^2 x(t) + \cdots + e(t) \qquad (6\text{-}44)$$

Therefore, the backshift operator furnishes easy notation for processes dependent on the past.

These autoregressive processes were explained in more detail in Section 6.6.2.

The **second element** of the ARMA process **is the MA, i.e. the moving average process**. In this theoretical process, a value of the time series $x(t)$ is a weighted average of the current value and one or more preceding values of the process. The moving average process expresses the actual time series value $x(t)$ as a linear combination of an actual random disturbance $e(t)$, i.e. deviation from any theoretical values of the series, and the weighted level of the recent disturbances. **The moving average process is a smoothed random process**.

One can imagine the observations $x(t)$ of the moving average time series model as a current disturbance $e(t)$ which is corrected or smoothed by the weighted mean of recent disturbances. The order of a moving average process is the number of recent disturbances used for the smoothing, i.e. the number of significant coefficients b_i. The coefficients $b_1, b_2, \ldots, b_n$ were computed by iteration with the criterion of the best fit.

A moving average process can be represented as:

$$x(t) = (b_0 + b_1 B^1 + b_2 B^2 + \cdots) e(t) \qquad (6\text{-}45)$$

and rewritten without the backshift operator as:

$$x(t) = b_0 e(t) + b_1 e(t-1) + b_2 e(t-2) + \cdots \qquad (6\text{-}46)$$

The value of the series at time t, $x(t)$, is a weighted average of the levels of the current and most recent random disturbances $e(t)$.

Imagine the path of a flying ball. This is a deterministic curve. But rain and wind are random disturbances which have to be smoothed and included in the resulting path.

The following theoretical example of a second order moving average process shall be demonstrated in Tab. 6-3.

Tab. 6-3. Theoretical example of a moving average process

	t_1	t_2	t_3	t_4	t_5	t_6
Disturbances $e(t)$	1	−2	−5	3	−2	5
Theoretical moving average time series $x(t)$			−5.3	2.2	−1.9	4.9

With the constants $b_1 = 0.2$ and $b_2 = 0.1$ the calculation is, e.g.

$$x(t_6) = 5 + (0.2 \cdot (-2)) + (0.1 \cdot 3) = 4.9$$

It is apparent that the moving average time series $x(t)$ consists of the smoothed values from the values of the disturbances $e(t)$.

The errors of any functional relationships may be such moving average processes. Then these errors depend not only on the goodness of fit at time t but also on the goodness of fit from recent estimations because of the moving average smoothing function.

The moving average is a process similar to exponential smoothing. The exponential smoothing method (see also Section 6.4.2) has exponentially decreasing coefficients of the recent values. In a MA model the single coefficients $b_1, b_2, ..., b_n$ were calculated by minimization of the sum of squared errors.

The ARMA method offers the possibility of combining both autoregressive and moving average processes. **The disturbance term from autoregressive processes may be described by a moving average**.

Stationary time series can be described by an ARMA process. The ARMA formula of a first-order autoregressive process and a first-order moving average is the following:

ARMA (1,1) model:

$$x(t) = a_1 B^1 x(t) + (1 + b_1 B^1) \cdot e(t) \qquad (6\text{-}47)$$

The result is a process which determines the values from the previous values (autoregressive process), and smoothes the errors using a moving average.

ARIMA modeling in contrast with the ARMA model, includes trend or seasonality of time series. For such series, the trend can be removed by one-step differencing. Seasonality may be removed by 12-step differencing. After differencing the time series one gets a stationary time series which can be described as an ARMA process.

The notation of ARIMA processes is:

ARIMA *(p,d,q)*

p – order of the autoregressive process
d – order of the integration, e.g. differencing
q – order of the moving average

ARIMA (1,1,1) for trend effects:

$$(1 - B^1)x(t) = a_1 B^1 x(t) + (1 + b_1 B^1) \cdot e(t) \qquad (6\text{-}48)$$

ARIMA (1,1,1) for seasonal effects:

$$(1 - B^{12})x(t) = a_1 B^1 x(t) + (1 + b_1 B^1) \cdot e(t) \qquad (6\text{-}49)$$

It is easy to interpret Eq. 6-49, difficult as it may look. The series for the moving average term equals a combination of the current disturbance and of the disturbance of one previous seasonal period. The original series, therefore, follows a model integrated from the ARMA model of the stationary series, see Eq. 6-47.

6.7.2 Application of ARIMA Models

6.7.2.1 Specification of ARIMA Models

The specification of ARIMA models is very expensive for the operator who analyzes time series. The first phase is the estimation of the order of three inherent processes, autoregression, integration, and moving average.

For composed trend and seasonal processes, the ARIMA trend and the ARIMA seasonal model were multiplied, e.g. both AR components as well as both integration components and both MA components. Then the notation is:

ARIMA *(p,d,q)(sp,sd,sq)*

p – order of the autoregressive process of the trend model
d – order of the integration of the trend model
q – order of the moving average of the trend model
sp – order of the autoregressive process of the seasonal model
sd – order of the integration of the seasonal model
sq – order of the moving average of the seasonal model

ARIMA (1,1,1)(1,1,1) for combined trend and seasonal effects:

$$(1 - B^1)(1 - B^{12})x(t) = a_{1t}B^1 a_{1s}B^{12}x(t) + (1 + b_{1t}B^1)(1 + b_{1s}B^{12}) \cdot e(t) \quad (6\text{-}50)$$

a_{1t} – autoregression coefficient for the trend component
a_{1s} – autoregression coefficient for the seasonal component
b_{1t} – moving average coefficient for the trend component
b_{1s} – moving average coefficient for the seasonal component.

The following practice is advised for determining these parameters:

Making a Series Stationary by Differencing – to Yield the Parameters *d* and *sd*

Removing trend:

$d = 1$
$$(1 - B^1)x(t) = x(t) - x(t - 1) \quad (6\text{-}51)$$

$d = 2$
$$(1 - B^1)(1 - B^1)x(t) = [x(t) - x(t - 1)] - [x(t - 1)] \quad (6\text{-}52)$$

238   6 Basic Methods of Time Series Analysis

Removing seasonality:

$sd = 0$

$$(1 - B^{12})x(t) = x(t) - x(t - 12) \qquad (6\text{-}53)$$

This differencing is continued until the time series is stationary (mean and variance are not dependent on time). Frequently, single time differencing, i.e. first order for the seasonal ARIMA model, is sufficient. Second order differencing is necessary for quadratic trends. Please note the loss of values after differencing (e.g. after first-order seasonal differencing twelve values will be lost).

Searching for the Order of the AR and the MA Process
The search for the right order of the AR and the MA elements requires the computation of the autocorrelation function and the partial autocorrelation function. Patterns of these functions give hints to the order of these processes. Here are some general rules (Figs. 6-21 to 6-23):

– Autoregressive processes have an exponentially decreasing autocorrelation function and one or more spikes in the partial autocorrelation function. The number of spikes in the partial autocorrelation function indicates the order of autoregression.

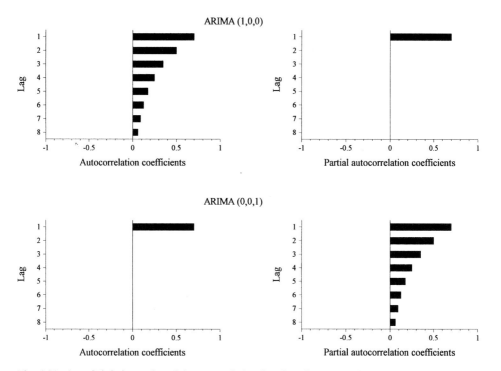

Fig. 6-21. Autocorrelation and partial autocorrelation functions for ARIMA(1,0,0) and ARIMA(0,0,1)

6.7 ARIMA Modeling   239

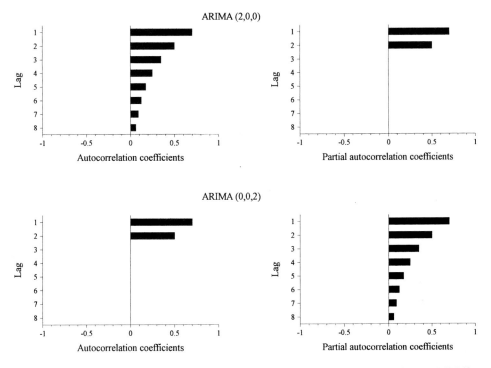

Fig. 6-22. Autocorrelation and partial autocorrelation functions for ARIMA(2,0,0) and ARIMA(0,0,2)

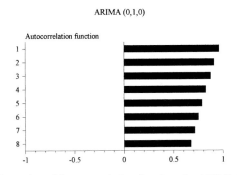

Fig. 6-23. Autocorrelation and partial autocorrelation functions for ARIMA(0,1,0)

− Moving average processes have one or more spikes in the autocorrelation function, the number indicates the order of the moving average process. The partial autocorrelation functions have an exponentially decreasing shape.
− Nonstationary series have an autocorrelation function with different shapes. In the case of a trend, the autocorrelation function shows some significant spikes and then

decreases rather quickly to nonsignificant values. Autocorrelation functions for seasonal changing series also demonstrate the seasonality.

In reality, these functions are more complex and the operator has to use the "trial and error" mode. Practical criteria which improve the likelihood of correct selection of the parameters of the ARIMA model are the autocorrelation and the partial autocorrelation function of the errors of the resulting ARIMA fit. If they do not have significant spikes the model is satisfactory.

Improving the Noncorrelation between the ARIMA Model Components
The autoregressive component, the seasonal autoregressive component, the moving average, and the seasonal moving average component should not have any correlation. Otherwise the model contains redundant components.

6.7.2.2 Application of the ARIMA Modeling to the Example Time Series

The nitrate time series of the drinking water reservoir is very complex and demonstrates the difficulty of identifying such models. A model with the shape **ARIMA (p,d,f) (sp,sd,sf)** shall be estimated.

Making a Series Stationary by Differencing – to Yield the Parameters d and sd
First, the series of the nitrate concentrations within the storage reservoir is made stationary in order to obtain the parameters d and sd for the trend and the seasonal ARIMA model. With one-time differencing at the differences 1, the series becomes stationary and the parameter d is set to unity (Fig. 6-24), but seasonal fluctuations are present. With one-time differencing of the original nitrate series at the difference 12, the seasonal fluctuations disappear, but the trend is present (Fig. 6-25). It is, therefore, necessary to include the seasonal ARIMA component in the model, the parameter sd is set to zero. The deduced possible model is ARIMA (?,1,?)(?,0,?).

Searching for the Order of the AR and the MA Process
Next, one calculates the autocorrelation function, ACF, (Fig. 6-16) and the partial autocorrelation function, PACF, (Fig. 6-17) from the original nitrate time series. The autocorrelation function leads to the following conclusions:

- An autoregression component is present because of the significant values of the autocorrelation function.
- A seasonal autocorrelation component is detected by high autocorrelation values at lags 12, 24, 36, …
- The die-away behavior of the autocorrelation function (the long declining shape of the ACF) hints at a trend, because the die-away of an autocorrelation function in a stationary time series would show an exponential function.

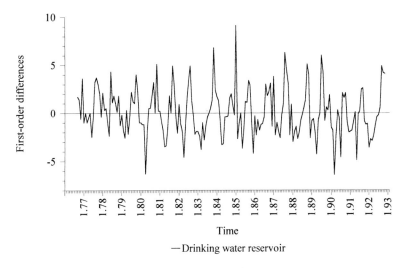

Fig. 6-24. First order differences from the example time series

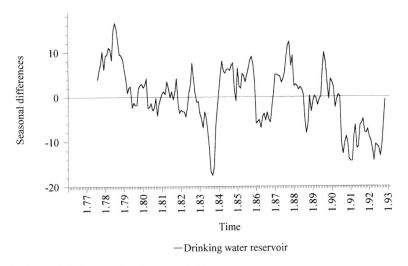

Fig. 6-25. Seasonal differences from the example time series

The PACF illustrates the order 1 for the AR component, but at this stage of estimation of the model it is unknown if the trend or the seasonal model follow the autoregression with the order of one. No moving average component can be found from the PACF. Deduced possible models are ARIMA (1,1,0)(1,0,0), ARIMA (0,1,0)(1,0,0), or ARIMA (1,1,0)(0,0,0).

These were the general conclusions from ACF and PACF of the time series for modeling ARIMA $(p,d,q)(sp,sd,sq)$. Now the second term of the multiplicative ARIMA model – the seasonal ARIMA component, ARIMA $(0,0,0)(sp,sd,sq)$ – must be estimated.

One of the single seasonal models deduced from the previous conclusion is ARIMA (0,0,0)(1,0,0). This model will be proved for its significance relating to the seasonal fluctuations of the time series. The model ARIMA (0,0,0)(1,0,0) (Tab. 6-4) confirms the high significance of the seasonal AR component. Therefore, sp is set to unity. The resulting standard error of the model is 6.35 mg L^{-1} NO_3^-. The resulting fit and the errors from ARIMA (0,0,0)(1,0,0) are demonstrated in Fig. 6-26.

Tab. 6-4. Model specifications of ARIMA(0,0,0)(1,0,0)

	Coefficients	Standard error of coefficients	Error probability
Seasonal autoregression, a_{1s}	0.632	0.062	0.000
Constant	30.67 mg L^{-1} NO_3^-	1.121 mg L^{-1} NO_3^-	0.000

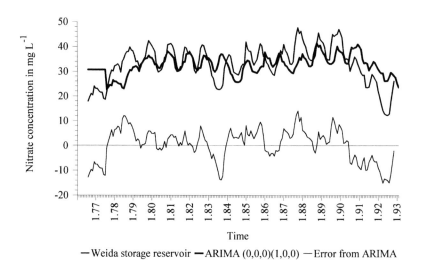

Fig. 6-26. Fit from ARIMA(0,0,0)(1,0,0)

These resulting errors must now be proved by autocorrelation function for seasonal effects. If the seasonal model is valid, no seasonal effects for autocorrelation function (Fig. 6-27), ACF, should be detectable.

The patterns of both ACF (Fig. 6-27) and PACF (Fig. 6-28) of the errors from the previous model are used to find the complete multiplicative trend and the seasonal ARIMA model. The following conclusions can be drawn:

- Seasonal effects from ACF of errors disappear. The seasonal model is sufficient.
- The die-away shape indicates a trend, d is set to unity.

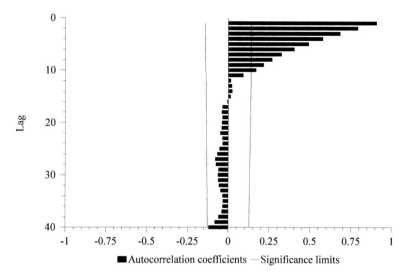

Fig. 6-27. Autocorrelation function of errors from ARIMA(0,0,0)(1,0,0)

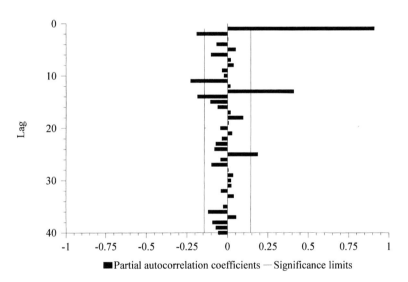

Fig. 6-28. Partial autocorrelation function of errors from ARIMA(0,0,0)(1,0,0)

- PACF of the errors indicates a first-order autoregression component, therefore the parameter p is set to unity. The spikes at lags 13 and 25 are a consequence of the multiplicative seasonal model.

The resulting model to be proved is ARIMA (1,1,0)(1,0,0), see Tab. 6-5 and Fig. 6-29.

Tab. 6-5. Model specifications of ARIMA(1,1,0)(1,0,0)

	Coefficients	Standard error of coefficients	Error probability
Autoregression, a_{1t}	0.343	0.069	0.000
Seasonal autoregression, a_{1s}	0.141	0.075	0.060
Constant	0.060 mg L^{-1} NO$_3^-$	0.285 mg L^{-1} NO$_3^-$	0.833

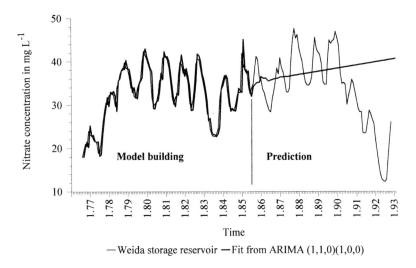

Fig. 6-29. Fit from ARIMA(1,1,0)(1,0,0)

After estimation this model contains a standard error of 2.27 mg L^{-1} NO$_3^-$. Both the autoregression component a_{1t} and the seasonal autoregression component a_{1s} are significant. The constant is not significantly different from zero.

The resulting model is:

$$(1 - B^1)x(t) = 0.343\, B^1 \cdot 0.141\, B^{12} x(t) \tag{6-54}$$

The ACF and PACF of the resulting errors from ARIMA (1,1,0)(1,0,0) do not show spikes (Figs. 6-30 and 6-31). This means they do not have significant autoregression or moving average components.

Improving the Noncorrelation between the ARIMA Model Components
Last but not least, it is necessary to check the correlation between the ARIMA components: the autoregression component and the seasonal autoregression component using the correlation matrix. In an adequate model, there should be no significant correlations between the single components.

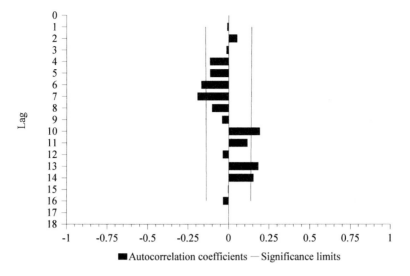

Fig. 6-30. Autocorrelation function of errors from ARIMA(1,1,0)(1,0,0)

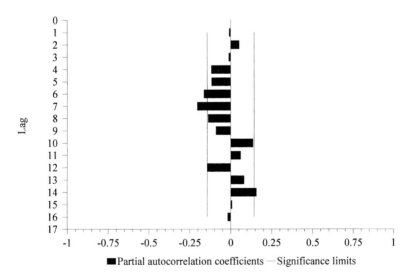

Fig. 6-31. Partial autocorrelation function of errors from ARIMA(1,1,0)(1,0,0)

In the example with the ARIMA model (1,1,0)(1,0,0), the correlation coefficient between the autoregression and the seasonal autoregression component is -0.19. This means they are not correlated with an error probability of 0.002. Therefore, this model is valid.

6.7.3 Forecasting with ARIMA

Forecasting is one of the most useful results of time series analysis. But with increasing distance from the present time, errors of forecasting will also increase. In general, forecasting without errors is only possible for an infinite-order autoregressive process. Then, future values may be calculated from preceding values without any errors. The opposite is a pure moving average process. Here, forecasting is not possible because of the random character of $e(t)$.

In ARIMA modeling, the order of the autoregressive component is frequently zero, one or sometimes two. Therefore, only **short forecasting intervals** are of any use. Disturbances in future values, normally smoothed by the moving average, were set to zero. The following example demonstrates this fact:

Assume the ARMA(1,1) model:

$$x(t) = 0.4\,x(t-1) + e(t) + 0.6\,e(t-1) \tag{6-55}$$

The forecasting formula is:

$$\begin{aligned}\hat{x}(t+1) &= 0.4\,x(t) + e(t+1) + 0.6\,e(t)\\ e(t+1) &= 0\end{aligned} \tag{6-56}$$

The forecast result is the following:

$$\begin{aligned}e.g.\quad \hat{x}(t+1) &= 0.4\,x(t) + 0.6\,e(t)\\ \hat{x}(t+2) &= 0.4\,\hat{x}(t+1)\end{aligned} \tag{6-57}$$

Fig. 6-29 demonstrates the satisfactory prediction for only one step, i.e. one month.

References

Berryman, D., Bobee, B., Cluis, D., Haemmerli, J.: Water Resour. Bull. 24 (**1988**) 545
Box, G.E.P., Jenkins, G.M.: Time Series Analysis, Forecasting and Control, Holden-Day, San Francisco, **1976**
Brockwell, P.J., Davis, R.A.: Time Series: Theory and Methods, Springer, New York, Berlin, Heidelberg, London, Paris, Tokyo, **1987**
Chatfield, C.: The Analysis of Time Series: An Introduction, 4th Ed., Chapman and Hall, London, **1989**
Cox, D.R.: Scand. J. Statist. 8 (**1981**) 93
Doerffel, K., Wundrack, A. in: Bock, R., Fresenius, W., Günzler, H., Huber, H., Tölg, G. (Hrsg.): Analytiker-Taschenbuch, Bd. 6, Akademie-Verlag, Berlin, **1986**, pp. 37
Doerffel, K., Küchler, L., Meyer, N.: Fresenius' J. Anal. Chem. 337 (**1990**) 802

Doerffel, K., Herfurth, G., Liebich, V., Wendlandt, E.: Fresenius' J. Anal. Chem. 341 (**1991**) 519

E.U.-TW: Directives of Council (15.7.1980) on the Quality of Water for Human Use (printed as an official paper of the European Community no. L 229/11–29 from 30.8.**1980**)

Einax, J., Geiß, S., Michaelis, W.: Fresenius' J. Anal. Chem. 350 (**1994**) 614

Fomby, T.B., Hill, R.C., Johnson, R.: Advanced Econometric Methods, Springer, New York, Berlin, Heidelberg, London, Paris, Tokyo, **1984**

Förster, E., Rönz, B.: Methoden der Korrelations- und Regressionsanalyse, Die Wirtschaft, Berlin, **1979**

Geiß, S., Einax, J.: Vom Wasser 78 (**1992**) 201

Geiß, S., Einax, J.: Chemom. Int. Lab. Syst. 32 (**1996**) 57

Geiß, S., Einax, J., Danzer, K.: Anal. Chim. Acta 242 (**1991**) 5

Johnson, T.R., Wijnberg, L.: Time Series Analysis of Hourly Average Air Quality Data, Proc. 74th Ann. Meet., Air Pollut. Control Assoc., Philadelphia, **1981**, 81–33.5

Kateman, G.: Chemometrics – Sampling Strategies in: Chemometrics and Species Identification, Topics in Current Chemistry, Vol. 141, Akademie-Verlag, Berlin, **1987**, pp. 43

Metzler, P., Nickel, B.: Zeitreihen- und Verlaufsanalysen, S. Hirzel, Leipzig, **1986**

Pandit, S.M., Wu, S.: Time Series and System Analysis with Applications, R. E. Krieger, Malabar, Florida, **1990**

Priestley, M.B.: Spectral Analysis and Time Series, Academic Press, London, **1981**

Sachs, L.: Angewandte Statistik, 7. Aufl., Springer, Berlin, **1992**

Schlittgen, R., Streitberg, B.H.J.: Zeitreihenanalyse, R. Oldenbourg, München, Wien, **1989**

SPSS: SPSS/PC+ Trends, SPSS Inc., **1990**

Stähle, L.: J. Pharmaceut. Biomed. Anal. 9 (**1991**) 671

Stock, H.-D.: Vom Wasser 57 (**1981**) 289

Terry, W.R., Kumar, A., Lee, J.B.: Time Series Analysis of Acid Rain Data collected in North-eastern United States, Proc. APCA Ann. Meet., 78th (Vol.1), 6A.3 (**1985**) 15

Tiao, G.C., Reinsel, G.C., Pedrick, J.H., Allenby, G.M., Mateer, C.L., Miller, A.J., Deluisi, J.J.: J. Geophys. Res., D: Atmos., 91 (D12), 13 (**1986**) 121

Trinkwasserverordnung – TrinkwV: Verordnung über Trinkwasser und über Wasser für Lebensmittelbetriebe vom 5.12.1990, Bundesgesetzblatt, Jahrgang **1990**, Nr. 66, Teil I, pp. 2612

Tümpling, W. von: Acta Hydrochim. Hydrobiol. 1 (**1973**) 477

Van Straten, G., Kouwenhoven, J.P.M.: Wat. Sci. Tech. Vol. 24, 6 (**1991**) 17

Weber, E.: Einführung in die Faktorenanalyse, Gustav Fischer, Jena, **1974**

Zetterqvist, L.: Statistical Methods for Analysing Time Series of Water Quality Data, Doctoral Dissertation, Lund University, Department of Mathematical Statistics, **1989**

Part B

Case Studies

Since the beginning of the industrial revolution, and especially in the last few decades, anthropogenic activities have accelerated many biogeochemical cycles. Subsequently environmental pollution by gaseous, particulate, and liquid and solid emissions has increased. Today, in some particularly highly polluted regions, pollutant concentrations are reached which considerably load the environment. Influences on human health have to be assumed. This generates the need for environmental monitoring to obtain qualitative and quantitative information about the kind and the degree of pollution, and its local and temporal changes. This knowledge can be used to assess the potential hazards for life, in order that the required measures can, if necessary, be taken.

The diversity of existing pollutants is large in respect both of the range of possible concentrations and the character of species present. For this reason it is definitely necessary to adapt the analytical process to the actual environmental problem.

In environmental chemistry in particular it cannot be the last step of the investigation to "produce" data tables on pollutant concentrations. Multivariate statistical methods offer a tool for the investigation of the multifactorial and complex events in the environment.

In the majority of studies pollution load data result from many, often unknown entry paths of noxious substances and from interactions among one another and the environmental compartments. In brief, the scope of application of chemometric methods is to extract the latent information from environmental data. In this section the power of multivariate data analysis shall be demonstrated for specific and typical examples of environmental investigation.

The main questions to be answered are:
- Can emission sources or dischargers be identified from complex combinations of measurable variables?
- Is it possible to characterize local or temporal loading states and their changes objectively? How we can illustrate groups or cluster associations among different samples?
- Can spatial distributions of environmental factors or pertubations be assessed?
- How representative are environmental samples in the specific case under investigation?
- Can interactions between environmental compartments be described by means of chemometric methods? Is it possible to predict a property of interest?

The main aim of this part of the book is to demonstrate the advantages of using multivariate statistical computations. The application of chemometric methods to the results of routine environmental monitoring and their relevant interpretation facilitates assertations concerning the identification of effective factors in the environment and the objective assessment of pollutant loading. These factors and loading states are either not accessible or are of only very restricted accessibility in current environmental monitoring.

The following case studies cannot and shall not give an overview of all possible applications of chemometric methods, but they may stimulate the reader into using chemometrics to solve environmental tasks and problems in his own working field.

7 Atmosphere

The present state of pollution of the atmosphere, especially the regional and semi-global pollution of its lower layer, is determined by routine environmental monitoring. Generally, the values of pollution levels, calculated as means, medians, or percentiles, are compared with legally fixed limits or thresholds. The widely scattered nature of those data does not give the possibility of drawing many conclusions by univariate methods (see the following examples). The methods of multivariate statistics should be able to extract more information because of the possibility of handling the whole data matrix.

Problems of environmental pollution by particulate emissions as a part of the complex complicated palette of environmental analytical problems will be covered in the centre of this section. Airborne particulates and their composition are affected by manifold influences. The main influences shall be ascertained. The power and the limits of multivariate statistics when applied to environmental problems under the compromising conditions of routine monitoring shall also be demonstrated.

7.1 Sampling of Emitted Particulates

The representativeness of samples is a very important premise for true assessment of the state of air pollution. The following example demonstrates the temporal frequency necessary for objective assessment of the impact of emissions of total suspended dust at one sampling location [GEISS et al., 1991]:

The element loadings of particulate emissions were investigated in the surroundings of a large ferrous metallurgical plant. The suspended dust was sampled by sucking the ambient air (1 m^3 h^{-1}) for 24 h through a membrane filter (Synpor) with a pore size of (0.85 ± 0.15) µm [ARBEITSMAPPE, 1984] at intervals of one week over a period of six months. The loaded filters were digested with concentrated nitric acid followed by addition of 30% hydrogen peroxide [ARBEITSMAPPE, 1984]. The concentrations of 17 elements (Al, Ba, Be, Ca, Cd, Cr, Cu, Fe, Mg, Mn, Na, P, Pb, Sr, Ti, V, and Zn) in the digestion solutions were determined by inductively coupled plasma-optical emission spectrometry.

It was necessary to answer the following question: Is the applied sampling frequency appropriate for characterization of the average impact of emissions at the sampling point over the time period under investigation?

252   7 Atmosphere

Because of the highly scattered temporal distributions of the individual loadings of the elements, the multivariate autocorrelation function was computed as described in Section 6.6.3. The results are demonstrated in Fig. 7-1.

Comparison with the highest possible values of a multivariate random correlation (see also Section 6.6.3) shows a correlation period of approximately two weeks and two days. This means that sampling of suspended dust at intervals of two weeks is sufficient for the characterization of the average impact of multielement emissions (for all investigated elements simultaneously) at that particular sampling point.

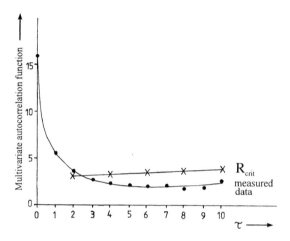

Fig. 7-1. Multivariate autocorrelation function of the sampling interval for suspended dust

7.2 Evaluation and Interpretation of Atmospheric Pollution Data

7.2.1 Chemometric Characterization of the Impact of Particulate Emissions in Monitoring Raster Screens

7.2.1.1 Problem and Experimental

Emissions of dust are crucial sources of pollution of the atmosphere by anthropogenic activities. The impact of emissions on territories is essentially determined by the amount of sedimented airborne particulate matter [KOMMISSION FÜR UMWELTSCHUTZ, 1976]. In routine monitoring the assessment of these loadings is usually conducted by determining the total sedimented airborne particulates (TSP) in monitoring raster screens and comparison with legally fixed thresholds. Commonly, the positions of the dust-sampling points were chosen on the lattice points of intersections of GAUSS-KRUEGER coordinate systems with an edge length of 1 km per 1 km [ARBEITSMAPPE, 1984; VDI 2119, 1972]. But the distance of the measuring points actually depends on the particular emission structure and the nature of the monitored territory.

The following example demonstrates the possibilities of chemometric characterization of particulate emissions in monitoring raster screens [EINAX and DANZER, 1989; EINAX, 1993]:

– The characteristic loadings from sedimented airborne particulates were determined at each sampling point of the screen by gravimetric determination of the dust which was collected in glass vessels with a defined opening area during an exposure time of about 30 days, by analogy with the BERGERHOFF method [VDI 2119, 1972].
– The elemental content of the dust samples was analyzed by means of optical emission spectrography by direct evaporation into a direct current arc. The analyses were performed in a PGS 2 plan-grating-spectrograph (Carl Zeiss Jena) with an UBI 2 arc excitement source (Carl Zeiss Jena) with total consumption of the dust samples, mixed with graphite powder, and the registration of the first order spectra on spectral photoplates. The specific analytical conditions are given in [EINAX and DANZER, 1989]. Measurement of the blackening was by means of a MD 100 microdensitometer (Carl Zeiss Jena). Calculation of the elemental concentrations was performed on the basis of triplicate analysis using the HARVEY method [HARVEY, 1947].

7.2.1.2 Analytical Results and Chemometric Interpretation

The amounts of 16 elements (see also Tab. 7-1) were determined with the above mentioned method of emission spectrographic analysis in a total of 170 samples of sedimented airborne particulates from three urban areas (Gera, Jena, and Greiz) in Thuringia (Germany) during one year of investigation.

The routine analytical method applied is relatively inaccurate in respect of the absolute values, but sufficiently accurate for determination of element correlations in the following multivariate statistical computations.

To demonstrate the accuracy, two dust and two soil reference materials were analyzed with the described method. The mean value of the correlation coefficients between the certified and the analyzed amounts of the 16 elements in the samples is $r = 0.94$. By application of factor analysis (see Section 5.4) the square root of the mean value of the communalities of these elements was computed to be approximately 0.84. As frequently happens in the analytical chemistry of dusts several types of distribution occur [KOMMISSION FÜR UMWELTSCHUTZ, 1985]; these can change considerably in proportion to the observed sample size. In the example described the major components are distributed normally and most of the trace components are distributed log-normally. The relative ruggedness of multivariate statistical methods against deviations from the normal distribution is known [WEBER, 1986; AHRENS and LÄUTER, 1981] and will be tested using this example by application of factor analysis.

7.2.1.2.1 Univariate Aspects

The total mean values of the investigated territories are shown in Tab. 7-1. It can be seen that the variation range of the elemental content amounts to three orders of magnitude. Territorial or even loading structures of the sampling points at the receptor sites cannot be recognized in the adequate univariate, or even bivariate graphical representations.

Tab. 7-1. Elemental content of sedimented airborne particulates as mean values over one year of investigation (relative to total sedimented airborne particulates TSP = 100) in Thuringia (Germany)

Element	Mean value	Minimum value	Maximum value
TSP	100	24.0	531.66
Si	36.05	0.44	224.41
Fe	14.83	1.47	83.82
Al	29.98	4.69	179.12
Ti	4.12	0.745	28.51
Mg	8.13	0.238	136.24
Pb	0.743	0.00396	5.42
Zn	0.715	0.0742	7.91
Cu	0.500	0.0378	1.70
V	0.158	0.0169	1.42
Mn	0.416	0.0666	5.37
Ni	0.0770	0.00275	0.412
Sn	0.0510	0.00386	0.365
Cr	0.0213	0.000689	0.132
Mo	0.0112	0.00151	0.0887
B	0.0717	0.00999	0.878
Ba	3.41	0.314	31.19

The representation of total sedimented particulates and of some elemental precipitation shows only very strongly scattered seasonal variation (Fig. 7-2); this can be interpreted as follows:

- The summer maximum of TSP (total sedimented particulates) results from the higher levels of disturbed geogenic material and secondary dust in the relatively arid summer season.
- The pattern of deposition of silicates is highly correlated with, and corresponds to, the TSP.
- Again, lead deposition shows a summer maximum which is due to emissions from motor vehicles. The described investigations took place when the majority of motor cars in the former GDR were using leaded petrol.
- Compared with that, the maximum deposition of aluminum occurs in the winter months. Aluminum can serve as an indicator element for lignite combustion, as thermo-analytical investigations of flue gases in lignite heating plants have shown [EINAX and LUDWIG, 1991].

– The other investigated elements do not show any discernible seasonal trends. Representations of single measuring points do not enable any conclusions to be made about temporal trends, because of the high variability of the impact of emissions.

The large variance of the elemental depositions, also demonstrated by the very uncertain temporal courses of the elemental deposition rate (Fig. 7-2), strongly limits visual inspection of the obtained data, the interpretation can be subjective only. Otherwise practically all simple correlation coefficients are significant. Both facts show that it seems to be useful to apply advanced statistical methods to attempt recognition of possible existing data structures which may enable the characterization of pollutant loading and the possible identification of emission sources.

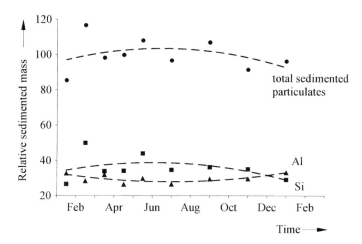

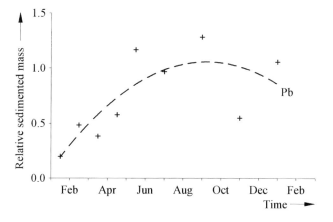

Fig. 7-2. Relative sedimented mass as a function of time for total mean values of different analytes in the investigated areas

7.2.1.2.2 Cluster Analysis

The principle of unsupervised learning consists in the partition of a data set into small groups to reflect, in advance, unknown groupings [VARMUZA, 1980] (see also Section 5.3). The results of the application of methods of hierarchical agglomerative cluster analysis (see also [HENRION et al., 1987]) were representative of the large palette of mathematical algorithms in cluster analysis.

In this passage we demonstrate that comparable results may also be obtained when other methods of unsupervised learning, e.g. the non-hierarchical cluster algorithm CLUPOT [COOMANS and MASSART, 1981] or the procedure of the computation of the minimal spanning tree [LEBART et al., 1984], which is similar to the cluster analysis, are applied to the environmental data shown above.

In the representation of the number of clusters obtained versus their similarity we see the following (Fig. 7-3):

– The best structuring of the data material can be obtained by use of the method of WARD. This fact was also shown for other cases of application reported in the literature [DERDE and MASSART, 1982].
– The multivariate structure of the impact of emissions changes continuously in the territory investigated.

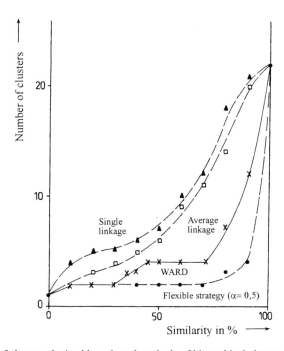

Fig. 7-3. Stability of clusters obtained by selected methods of hierarchical cluster analysis

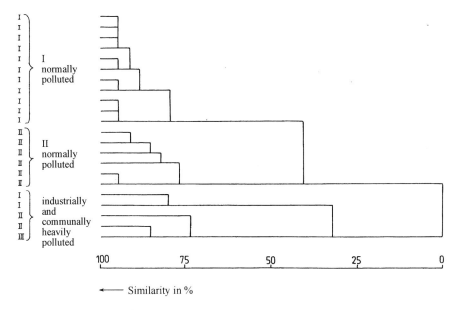

Fig. 7-4. Dendrogram for different emission impact monitoring raster screens for one month (Cluster algorithm according to WARD)

The dendrogram of different emission impact monitoring raster screens computed by means of the cluster algorithm of WARD is represented in Fig. 7-4.

The moderately polluted measuring points of the urban emission impact monitoring raster screen I (town Gera) are clearly separated from those of the town II (Jena). The highly polluted regions of all investigated territories (I, II, and III (Greiz)) are contrasted with those regions where the pollution is normal.

With regard to the monitoring of the impact of emissions cluster analysis proves a useful tool for recognition of particularly highly polluted measuring points and accordingly gives the first hints of possible emission sources.

It should be remarked that the application of CA is not limited to the description of the impact of particulate emissions. BIRRING and HASWELL [1994] describe the use of both CA and PCA for the identification of different sources of biogases.

Certainly it is possible to apply also other display methods for the visualization of such complex environmental data, as particulate emissions. TREIGER et al. [1993; 1994] describe the study of different aerosol samples by nonlinear mapping of electron probe microanalysis data. Different interpretable groups of chemical elements which determine the composition of aerosol samples can be obtained. More recent work by WIENKE and HOPKE [1994] and WIENKE et al. [1994] discuss the combination of different chemometric techniques for better graphical representation of aerosol particle data. The authors use receptor modeling with a minimal spanning tree combined with a neural network.

7.2.1.2.3 Multivariate Analysis of Variance and Discriminant Analysis, and PLS Modeling

Multivariate Analysis of Variance and Discriminant Analysis

If values for the impact of emissions from different monitoring raster screens are submitted, or if differently loaded areas can be assumed from a knowledge of the emission structure and the specific territorial situation, it is obvious that differences in the impact of emissions can be classified in a multivariate manner.

In the beginning of this section we will test if it is possible to classify the different impacts of emission levels in the three urban territories investigated by means of the different compositions of the sampled sedimented dusts. This will be followed by application of the method of multivariate analysis of variance and discriminant analysis (MVDA), as a very common method of supervised learning, to the data set of the three different *a priori* classes (towns). The mathematical fundamentals of MVDA are described in detail in Section 5.6. Fig. 7-5 shows the results from MVDA of different emission impact monitoring raster screens in the plane of the computed discriminant functions. The separation lines correspond to the limits of discrimination for the highest probability. These results prove that good separation of the three investigated territories is possible by means of MVDA. The misclassification rate amounts to 15.3%, as compared to the error of a randomly correct reclassification of 48.9%. But the scattering radii of the 5% risk of error of the multivariate analysis of variance, which overlap considerably, show also that the differences in the multidimensional structure of the impact of emissions are only small.

From the only incomplete separation of the classes and the error of discrimination, which is not very small, the critical reader may raise the question: is MVDA really a powerful chemometric tool for the distinction of slight environmental differences?

To give a positive answer to this question the following example from another environmental case study will be demonstrated.

In a small territory of the Thuringian Forest the water quality of five closely situated springs was monitored over a period of 14 weeks. The geological background in the investigated area is very homogeneous and different emission influences or defined dischargers can be excluded. Eight parameters (conductivity, total hardness, pH, and the concentrations of calcium, magnesium, chloride, nitrate, and sulfate) were analyzed in the spring waters, sampled weekly over a period of 14 weeks. No significant differences between the springs or changes over the time under investigation could be found by means of univariate tests. By application of MVDA with the *a priori* assumption of the existence of five different classes (five springs, each occupied with 14 objects) complete separation of the springs was possible. To illustrate this result, and with it the discriminative power of MVDA, the scores of the two strongest discriminant functions by separation of the different spring waters are demonstrated in Fig. 7-6. It is apparent that the separation of the individual classes is considerably better. In this case the error of the reclassification amounts to 0%.

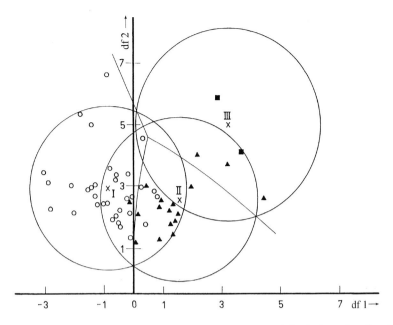

Fig. 7-5. Plot of the scores of discriminant function df 2 vs. scores of discriminant function df 1 of the different emission impact monitoring raster screens I, II and the emission impact sampling point of town III [(o) town I, (▲) town II, (■) town III]. (The circles correspond to the 5% risk of error of the MANOVA)

		Result of discrimination			
		Discriminated samples in class			Misclassification rate
		I	II	III	
Given	I	29	4	0	4/33
classes	II	1	13	3	4/17
	III	0	0	2	0/2

This example demonstrates that differences between the impacts of particulate emissions are essentially smaller because of the high variability, inhomogeneity, and mobility of the atmosphere than in the discussed case study of the hydrosphere, which is an example of an environmental medium largely uninfluenced by human civilization.

The extent to which it is possible to classify differences between the impact of emissions within one monitoring raster screen, and accordingly to detect heavily loaded areas, will be the subject of further investigation.

A model of three loaded areas (Fig. 7-7) has therefore been constructed from a knowledge of the emission pattern but also from the particular territorial situation in the region investigated. The results from MVDA of the impact data in the summer months are represented in Fig. 7-8. The isolated class of the relatively highly polluted center clearly contrasts with the others. The loaded areas A and B are more similar to one another, but they are well reclassified, as the small error of discrimination of 10.0% proves.

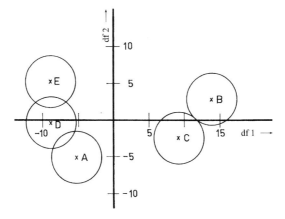

Fig. 7-6. Plot of the scores of discriminant function df 2 vs. scores of discriminant function df 1 for several spring waters (A-E) from one hydrogeological origin. (The circles correspond to the 5% risk of error of MANOVA)

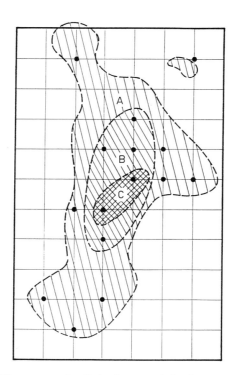

Fig. 7-7. Sketch of the different areas of pollution in one emission impact monitoring raster screen: A – slightly polluted; B – moderately polluted; C – heavily polluted

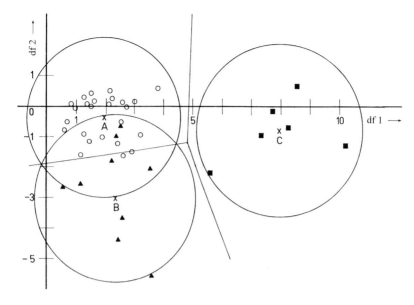

Fig. 7-8. Plot of the scores of discriminant function df 2 vs. scores of discriminant function df 1 of one emission impact monitoring raster screen and differently loaded areas during the summer months; areas of pollution: (o) A, (▲) B, (■) C. (The circles correspond to the 5% risk of error of MANOVA)

		Result of discrimination			
		Discriminated samples in class			Misclassification rate
		A	B	C	
Given	A	23	2	0	2/25
classes	B	2	7	0	2/9
	C	0	0	6	0/6

On the other hand the results from MVDA in the winter months show a more regular structure of the impact of emissions (Fig. 7-9), as it proved by the greater overlapping of the scattering radii of the 5% risk of error for the three loaded areas.

The degree of discrimination was less than in the summer months, with a corresponding error of 20.9%. The higher similarity of the classes A and B results, above all, from stronger and more similar influences of communal heating plants on the impact of emissions in these two loaded areas during the winter period.

The following investigation has also been conducted to detect the influence of a regular emitter on the area under investigation. A representative reference dust sample was taken in the stack of a big industrial heating plant, located in the heavily loaded area C (see Fig. 7-7). The dust sample was analyzed in a manner analogous with the sedimented airborne particulates. The obtained data vector was added as a test data set to the above described classification model. The emitted dust from the stack showed the high-

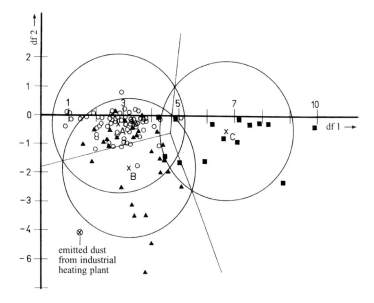

Fig. 7-9. Plot of the scores of discriminant function df 2 vs. scores of discriminant function df 1 of one emission impact monitoring raster screen and different loaded areas during the period when buildings were heated; areas of pollution: (o) A, (▲) B, (■) C. (The circles correspond to the 5% risk of error of MANOVA)

	Result of discrimination			
	Discriminated samples in class			Misclassification rate
	A	B	C	
Given A	61	4	1	5/66
classes B	15	15	0	15/30
C	1	2	11	3/14

est similarity to that sampled in the moderately loaded area B. The following conclusions can be drawn:

- Because of the extreme height of the stack an emission influence is not detectable in the surrounding territory C and also not in the more distant area A. Results from investigations on the spreading of sedimenting dusts from power plants are in accordance with this chemometrically obtained explanation [BOHN, 1979].
- The dust emitter influences the pollutant loading in area B.
- The relative dissimilarity between the measuring value vector of the reference emission dust and the centroid of class B has three origins:
  – The emitted dust is a total dust, whereas the sedimented dust has been sampled in some distance from the point of emission. Some authors refer to differences in the elemental contents of these two types of dust [BONTRON and PATTERSON, 1983; KLAMM, 1986].

- During the process of transmission through the atmosphere a change in the chemical composition of the dust takes place, basically as a result of meteorologically induced chemical and physical processes.
- Further components, e.g. the elemental pattern of other emission sources or diffuse emitters (for instance secondary dust) are also contained in the sedimented dust of area B.

PLS Modeling

Another possibility of finding relationships between the impact of emissions in a territory and existing emission sources is the use of PLS modeling. For the above discussed case PLS modeling between the data matrix of the pollutant load in territory B and the data vector for the composition of the emitted dust was performed according to the mathematical basis described in Section 5.7.2. The elemental compositions both of the emitted dust and the impact of emissions were normalized to their concentrations, thus giving a uniform data basis.

Both in the winter and in the summer months the relationship between emission and pollutant loading can be reflected by the following elements in decreasing order of their influence:

$$Al > Fe > Mg > Si > Ti$$

The other investigated elements were less significant, which means that the impact of emissions resulting from the heating of buildings is determined by the above elements, particularly aluminum which has the highest weight in the describing t-vector.

Thermoanalytical investigations of sedimented airborne particulates [EINAX and LUDWIG, 1991] confirm experimentally the chemometrically found interpretation that aluminum can serve as an indicator element for lignite combustion. Thermogravimetric analysis of mixed samples of sedimented dusts detect a loss of mass at a temperature of 714 °C; this can be interpreted as dehydration of aluminosilicates. This loss of mass exhibits a well defined summer minimum and a strong winter maximum. These findings also correspond to the results from factor analysis (see Section 7.2.1.2.4).

At this point it should be remarked that multivariate regression with latent variables is a useful tool for describing the relationship between complex processes and/or features in the environment. A specific example is the prediction of the relationship between the hydrocarbon profile in samples of airborne particulate matter and other variables, e.g. extractable organic material, carbon preference index of the n-alkane homologous series, and particularly mutagenicity. The predictive power was between 68% and 81% [ARMANINO et al., 1993]. VONG [1993] describes a similar example in which the method of PLS regression was used to compare rainwater data with different emission source profiles.

The application of MVDA and PLS modeling to results of investigations of the impact of particulate emissions enables the following conclusions to be drawn:

- MVDA is a useful chemometric tool for detection and classification of differences in the multivariate structure of the impact of emissions on different urban territories.

- MVDA is highly suitable for classifying differently loaded areas inside one emission impact monitoring raster screen. By this means the *a priori* class division, which is given by a knowledge of the structure of emission sources and the particular territorial situation, can be corroborated.
- The territorial differences between areas affected by particulate emissions are relatively small in comparison to the relationships between other environmental media, which also vary from measuring point to measuring point, so that the given classes may be discriminated relatively well for the highest probability, but mostly they do not exist in isolation.
- It is possible to detect the influence of emission sources in defined territories by application of MVDA.
- A loading hypothesis resulting from a knowledge of the emission structure can be proved by emission sampling, followed analysis and application of MVDA. This means that emitters which pollute the environment can be identified.
- PLS modeling is a very powerful chemometric tool for detection and quantification of the relationships between emission and their impact on the environment. The chemometric solution could be confirmed by independent thermoanalytical investigations.

7.2.1.2.4 Factor Analysis

The aim of the application of factor analysis (FA) to environmental problems is to characterize the complex changes which occur to all the features observed in partial systems of the natural environment. These common factors explain the complex state of the environment more comprehensively and causally and so enable extraction of the essential part of the information contained in a set of data.

The mathematical fundamentals of FA (see also [WEBER, 1974; 1986; MALINOWSKI, 1991]) are described in detail in Section 5.4.

The exclusive consideration of common factors seems to be promising, especially for such environmental analytical problems, as is shown by the variance splitting of the investigated data material (Tab. 7-2). Errors in the analytical process and feature-specific variances can be separated from the common reduced solution by means of estimation of the communalities. This shows the advantage of the application of FA, rather than principal components analysis, for such data structures. Because the total variance of the data sets has been investigated by principal components analysis, it is difficult to separate specific factors from common factors. Interpretation with regard to environmental analytical problems is, therefore at the very least rendered more difficult, if not even falsified for those analytical results which are relatively strongly affected by errors.

As above represented it is necessary to test the ruggedness of FA against deviations from the normal distribution; FA was, therefore, conducted on the basis of the SPEARMAN rank correlation coefficient [WEBER, 1986]. The obtained factor loading matrix

Tab. 7-2. Origins of variances in the elemental content of sedimented airborne particulates

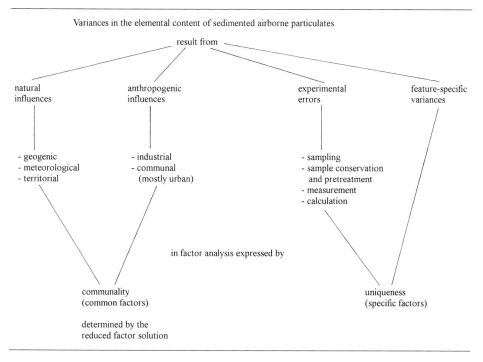

is qualitatively the same as the result of FA based on the normal linear correlation coefficient.

The basis of the application of FA was a data matrix containing 17 features of 52 sedimented airborne particulate samples collected during a period when buildings were being heated. According to Fig. 7-10 the application of the scree plot [CATTELL, 1966] indicates four common factors.

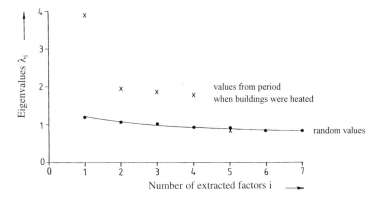

Fig. 7-10. Scree plot for the determination of the number of common factors

The interpretation of the obtained four factors according to Tab. 7-3 can be proved by the following discussion of the factor scores computed by means of the BARTLETT estimation [JAHN and VAHLE, 1970].

Tab. 7-3. Common factors of the impact of particulate emissions during a period when buildings were being heated

Factor	Determinable variance in %	High factor loadings for	Main origin
1	34.3	TSP, Si, Mg, Pb, Ti, Mn, Fe	Raised soil-derived material and secondary dust
2	17.2	Al, V, Ti, TSP, Ni	Industrial and communal heating
3	16.5	Cu, Mn, Cr, Sn, Mo, Mg	Emissions from metallurgy and metal working
4	15.9	Zn, Ni, Ti	Ubiquitous elements washed out by precipitation

Graphical representation of the scores of these factors characterizing differences between the impact of emissions on the territories investigated best enables territorial clustering of the analyzed dusts (Fig. 7-11). Fig. 7-12 shows the relative fuzzy resolution of temporal changes in the impact of these emissions.

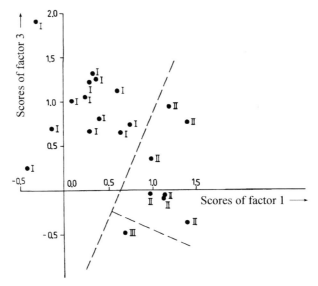

Fig. 7-11. Representation of the scores of factor 3 vs. scores of factor 1 for territorial separation of the sampling points during the period when buildings were being heated (I, II, and III – different towns)

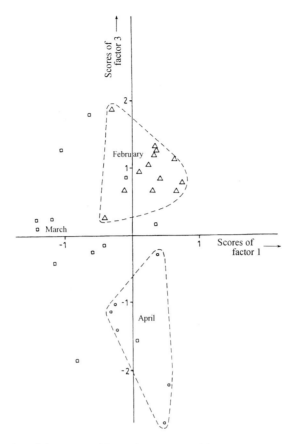

Fig. 7-12. Representation of the scores of factor 3 vs. scores of factor 1 for the detection of temporal separation during the period when buildings were being heated

FA was accomplished from the data set of sedimented airborne particulates sampled in the summer. The factor loading matrix is demonstrated and discussed in Tab. 7-4.

The absence of the heating factor and the wash-out factor should be emphasized; in addition, industrial emissions can be split into two sources.

By means of FA four common factors could be extracted during the period when buildings are heated, and three common factors in the summer months. Interpretation of the factor loading matrices gives a plausible explanation of the main emission sources and their seasonal variations in the investigated territories. It could be shown that a main cause of particulate levels is the raised geogenic material and secondary dust, with 34.3% of determinable variance during the period when buildings are heated and with 27.6% of determinable variance in the summer months. During the period of heating the results from FA show the influence of combustion of lignite for heating in the investigated territories, expressed in a common factor with 17.2% of the determinable var-

Tab. 7-4. Common factors of the impact of particulate emissions during the summer

Factor	Determinable variance in %	High factor loadings for	Main origin
1	27.6	Mn, Al, Si, TSP, Fe, Pb, Ti	Raised soil-derived material and secondary dust
2	28.5	Cr, V, Ti, B, Mo, Ni, Fe, TSP, Mg	Emissions from ferrous metallurgy
3	10.4	Mg, Zn, Sn, Mo, Cu	Emission from nonferrous industry

iance. The other factors are caused by emission of ferrous and nonferrous metallurgy and metal-working.

The results and their interpretation are comparable with other published results (for instance [HOPKE et al., 1976; VAN ESPEN and ADAMS, 1983; KEIDING et al., 1986; VONG, 1993]), but this is only a plausible explanation of a mathematical hypothesis. The graphical representation of the computed factor scores can verify this hypothesis. As demonstrated in the literature [EINAX and DANZER, 1989], representation of the scores of the main factors enables complete territorial separation of the investigated emission impact monitoring raster screens but only a relatively fuzzy temporal separation.

Using multivariate statistics it is, therefore, possible to identify not only heavily polluted regions but also regions which are not subject to pollution from specific sources.

The advantages of the application of FA to the discussed environmental analytical problems are:
- In the complex phenomenon "environment", particularly in the atmosphere, the complex causes which exist can be recognized by the application of FA.
- The elimination of experimental errors and variances specific to the features in the course of FA, enables description of the investigated environmental problem by means of a model of common factors.
- The hypothesis of causes, by interpretation of the factor loading matrix, can be proved by computation of the factor scores and their graphical representation.
- This representation enables successful territorial structuring of the effect of emissions by simultaneously emphasizing the polluted measuring points.

7.2.1.3 Conclusions

The results of spectrographic analysis of sedimented airborne particulates collected during routine monitoring of the environment show high, environmentally caused, variability. Objective assessment of loading states, and still less of emission sources, can, therefore hardly be achieved by visual inspection of the data or by application of simple statistical methods, whereas the techniques of multivariate data analysis enable comprehensive and causally explicable demonstration of complex and intricate environmental data sets, with the result that their latent information content is recognizable. In each case knowledge of the emission structure and specific territorial conditions in the region investigated is necessary for the useful interpretation of the chemometric solutions obtained.

Commonly the compromising conditions of routine environmental monitoring lead to restrictions on the accuracy and the precision of sampling and analysis. The purpose of this section is to show that under these conditions multivariate statistical methods are a useful tool for qualitative extraction of new information about the degree of stress of the investigated areas, and for identification of emission sources and their seasonal variations. The results represented from investigation of the impact of particulate emissions can, in principle, be transferred to other environmental analytical problems, as described in the following case studies.

7.2.2 Chemometric Characterization of the Temporal Course of the Impact of Particulate Emissions at One Sampling Location

The results of the chemometric investigation of particulate contamination in monitoring raster screens (discussed in detail in Section 7.2.1.2.4) show the limitations of the interpretation of temporal courses of the impact of such emissions; this is easy to understand because such temporal changes are averaged over many sampling points in a large territory.

In this section the possibility and the power of chemometric methods for detection of the temporal changes in the levels of the impact of particulate emissions will be studied by discussion of two examples of results from investigations of the impact of emission at one sampling point [EINAX et al., 1991; 1994].

7.2.2.1 First Example

7.2.2.1.1 Experimental

At one sampling point near the Technical University of Košice (Slovakia) (see Fig. 7-13) sedimented airborne particulate matter was sampled by the above described BERGERHOFF method (see also Section 7.2.1.1) over a period of two and a half years. Tab. 7-5 shows the analytical methods applied for the determination of the elemental composition of the dust samples.

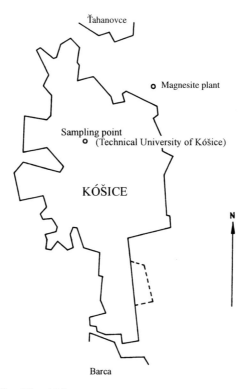

Fig. 7-13. Sketch of Kóšice (Slovakia)

Tab. 7-5. Analytical methods applied for the determination of the elemental composition of the dust samples

Applied analytical method	Analyzed elements
Flame atomic absorption spectrometry	Ba, Ca, K, Na, Sr
X-ray fluorescence spectrometry	Al, Ca, Fe, Mg, Si
Optical emission spectrography	Ag, B, Ba, Co, Cr, Cu, Mn, Mo, Ni, Pb, Sb, Sn, Sr, Ti, V, Zn
Differential pulse anodic stripping voltammetry	Pb, Zn (in lower concentration)
Potentiometry with ion-sensitive electrodes	F, I, S

7.2.2.1.2 Data Preparation and Univariate Aspects

The elements Co, Mo, and Sb were excluded from the determined 27 features (26 elements and total sedimented particulates) because their values were below the detection limits. The relative frequencies of the four main wind directions at the sampling point were taken into consideration in the multivariate statistical computations as an important factor influencing the impact of the emissions. Tab. 7-6 illustrates the expected strong variations of the analyzed features. The measured variations amount to two orders of magnitude during the duration of the investigation. The fluctuations in the East Slovakian region are lower than in Germany (see Section 7.2.1.2). An important reason is the greater uniformity of the impact of particulate emissions in the more continental climate in this region.

Tab. 7-6. Mean, minimum, and maximum values of element precipitation at the sampling point in Kóšice (Slovakia) in kg km^{-2} year^{-1}

Feature	Mean value	Minimum value	Maximum value
TSP*	142.13	84.5	223.3
Al	5269.26	2570.0	9770.0
Ag	1.410	0.02	14.8
B	11.067	3.1	51.4
Ba	370.53	185.9	790.0
Ca	2358.53	783.9	8040.0
Cr	15.089	6.9	32.2
Cu	35.452	7.5	92.5
Fe	12217.03	6260.0	27680.0
K	385.54	137.4	813.6
Mg	34126.29	17310.0	54230.0
Mn	162.95	56.0	430.8
Na	419.77	102.2	1310.0
Ni	9.241	3.7	22.7
Pb	63.711	24.1	242.8
Si	17968.14	9550.0	42940.0
Sn	4.867	0.7	12.5
Sr	10.80	3.4	31.3
Ti	504.08	122.5	1190.0
V	17.892	4.8	38.1
Zn	415.38	148.0	973.1
F	203.06	3.0	440.1
I	327.00	29.1	1650.0
S	5777.90	887.4	14370.0

* in t km^{-2} year^{-1}

7.2.2.1.3 Combination of Cluster Analysis, and Multivariate Analysis of Variance and Discriminant Analysis

Results from hierarchical agglomerative cluster analysis according to the algorithm of WARD (see Section 5.3) are illustrated as a dendrogram in Fig. 7-14. Distinction of the months in which the heating of buildings has a large influence from the summer months is clearly demonstrated. November and December of the second year of the in-

272   7 Atmosphere

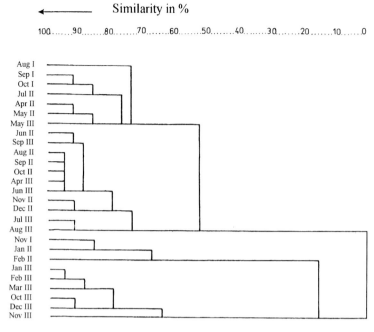

Fig. 7-14. Dendrogram of the hierarchical agglomerative cluster analysis according to WARD

vestigation were classified as summer months because of the small influence of building heating resulting from relatively high temperatures in this period.

Because of the qualitative character of classification results obtained by cluster analysis, it is useful to combine this unsupervised technique with a supervised method, here MVDA. To accomplish MVDA it becomes necessary to change the structure of the original data set in such a manner that the required matrix inversion becomes possible. This means that the number of objects (months) must be higher than the number of features. Sixteen elements and the four wind directions were, therefore, selected. These features have high factor loadings in the results from FA, i.e. they determine a large part of the variance-covariance of the data space. One discriminant function with the optimum separation set silicon, nickel, and iron results from MVDA. The result of discrimination is:

	Discriminated samples in class	
	A	B
Given class A (with low influence of building heating)	16	2
B (with large influence of building heating)	1	8

The error of discrimination is 11.1%; the error of a randomly correct reclassification amounts to 44.4%.

7.2.2.1.4 Factor Analysis

The first result of FA according to Section 5.4, the matrix of factor loadings, is represented in Tab. 7-7.

Tab. 7-7. Matrix of factor loadings of the common factor analytical solution of sedimented airborne dusts from Kóšice (factor loadings < 0.450 are set to 0 for greater clarity)

Features	Loadings of factor					Communality h_i^2
	1	2	3	4	5	
TSP	0.871	0	0	0	0	0.872
Al	0	0	0	0.762	0	0.666
Ag	0	0	0	0	0	0.239
B	0	0	0	0	0.754	0.840
Ba	0	0	0	0.717	0	0.770
Ca	0.837	0	0	0	0	0.739
Cr	0	0	0	0	0.805	0.730
Cu	0	0	0	0.618	0.504	0.665
Fe	0	0	0	0.719	0	0.637
K	0.780	0	0	0	0.573	0.950
Mg	0.629	0	0	0	0	0.666
Mn	0.766	0	0	0	0	0.688
Na	0.718	0	0	0	0	0.866
Ni	0.732	0	0	0	0	0.762
Pb	0	0.700	0	0	0	0.551
Si	0	0.877	0	0	0	0.959
Sn	0	0	0	−0.559	0	0.590
Sr	0.912	0	0	0	0	0.883
Ti	0	0.609	0	0.607	0	0.796
V	0.592	0	0	0.525	0	0.674
Zn	0	0	0	0.734	0	0.832
F	0	0	0	0.652	0	0.629
I	0	0	0	0	0	0.158
S	0.789	0	0	0	0	0.709
North wind	0	0	0.960	0	0	0.946
East wind	0	0	0	0.450	0	0.479
West wind	0	0	0	0	0	0.281
South wind	0	0	−0.958	0	0	0.943

The features Ag and the west wind direction have very low communalities in the common factor solution, i.e. their variance does not correlate with the variance of the other determined features. Origins may be both errors in the determination of these features and the features' specific variances. These features should not, therefore, be interpreted together with the factor analytical solution. The most weighted factors may be interpreted as:

Factor 1: Influence of domestic and industrial heating, especially the combustion of salt- and sulfur-containing lignite in Kóšice

Factor 2: Increased soil-derived material (large amounts of variance for the main soil-derived elements silicon and titanium) and raised secondary dust (characterized by lead, from traffic emissions enriched in the upper soil layer [EWERS and SCHLIPKÖTER, 1984])

Factor 3: North-south wind-factor (resulting from the north-south extension of the valley of Kóšice); no detectable interaction with element precipitates; no remarkable emission sources exist either in the north or in the south

Factor 4: Mainly originating from emission by the magnesite plant; positive interaction with the east-wind direction according to the position of the plant (Fig. 7-13)

The results from the second part of FA, the factor scores, are represented in Fig. 7-15 as a function of time. The graphical representation shows clear seasonal courses for the factors 1 and 2 with maxima in the winter and minima in the summer. The explanation for factor 1 is obvious (influences of building heating). The winter maxima for factor 2 can be interpreted in terms of the greater amount of soil-derived material in the winter and early spring months with relative low precipitation. The scores of factor 4 show a significant increasing trend proved by application of the CUSUM technique [DOERFFEL, 1990] (see also Section 6.4.3). During the investigation period the production of the magnesite plant was increased by nearly 12%. The local minimum of the scores of factor 4 at the beginning of year III is a result of a general repair of the plant.

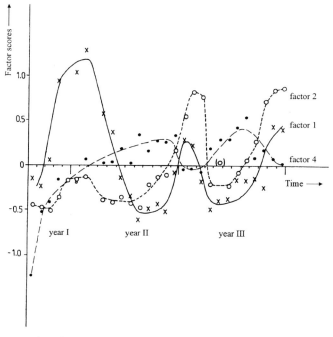

Fig. 7-15. Representation of the time-dependence of the factor scores

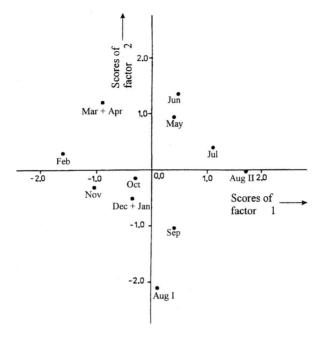

Fig. 7-16. Representation of the scores of factor 2 vs. scores of factor 1 for one year of the investigation

The seasonal course of the impact of particulate emissions becomes clearer, if the data from one year of the investigation are studied by FA (Fig. 7-16). The results of only a part of the data matrix demonstrate that even when the rule of HORST is not considered, interpretable solutions can still be obtained. The rule of HORST demands that the number of objects should be at least three times the number of features [WEBER, 1986].

This fact is of relevance for environmental analysis, since it is often objectively impossible (time of sampling, experimental expenditure, etc.) to obtain the required data structure.

7.2.2.2 Second Example

This second example demonstrates the possibility of investigating long range transport of particulate emissions by application of chemometric methods [EINAX et al., 1994].

7.2.2.2.1 Experimental

Aerosol samples were taken weekly over a period of one year from May 1984 to June 1985 on the North Frisian island of Pellworm (Germany) which is not polluted by defined emitters. The high volume sampler used was equipped with a five-stage slotted

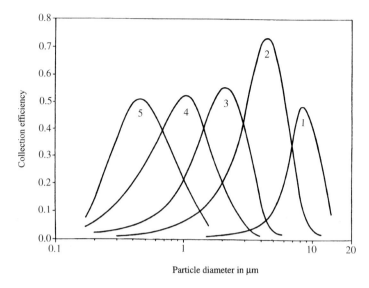

Fig. 7-17. Effective collection efficiency of the five-stage slotted cascade impactor

cascade impactor. The effective collection efficiencies are shown in Fig. 7-17. The maximum collection efficiencies were reached at a particle diameter of 8.5 µm for the first fraction, 4.6 µm for the second fraction, 2.1 µm for the third fraction, 1.0 µm for the fourth fraction, and 0.45 µm for the fifth fraction.

The loaded filters were digested with nitric acid under pressure at 160 °C. The quantitative determination of the concentrations of 23 trace elements (As, Ba, Ca, Cd, Cr, Cu, Fe, K, Mn, Mo, Ni, Pb, Rb, S, Sb, Se, Sn, Sr, Ti, V, Y, Zn, and Zr) was performed by total-reflection X-ray fluorescence analysis (TXRF) after addition of an internal Co standard [STÖSSEL, 1987; MICHAELIS, 1988]. Further details of the sampling method and the trace analysis were given in [MICHAELIS and PRANGE, 1988].

7.2.2.2.2 Multivariate Autocorrelation Analysis

When applying multivariate autocorrelation analysis to this multivariate problem (for mathematical fundamentals see Section 6.6.3) two questions should be answered:

- How does the relationship between the deposition events for the different fractions depend on time?
- Which is the required optimum sampling frequency of the different fractions for the representative assessment of the impact of the emissions?

The multivariate autocorrelation function (MACF) was computed for each particle-size fraction of the data set, consisting in the concentrations of 23 elements in aerosol samples taken weekly over a period of 60 weeks.

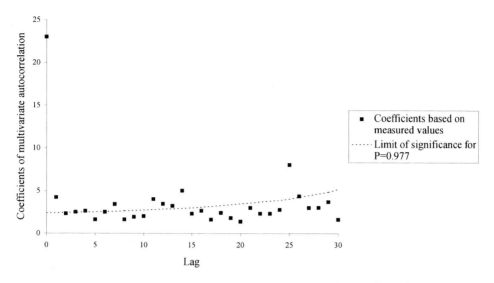

Fig. 7-18. Multivariate autocorrelation function for the first fraction (d_{medium} = 8.5 µm)

In the first fraction (d_{medium} = 8.5 µm) significant autocorrelation can be found after a lag of one week only (Fig. 7-18). Because of the rapid deposition of these large particles there is no longer-term correlation with impact of the emissions.

In comparison with Fig. 7-18 the MACF for the fifth fraction (d_{medium} = 0.45 µm) (Fig. 7-19) is less scattered and the duration of correlation is definitely longer – significant multivariate correlation with the impact of the emissions can be found, for up to 12 weeks.

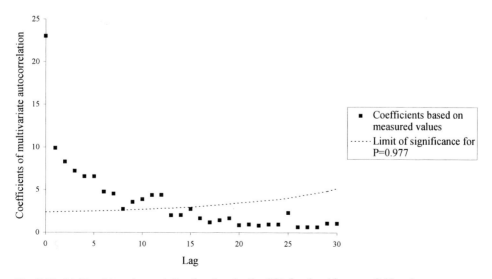

Fig. 7-19. Multivariate autocorrelation function for the fifth fraction (d_{medium} = 0.45 µm)

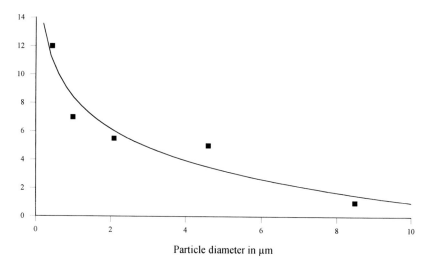

Fig. 7-20. Duration of multivariate autocorrelation as a function of particle size

This tendency of increasing duration of correlation with decreasing particle size is illustrated in Fig. 7-20. This means, that the impact of smaller particles is more uniform than that of larger particles, which are deposited relatively quickly. Thus it is possible to characterize the long-term impact level of smaller airborne dust particulates.

7.2.2.2.3 Factor Analysis

FA was next applied to identify and to characterize sources of the airborne particulates in the different particle sizes. The use of FA for solving this task is described in the literature, e.g. for total suspended dust and for total sedimented airborne particulates [KEIDING et al., 1987; EINAX and DANZER, 1989]. For the present problem the question is: is it possible to obtain plausible and interpretable solutions by applying FA to each particle size fraction if they are very similar in their composition?

By application of FA two common factors were extracted from the data set of each particle size fraction. These two factors describe the main part of total variance of the data set in the range 65 to 86%. The factors and their high loading elements are illustrated in Tab. 7-8 and 7-9. They can be interpreted as sea spray and anthropogenic factor.

The first factor of the fraction with the largest particles comprises Ca, Rb, Ti, Mn, Fe, K, Y, Ba, Se, and Zr. Qualitative comparison with the average composition of sea water (Tab. 7-10) shows that the origin of this first factor is probably sea spray. The second factor of the first particle size fraction summarizes the elements Pb, Zn, As, S, V, Sn, Sb, and Cr. These elements indicate anthropogenic influence. Although sulfur exists at a higher concentration in sea water (Tab. 7-10) the high factor loading for sulfur in the anthropogenic factor is obviously caused by emissions of sulfur diox-

Tab. 7-8. Factorial pattern for the sea spray factor

Fraction	Elements, arranged in decreasing order of their factor loadings	Eigenvalue	Part of total variance in %
1	Ca, Rb, Ti, Mn, Fe, K, Y, Ba, Sr, Zr	8.06	35.0
2	Rb, Y, Ti, Ca, Zr, Ba, Sr, Fe, Mn, K	9.91	43.1
3	Ca, Sr, Ti, Zr, Ba, Rb, Fe, Y, K, Mn	11.44	49.7
4	Ca, Ti, Zr, Sr, Fe, Ba, Rb, Mn, Cr, K, As, Zn	9.71	42.2
5	Ca, Sr, Ti, Zr	4.57	19.9

Tab. 7-9. Factorial pattern for the anthropogenic factor

Fraction	Elements, arranged in decreasing order of their factor loadings	Eigenvalue	Part of total variance in %
1	Pb, Zn, As, S, V, Sn, Sb, Cr	6.79	29.5
2	Pb, Sb, Mo, Zn, Cr, Sn, Se, S, V, As	8.58	37.3
3	Sb, Mo, Sn, Pb, S, V, Zn, Cu, Cr, As, Cd, Se	8.40	36.5
4	Se, Mo, S, V, Sb, Pb, Sn, Cd, Cu, Zn, As, Cr	9.74	42.3
5	Pb, Cu, Sb, Zn, Se, As, Mn, Mo, Cd, Rb, Sn, Cr, Fe, Y, V, S	13.10	57.0

Tab. 7-10. Medium composition of sea water in mg L^{-1} [DEMAYO, 1992]
(analyzed elements in bold letters)

O $8.57 \cdot 10^5$	F $1.3 \cdot 10^0$	**Ni $5.4 \cdot 10^{-3}$**	Y $3.0 \cdot 10^{-4}$
H $1.08 \cdot 10^5$	Ar $6.0 \cdot 10^{-1}$	**As $3.0 \cdot 10^{-3}$**	Co $2.7 \cdot 10^{-4}$
Cl $1.90 \cdot 10^4$	N $5.0 \cdot 10^{-1}$	**Cu $3.0 \cdot 10^{-3}$**	Ne $1.4 \cdot 10^{-4}$
Na $1.05 \cdot 10^4$	Li $1.8 \cdot 10^{-1}$	**Sn $3.0 \cdot 10^{-3}$**	Cd $1.1 \cdot 10^{-4}$
Mg $1.35 \cdot 10^3$	**Rb $1.2 \cdot 10^{-1}$**	U $3.0 \cdot 10^{-3}$	W $1.0 \cdot 10^{-4}$
S $8.85 \cdot 10^2$	P $7.0 \cdot 10^{-2}$	Kr $2.5 \cdot 10^{-3}$	**Se $9.0 \cdot 10^{-5}$**
Ca $4.00 \cdot 10^2$	I $6.0 \cdot 10^{-2}$	**Mn $2.0 \cdot 10^{-3}$**	Ge $7.0 \cdot 10^{-5}$
K $3.80 \cdot 10^2$	Ba $3.0 \cdot 10^{-2}$	V $2.0 \cdot 10^{-3}$	Xe $5.2 \cdot 10^{-5}$
Br $6.50 \cdot 10^1$	In $2.0 \cdot 10^{-2}$	Ti $1.0 \cdot 10^{-3}$	Cr $5.0 \cdot 10^{-5}$
C $2.80 \cdot 10^1$	Al $1.0 \cdot 10^{-2}$	Cs $5.0 \cdot 10^{-4}$	Th $5.0 \cdot 10^{-5}$
Sr $8.10 \cdot 10^0$	**Fe $1.0 \cdot 10^{-2}$**	Ce $4.0 \cdot 10^{-4}$	Ga $3.0 \cdot 10^{-5}$
B $4.60 \cdot 10^0$	**Mo $1.0 \cdot 10^{-2}$**	Sb $3.3 \cdot 10^{-4}$	Hg $3.0 \cdot 10^{-5}$
Si $3.00 \cdot 10^0$	**Zn $1.0 \cdot 10^{-2}$**	Ag $3.0 \cdot 10^{-4}$	**Pb $3.0 \cdot 10^{-5}$**

ide into the atmosphere. The element Ni has no significant loading in either of the common factors, which means that the origin of the variance is feature-specific. The reason is probably contamination of the dust with nickel during the sampling process.

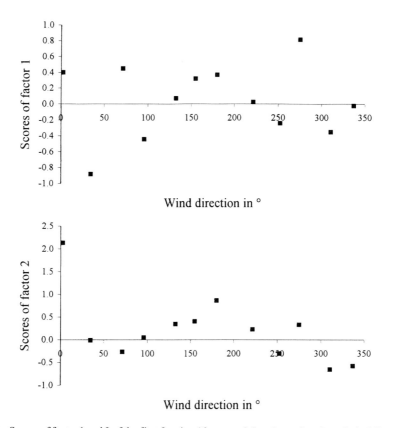

Fig. 7-21. Scores of factor 1 and 2 of the first fraction (d_{medium} = 8.5 µm) as a function of wind direction

To investigate the influence of wind direction, the factor scores for each fraction were averaged within a sector of 30°. The graphical representation of the scores of both factors (computed from the data set of the first fraction) versus the angle of wind direction is very noisy (Fig. 7-21) and does not enable any conclusions to be drawn on the location of these emission sources. This result is in good agreement with the result from multivariate autocorrelation analysis of the first fraction.

The results from FA for the other, smaller particle size fractions are qualitatively comparable; the extracted factors have a similar pattern (Tab. 7-8 and 7-9). The dependence of the eigenvalues for the extracted factors on the particle diameter (Fig. 7-22) illustrates that the part of common variance described by the sea spray factor decreases with decreasing particle size and that the part of common variance described by the anthropogenic factor increases continuously with decreasing particle size.

The number of elements in the anthropogenic factor also grows with decreasing particle size. The main elements are Pb, Sb, and Zn; frequent elements are Mo, As, Sn, Cu, V, and S. The distance of Pellworm from any industrial emission sources does not enable further differentiation of the impact of the emissions without a key pattern derived

7.2 Evaluation and Interpretation of Atmospheric Pollution Data   281

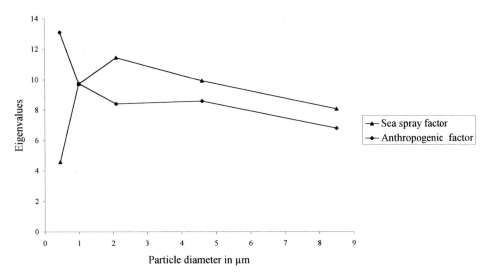

Fig. 7-22. Eigenvalues of the extracted factors as a function of particle size

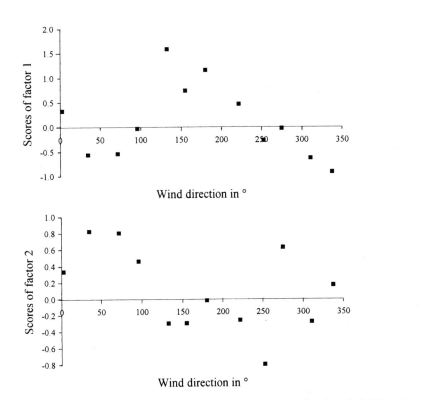

Fig. 7-23. Scores of factors 1 and 2 of the fifth fraction (d_{medium} = 0.45 μm) as a function of wind direction

282   7 Atmosphere

from the emission profiles of different industries. In summary, the anthropogenic influence predominates in the impact of emissions of small airborne particulate matter.

In accordance with this fact and also with the result from multivariate autocorrelation analysis, the factor scores for smaller particles depend on wind direction. This dependence is illustrated by the example of the fifth fraction of particles in Fig. 7-23. The factor scores of the first, anthropogenic, factor have a broad maximum in the range of 130–180°. Comparison with the frequency distribution of wind direction in the time interval under investigation (Fig. 7-24) shows that the direction in which the scores of this anthropogenic factor have a maximum (Fig. 7-23) does not correspond with the most frequent wind direction (240–330°). This maximum of factor scores in the range of 130–180° indicates the influence of industrial and communal emissions in the conurbations of Bremen and Hamburg.

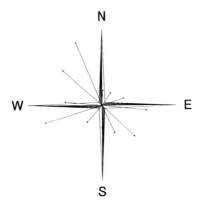

Fig. 7-24. Relative frequencies of the wind direction at the sampling point during the period of the investigation

The scores of the sea spray factor (factor 2 in Fig. 7-23) have a maximum in a wind direction in the range of 30–70°. A plausible explanation is that the wind blowing from 30–70° is directed against the main direction of the sea waves and captures more sea spray than winds from other directions.

References

Ahrens, H., Läuter, J.: Mehrdimensionale Varianzanalyse, 2. Aufl., Akademie-Verlag, Berlin, **1981**

Arbeitsmappe der Staatlichen Hygieneinspektion – Teil Lufthygiene: Immissionsmessung: Bestimmung des Staubniederschlages (Sedimentationsstaub) in der atmosphärischen Luft, 5. Lf., Kap.-Nr. 9, Lfd. Nr. 2, Staatsverlag der DDR, Berlin, **1984**

Armanino, C., Forina, M., Bonfanti, L., Maspero, M.: Anal. Chim. Acta 284 (**1993**) 79

Birring, K.S., Haswell, S.J.: Anal. Proc. 31 (**1994**) 201

Bohn, U. in: Kommission für Umweltschutz beim Präsidium der Kammer der Technik (Ed.): Meßtechnik und meteorologische Aspekte zur Luftüberwachung, Reihe "Technik und Umweltschutz", Bd. 21, Deutscher Verlag für Grundstoffindustrie, Leipzig, **1979**, pp. 130

Bontron, C.F., Patterson, C.C.: Geochim. Cosmochim. Acta 47 (**1983**) 1355

Cattell, R.B.: Multivariate Behavioral Research 1 (**1966**) 245

Coomans, D., Massart, D.L.: Anal. Chim. Acta 133 (**1981**) 225

Demayo, A. in: Lide, D.R. (Ed.): CRC Handbook of Chemistry and Physics, 73rd Ed., CRC Press, Boca Raton, Ann Arbor, London, Tokyo, **1992**, sec. 14, p. 10

Derde, M.P., Massart, D.L.: Fresenius' Z. Anal. Chem. 313 (**1982**) 484

Doerffel, K.: Wiss. Z. TH Leuna-Merseburg 32 (**1990**) 66

Einax, J.: Z. Umweltchem. Ökotox. 5 (**1993**) 45

Einax, J., Danzer, K.: Staub-Reinhalt. Luft 49 (**1989**) 53

Einax, J., Ludwig, W.: Staub-Reinhalt. Luft 51 (**1991**) 309

Einax, J., Danzer, K., Matherny, M.: Int. J. Environ. Anal. Chem. 44 (**1991**) 185

Einax, J., Geiß, S., Michaelis, W.: Fresenius' J. Anal. Chem. 348 (**1994**) 1748

Ewers, U., Schlipköter, H.-W. in: Merian, E. (Ed.): Metalle in der Umwelt, Verlag Chemie, Weinheim, Deerfield Beach/Florida, Basel, **1984**, pp. 351

Geiß, S., Einax, J., Danzer, K.: Anal. Chim. Acta 242 (**1991**) 5

Harvey, C.E.: A Method for Semiquantitative Spectrographic Analysis, A.R.L. Glendale, California, **1947**

Henrion, A., Henrion, R., Urban, P., Henrion, G.: Z. Chem. 27 (**1987**) 56

Hopke, P.K., Gladney, E.S., Gordon, G.E., Zoller, W.H., Jones, A.G.: Atmos. Environ. 10 (**1976**) 1015

Jahn, W., Vahle, H.: Die Faktorenanalyse und ihre Anwendung, Verlag Die Wirtschaft, Berlin, **1970**

Keiding, K., Jensen, F.P., Heidam, N.Z.: Anal. Chim. Acta 181 (**1986**) 79

Keiding, K., Sørensen, M.S., Pind, N.: Anal. Chim. Acta 193 (**1987**) 295

Klamm, K.: Staub-Reinhalt. Luft 46 (**1986**) 116

Kommission für Umweltschutz beim Präsidium der Kammer der Technik (Ed.): Luftreinhaltung in der Industrie, Reihe "Technik und Umweltschutz", Bd. 15, Deutscher Verlag für Grundstoffindustrie, Leipzig, **1976**

Kommission für Umweltschutz beim Präsidium der Kammer der Technik (Ed.): Umwandlung und Ausbreitung von Luftschadstoffen, Reihe "Technik und Umweltschutz", Bd. 30, Deutscher Verlag für Grundstoffindustrie, Leipzig, **1985**

Lebart, L., Morineau, A., Fenelon, J.-P.: Statistische Datenanalyse, Methoden und Programme, Akademie-Verlag, Berlin, **1984**

Malinowski, E.R.: Factor Analysis in Chemistry, 2nd Ed., Wiley, New York, Chichester, Brisbane, Toronto, Singapore, **1991**

Michaelis, W. in: van Dop, H. (Ed.): Air Pollution Modelling and its Application, VI. Plenum, New York, **1988**, pp. 61

Michaelis, W., Prange, A.: Nucl. Geophys. 2 (**1988**) 231

Stößel, R.-P.: Untersuchungen zur Naß- und Trockendeposition von Schwermetallen auf der Insel Pellworm, Dissertation, Universität Hamburg, **1987**

Treiger, B., Van Malderen, H., Bondarenko, I., Van Espen, P., Van Grieken, R.: Anal. Chim. Acta 284 (**1993**) 119

Treiger, B., Bondarenko, I., Van Espen, P., Van Grieken, R., Adams, F.: Analyst 119 (**1994**) 971

Van Espen, P., Adams, F.: Anal. Chim. Acta 150 (**1983**) 153

Varmuza, K.: Pattern Recognition in Chemistry, Springer, Berlin, Heidelberg, New York, **1980**, pp. 92

VDI Kommission Reinhaltung der Luft (Ed.): VDI 2119 Bl. 2: Messung partikelförmiger Niederschläge: Bestimmung des partikelförmigen Niederschlags mit dem Bergerhoff-Gerät (Standardverfahren), Juni **1972**

Vong, R.J.: Anal. Chim. Acta 277 (**1993**) 389

Weber, E.: Einführung in die Faktorenanalyse, Gustav Fischer, Jena, **1974**

Weber, E.: Grundriß der biologischen Statistik, 9. Aufl., Gustav Fischer, Jena, **1986**

Wienke, D., Hopke, P.K.: Anal. Chim. Acta 291 (**1994**) 1

Wienke, D., Gao, N., Hopke, P.K.: Environ. Sci. Technol. 28 (**1994**) 1023

8 Hydrosphere

8.1 Sampling in Rivers and Streams

8.1.1 Sampling Strategies in Rivers

The question of the representativeness of sampling and the corresponding sampling plan is of great importance for the characterization of the state of pollution of a river. At present sampling procedures in routine monitoring are fixed by obligatory standards for water analysis (for example [FACHGRUPPE WASSERCHEMIE IN DER GESELLSCHAFT DEUTSCHER CHEMIKER, since 1960; INSTITUT FÜR WASSERWIRTSCHAFT, 1986]). Whereas this standardization enables comparison of the results, conclusions about the representativeness of the results cannot be made.

The scope of this section is to demonstrate the advantages of multivariate statistical methods for the interpretation of multidimensional changes in the state of pollution of a river and to draw conclusions about the strategy to be used for sampling a river [GEISS et al., 1989; GEISS and EINAX, 1992]. Chemometric methods, like FA and MVDA, will be applied to find answers to the following questions:

– Which of the analyzed components of the water give a true indication of essential changes in the state of pollution of a river at a given sampling point?
– Are the sampling times or the sampling points perpendicular to the direction of flow to likely to lead to different results for the levels of the analyzed components?

8.1.1.1 Experimental

For identifying inhomogeneities perpendicular to the direction of flow and temporal load changes at one sampling location, samples were taken at five equidistant points at the times 5 a.m., 11 a.m., 5 p.m., and 11 p.m.. The sampling procedure was carried out using a plastic vessel, with a cover, at a depth of 10 cm below the river surface to guarantee sampling without contamination. The vessel and the plastic bottles used for storage were conditioned with 0.1 mol L^{-1} nitric acid [SALBU et al., 1985]. After membrane filtration (d_{Pore} = 0.45 µm) the samples were stabilized by addition of nitric acid to pH = 1. The concentration of iron was determined by flame AAS, those of chromium, copper, and nickel by AAS with electrothermal atomization. The method of standard addition was used to minimize systematic errors [WELZ, 1986]. The concentrations of suspended material were determined gravimetrically and those of calcium, magnesium, and carbonate by volumetry. The ammonium concentration was analyzed photometrically.

8.1.1.2 Interpretation of the Results

Multivariate statistical methods should be preferred for evaluating such multidimensional data sets since interactions and resulting correlations between the water compounds have to be considered. Fig. 8-1, which shows the univariate fluctuations in the concentrations of the analyzed compounds, illustrates the large temporal and local variability. Therefore in univariate terms objective assessment of the state of pollutant loading is hardly possible.

8.1.1.2.1 Factor Analysis

FA was applied to a matrix of 20 samples and 11 analyzed water components to find the main influences on changes in the pollutant loading of the river. Two factors with the highest eigenvalues (Tab. 8-1) describe 55.0% of the determinable variance. The largest part of the daily variance is caused by factor 1. The graphical representation of the factor scores (Fig. 8-2) demonstrates distinction between the samples at different sampling times by their chromium, iron, and suspended material content. The samples at 11 a.m. and 5 p.m. are closely similar. Neither factor 1 nor factor 2 can separate the different sampling points. This means that local inhomogeneities perpendicular to the direction of flow are not recognizable by the application of FA.

Tab. 8-1. Factor loading matrix

Factor	Part of determinable variance in %	Components with factor loadings > 0.7	Communality h_i^2
1	36.2	Cr	0.93
		Fe	0.90
		Suspended material	0.97
2	18.8	Ni	0.60
		SO_4^{2-}	0.44

8.1.1.2.2 Multivariate Analysis of Variance and Discriminant Analysis

The scope of the application of MVDA is to prove multivariate differences between samples taken at different sampling times and points. Therefore the samples at one

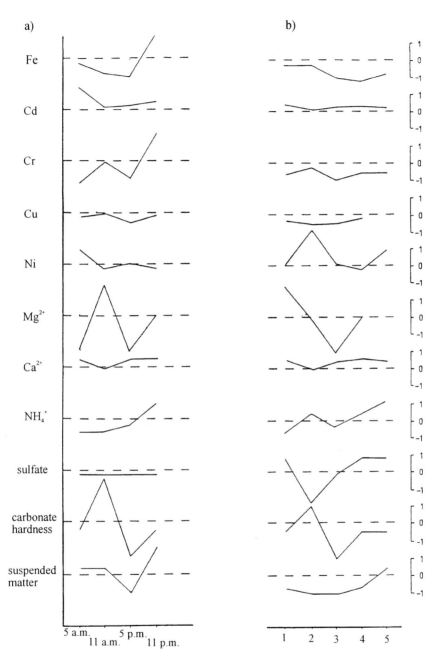

Fig. 8-1. Relative changes in the concentrations of the analyzed water components.
(a) Different sampling times at sampling point 3, (b) different sampling points at 5 p.m.

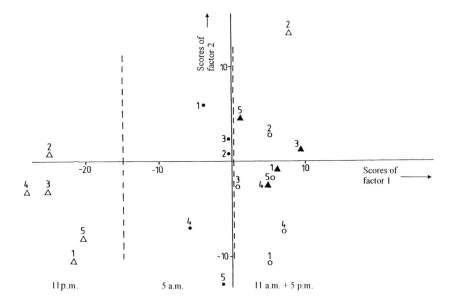

Fig. 8-2. Representation of the scores of factor 2 vs. the scores of factor 1 for the sampling points at different times

sampling time and sampling point have to be considered as one class, and the differences between their mean values are tested. Four classes result from the analysis of the sampling times which are always represented by five objects. The results of MANOVA are demonstrated in Tab. 8-2.

The MANOVA enables significant class separation with a multivariate scaled separation measure of 330.9. The sampling times 5 a.m. and 11 p.m. are well separable from the times 11 a.m. and 5 p.m. by the optimum separation set which consists in the features suspended material, iron, magnesium, nickel, and copper. The result of discriminant analysis is shown in the plane of the two strongest discriminant functions (Fig. 8-3).

The analysis of the five sampling points, each of which is occupied with four objects, demonstrates the very small differences in pollutant loading perpendicular to the direction of flow of the river (Fig. 8-4). Consideration of the optimum separation set, sulfate, ammonium, and chromium, results in a multivariate scaled separation measure of only 15.9. A significant difference in the simultaneous comparison exists for sampling points 2 and 5 only.

These multivariate statistical methods consider relative pollution changes or relationships of variances, because the basis of the computations is the matrix of the correlation coefficients. The absolute values of the concentration changes are not considered. Therefore conclusions regarding the actual state of pollution can only be drawn with respect to the actual data.

Tab. 8-2. Analysis of variance of the sampling times

Comparison of classes	Risk α of error of the first kind for	
	Single comparison	Simultaneous comparison
5 a.m.–11 a.m.	0.00*	0.00*
5 a.m.– 5 p.m.	0.00*	0.00*
11 a.m.– 5 p.m.	0.20	0.92
5 a.m.–11 p.m.	0.00*	0.00*
11 a.m.–11 p.m.	0.00*	0.00*
5 p.m.–11 p.m.	0.00*	0.00*

* indicates significance

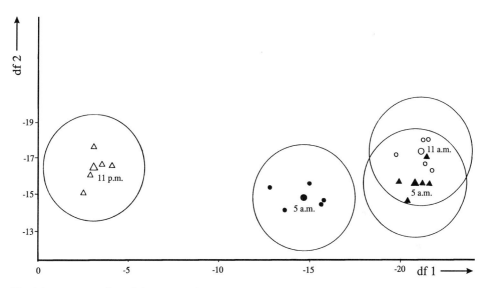

Fig. 8-3. Representation of the scores of the first two discriminant functions for classification of the sampling times

8.1.1.3 Conclusions

Both FA and MVDA are powerful instruments for differentiating complex environmental data sets on the basis of their variances. Samples at different sampling times or points can be separated at a defined statistical level. To assess the relationships between temporal and/or local pollution changes under consideration of ecotoxicologial aspects the absolute values of the changes must also be determined. To elucidate the size of changes in pollutant loading the scaled class mean and the ranges of class means for the optimum separation set features are represented in relation to the total mean for

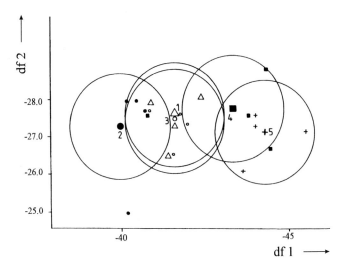

Fig. 8-4. Representation of the scores of the first two discriminant functions for classification of the sampling points

Tab. 8-3. Class means of the features of the optimum separation set for differentiation of the sampling times

Analyte	Classes				Range in %
	5 a.m.	11 a.m.	5 p.m.	11 p.m.	from total mean
Suspended material	1	0.72	0.63	2.16	136.0
Iron	1	0.72	0.76	1.26	56.9
Magnesium	1	1.02	1.00	1.01	1.9
Nickel	1	0.86	1.21	0.79	43.0
Copper	1	1.37	0.54	0.65	93.3

Tab. 8-4. Class means of the features of the optimum separation set for differentiation of the sampling points

Analyte	Classes					Range in %
	1	2	3	4	5	from total mean
Sulfate	1	0.97	0.99	1.02	1.02	4.5
Ammonium	1	0.92	1.02	1.15	1.32	43.3
Chromium	1	1.16	0.95	0.97	0.97	20.7

class differentiation (Tab. 8-3 and 8-4). The ranges show relatively high variations within a day, but only small variations perpendicular to the direction of flow.

Regarding the sampling strategy in routine monitoring of rivers and streams the following conclusions are possible:

- By application of multivariate statistical methods for consideration of the overall environmental situation of a river it is possible to optimize the sampling strategy.
- In the case investigated the necessity for time-dependent sampling was demonstrated because the daily loading changes are of the same order of magnitude as the variation along the river.
- The changes of concentration perpendicular to the direction of flow are so small that they do not have to be considered in routine monitoring, i.e. sampling is possible from the river bank, which means that expenditure is less.
- For exact determination of the concentration of selected water components analysis should be performed on a composite sample from several sampling points perpendicular to the direction of flow.
- When planning a sampling program for a whole river it is necessary to optimize the sampling strategy for each sampling point taking into account the overall environmental condition of the river.

8.1.2 Representative Sampling Distances along a River

The representative assessment of the pollution situation for a river section is possible if the samples taken at two neighbouring sampling points are correlated. This can be proved by calculation of the traditional univariate correlation coefficient, if the representativeness of only one parameter is of interest. To assess multivariate loads representatively it is necessary to compute the multivariate correlation coefficient (see Section 6.6.3) and to prove its significance [GEISS and EINAX, 1992]. The multivariate correlation coefficients for the investigated heavy metals cadmium, chromium, copper, iron, nickel, and zinc are represented in comparison to the highest value of a random correlation coefficient with a given probability of an error of first kind $\alpha = 0.05$ in Fig. 8-5. It is obvious that the distance between the sampling points is small enough for representative characterization of heavy metal pollution in the river section under investigation.

Computation of the multivariate correlation coefficients of some "classical" water parameters, like nitrate, ammonium, conductivity, total hardness, dissolved oxygen, biological oxygen demand, and chemical oxygen demand, gives a quite different result (Fig. 8-6). The distances between the sampling points from river-kilometer 362 to 263 are too great for representative characterization of the pollution. This is because some large storage reservoirs are located in this river section. Because of the long duration time in these storage reservoirs chemical and biochemical processes in the water body change the concentrations of the investigated components and the collective parameters considerably. Thus representative sampling in the sense of significant correlations between the sampling points in the section between the barrages is not possible.

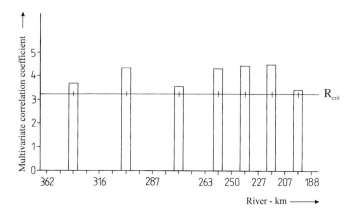

Fig. 8-5. Multivariate correlation coefficients between the sampling points on the river Saale for the heavy metals cadmium, chromium, copper, iron, nickel, and zinc

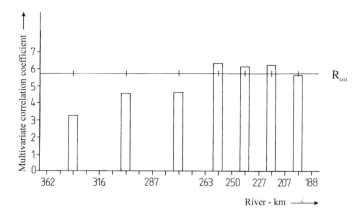

Fig. 8-6. Multivariate correlation coefficients between the sampling points on the river Saale for the water components nitrate and ammonium and the collective parameters conductivity, total hardness, dissolved oxygen, biological oxygen demand, and chemical oxygen demand

8.2 Analytical and Chemometric Investigations of Polluted River Waters and Sediments

8.2.1 Problem and Experimental

Changes in pollutant loading along a river originate geogenically, anthropogenically, and biochemically. They result from changes in the geological background, communal and industrial discharges, and transformations of water components by microorganisms.

8.2 Analytical and Chemometric Investigations of Polluted River Waters and Sediments   293

Sediment analyses are useful for characterization of pollution over a long period [MÜLLER, 1981]. Assessment of the state of a river and of the interactions between the components can be made by application of multivariate statistical methods only, because the strongly scattering territorial and temporal courses [FÖRSTNER and MÜLLER, 1974; FÖRSTNER and WITTMANN, 1983] are not compatible with many univariate techniques. FA shall serve as a tool for the recognition of variable structures and for the differentiated evaluation of the pollution of both river water and sediment [GEISS and EINAX, 1991; 1992].

In 140 water samples from the river Saale, sampled from 1986 to 1988 according to the technique described in Section 8.1.1.1, the heavy metals iron and zinc were determined using flame AAS and lead, cadmium, chromium, cobalt, copper, and nickel by AAS with electrothermal atomization in the soluble fraction (particle diameter <0.45 μm). The sampling points, located in Thuringia (Germany), are illustrated in Fig. 8-7. The method of standard addition, with three additions, was used to minimize matrix effects. The components ammonium, chloride, magnesium, nitrate, nitrite, phosphate, oxygen,

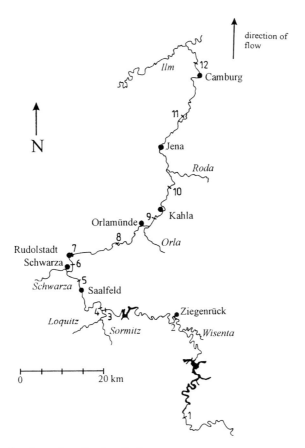

Fig. 8-7. Sampling locations along the river Saale

oxygen saturation, suspended material and the total parameters biological oxygen demand, chemical oxygen demand, and conductivity, pH and temperature were analyzed according to [INSTITUT FÜR WASSERWIRTSCHAFT, 1986].

The sediments were sampled over an area of 1 m² at the same locations as the water samples. After drying and sieving the fine-grained (particle diameter <63 µm) and coarse (particle diameter >63 µm) particles were digested separately in 1 mol L^{-1} nitric acid and 30% hydrogen peroxide.

8.2.2 Results from River Water Investigations, and their Interpretation

The mean values of the heavy metal concentrations in the river Saale are demonstrated in Tab. 8-5.

The interpretation of the mean values is severely limited because they do not reflect short time changes and extreme values.

Tab. 8-5. Mean heavy metal concentrations in the river Saale (1986–1988) in µg L^{-1}

River-km	Sampling point	Cd	Cr	Cu	Fe	Ni	Zn
362	Harra	0.77	9.7	6.5	494	8.9	82.7
316	Ziegenrück	1.28	6.5	4.2	393	10.7	85.0
287	Eichicht	1.32	6.3	4.15	211	11.9	84.0
263	Rudolstadt	0.91	5.9	7.5	413	11.0	250
250	Uhlstädt	0.97	7.9	6.4	333	9.4	188
227	Rothenstein	1.20	5.8	6.5	438	10.9	193
207	Porstendorf	0.80	7.2	6.8	856	14.1	194
188	Camburg	0.82	8.4	7.3	557	12.7	194

8.2.2.1 Scaling in Factor Analysis for River Assessment

Optimum scaling of the original data is a task of high importance when FA is to be applied. Normally autoscaling is accomplished during the computation of the matrix of correlation coefficients [WEBER, 1986]. This means that all autoscaled features have a mean value of 0 and a standard deviation of 1 for all samples. When FA is applied to concentrations of water components such scaling yields false results because the concentrations depend on the volume of water and consequently on the volumetric flow of the river. Different flow situations are not comparable. One solution is the scaling of the values in relation to the volumetric flow at different sampling points, so that the calculations are based on the loads. The autoscaling of the feature concentration for each day of sampling, i.e. for one specific flow situation, seems to be the better way. A comparison of both methods is demonstrated in Tab. 8-6.

Tab. 8-6. Comparison of scaling possibilities

Scaling of concentrations	Scaling of loads

• consideration of relative concentrations
• decreasing of relative concentrations
  – fast by dilution
  – slow by deposition to the sediment
• concentrations, originating from industry
  – separation of dischargers
⇒ enlightenment on the nature of interactions between the water components
⇒ expression of ecotoxic concentrations
	• consideration of mass flows (loads)
	• decreasing of loads
	  – fast owing to withdrawal of large volumes of water
	  – slow by deposition to the sediment
	• loads, originating from industry
	  – largely equal for all samples from one sampling point
	⇒ unification in one factor
	⇒ expression of total load

8.2.2.2 Application of Factor Analysis to Samples Taken from the River Saale in the Summer

To obtain detailed information of the total pollution of the Saale it is necessary to separate the vegetation periods in the process of data pretreatment. Essential factors in the river, e.g. the extent of plant growth, the activity of microorganisms, and the oxygen concentration, strongly depend on the season. A detailed discussion is given in the literature [GEISS, 1990; GEISS and EINAX, 1991].

The application of FA seems to be a useful tool for the interpretation of the environmental analytical results obtained. Thus, GRIMAULT and OLIVE [1993] found that the application of FA yields to more interpretable results than the use of PCA.

FA was performed for the above described water components with autoscaling of the features for each day of sampling during the summer period. The obtained factor loadings and communalities are demonstrated in Tab. 8-7.

The representation of the factor scores as a function of river-kilometers (Fig. 8-8) illustrates the truth of the following interpretation of the factors. Factor 1 contains the hardness components and may be explained mainly by geogenic influences. Factor 2 contains the oxygen content and the negatively correlated oxygen demand. The low oxygen concentration at Harra and Eichicht results from the biological oxidation of organic pollutants at Harra and from the barrage drain in Eichicht, where the water is not yet saturated with oxygen. The third factor illustrates pollution with copper, zinc, sulfate, and temperature arising from slate quarries along the small Sormitz tributary and discharges from the galvanic industry in and directly downstream of the town of Rudolstadt. The increase in temperature and in the sulfate concentration results from discharges from a chemical plant. Factor 4 describes a part of the activity of microorganisms, which is observable only in the summer. Low oxygen concentrations result from high values of biological oxygen demand; this enables nitrification (negative correlations between nitrate and nitrite concentrations). The fifth factor corresponds to the

Tab. 8-7. Matrix of factor loadings for the Saale in the summer
(factor loadings <0.50 are set to 0 for a greater clarity)

Component	Factor 1	2	3	4	5	Communality h_i^2
Cd	0	0	0	0	0.550	0.320
Cr	0	0	0.500	0	0	0.421
Cu	0	0	0.613	0	0	0.438
Fe	0	0	0	0	0	0.452
Ni	0	0	0	0	0	0.263
Zn	0	0	0.615	0	0	0.745
Temperature	0	0	0.719	0	0	0.587
pH	0.754	0	0	0	0	0.726
Suspended material	0	0	0	0	0.607	0.496
Chloride	0	−0.530	0	0	0.519	0.693
Sulfate	0.588	0	0.602	0	0	0.743
Nitrate	0	0	0	−0.766	0	0.823
Nitrite	0.514	0	0	0.516	0	0.719
Phosphate	0.762	0	0	0	0	0.642
Ammonium	0	0	0	0	0	0.340
Ca	0.588	0	0	0	0	0.481
Mg	0.821	0	0	0	0	0.771
Conductivity	0.873	0	0	0	0	0.906
Total hardness	0.922	0	0	0	0	0.872
Carbonate hardness	0.853	0	0	0	0	0.841
O_2	0	0.908	0	0	0	0.872
O_2-saturation	0	0.909	0	0	0	0.852
Biological oxygen demand	0	0	0	0.730	0	0.819
Chemical oxygen demand	0	−0.666	0	0.580	0	0.842
Eigenvalues	5.776	3.153	2.630	2.198	1.908	

communal discharge. The town Rudolstadt stands out because of the absence of sewage treatment. Further downstream the situation improves; the slight discharge from the Jena sewage treatment plant is remarkable.

The application of FA to another, also strongly polluted river, the Weiße Elster, demonstrates that generalization of the above obtained results is possible [GEISS and EINAX, 1991]. FA is, therefore, a powerful tool for:

– characterization of anthropogenic and geogenic loads
– identification of substantial dischargers
– detection of interactions between water components

Furthermore, it should be hinted that FA can also be used to select analytical variables for optimization of laboratory efforts in future groundwater or river water studies [ANDRADE et al., 1994].

8.2 Analytical and Chemometric Investigations of Polluted River Waters and Sediments   297

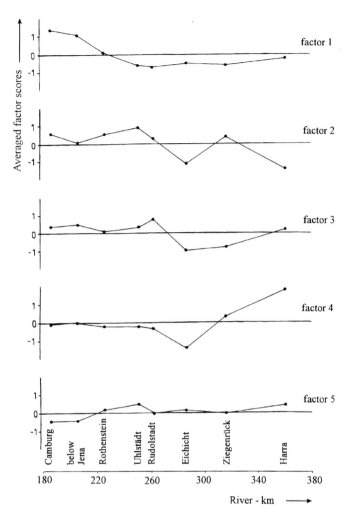

Fig. 8-8. Representation of the scores of the factors for the river Saale as a function of the distance of flow

8.2.3 Chemometric Description of River Sediments

From the FA of 12 sediment samples and 11 determined features, measured in three replicates, three factors result which describe 83.7% of total variance (Tab. 8-8).

The scores of the first factor increase remarkably after discharge from the Sormitz tributary because of the entry of Co- and Ni-containing fine-grained slate particles. In comparison, no significant Co- or Ni-enrichment is detectable in the flowing river water (see Fig. 8-8). The slate particles are only very poorly soluble and a strong dilution effect takes place after the discharge of the Sormitz into the Saale. These results

Tab. 8-8. Matrix of factor loadings of Saale sediments (factor loadings <0.50 are set to 0 for a greater clarity)

Element	Factor 1	2	3	Communality h_i^2
Ca	−0.535	0	0.514	0.807
Cd	0	0.974	0	0.975
Co	0.978	0	0	0.984
Cr	0	0.982	0	0.977
Cu	0	0	0.728	0.880
Fe	0.725	0	0	0.984
Mg	0	0	0.855	0.824
Mn	0	0	0	0.954
Ni	0.886	0	0	0.982
Pb	0	0	0.725	0.956
Zn	0	0.522	0.663	0.842
Eigenvalues	4.481	2.768	1.896	

demonstrate the hazard of sediment analyses for the characterization of the pollution of river waters.

Factor 2 results from the discharge of chromium by the leather industry and corresponds to the element pattern in the water. Factor 3, of geogenic origin, is caused by inflows from some small tributaries (Sormitz, Loquitz, Schwarza).

Some case studies on the investigation of anthropogenic loading of river sediments by organic compounds are given in the literature. For example, WENNING et al. [1993] describe the chemometric analysis, by means of PCA, of potential sources of polychlorinated dibenzo-p-dioxins and dibenzofurans in surface sediments.

8.3 Speciation of Heavy Metals in River Water Investigated by Chemometric Methods

Nowadays modern instrumental analytical methods enable largely precise and accurate determination of total concentrations of heavy metals in environmental compartments. The assessment of toxicologically relevant levels in river water on the basis of total concentrations is connected with the following:

– The humantoxic and ecotoxic effects of a metal depend strongly on its binding form [STUMM and KELLER, 1984]. Speciation of the heavy metal binding forms in the environmental compartments hydrosphere, atmosphere, and pedosphere is of considerable importance [LUND, 1990]. In most cases the individual species have to be separated before analysis [FUCHS and RAUE, 1981; FRIMMEL and SATTLER, 1982].

- The concentrations of heavy metals both in river water and in river sediment are strongly changed by deposition-remobilization processes. The deterministic modeling of the transition between both environmental compartments is severely limited by the complex chemical, physical, and biochemical processes.

The following examples demonstrate the contribution of chemometric methods to the differentiated evaluation of element trace analyses in rivers [EINAX and GEISS, 1994].

8.3.1 Comparing Investigations on Sediment Loadings by Means of Chemical Differentiation and Multivariate Statistical Data Analysis

8.3.1.1 Experimental

The sampling conditions were described in detail in Section 8.2.1. Sediments were sampled from seven locations on the river Weiße Elster in Thuringia (Germany) over a distance of 83 km (Fig. 8-9). The sediment samples were taken on both sides of the river for a better representativeness.

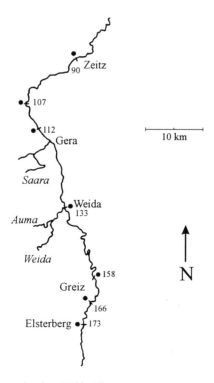

Fig. 8-9. Sampling locations at the river Weiße Elster

After drying and sieving the fine-grained fraction (particle diameter <63 μm) was split. One part was sequentially leached according to TESSIER [TESSIER et al., 1979] to give the following fractions:

- cationic-exchangeable
  (extraction with 1 mol L^{-1} sodium acetate solution at pH = 8.2)
- carbonate-bound
  (extraction with 1 mol L^{-1} acidic sodium acetate solution at pH = 5)
- easily-reducible
  (extraction with 0.04 mol L^{-1} NH_2OH/HCl in 25% acetic acid)
- organically bound
  (extraction with 0.02 mol L^{-1} HNO_3 and 30% H_2O_2)
- residual fraction
  (digestion with a mixture of HNO_3 and HF under pressure)

The total concentrations were determined after digestion with a mixture of HNO_3 and HF [SCHRAMEL et al., 1987]. The concentrations of the extracted heavy metals were determined using atomic absorption spectroscopy.

8.3.1.2 Results of Sequential Leaching

The applied leaching procedure [TESSIER et al., 1979] shows some lack of precision resulting from nonselective extraction, nonexactly definable phases in the environmental river compartment, and possible readsorption after the leaching steps [KHEBOIAN and BAUER, 1987; SAGER et al., 1990; TESSIER and CAMPBELL, 1990]. No leaching sequence yet exists for definite discrimination of different metal species.

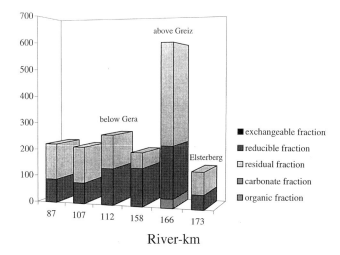

Fig. 8-10. Binding forms of cadmium in the sediment of the river Weiße Elster

8.3 Speciation of Heavy Metals in River Water Investigated   301

In the case under investigation cadmium is extensively bound to the easily-reducible and residual fraction (Fig. 8-10). The easily-reducible binding form results possibly from anthropogenic cadmium discharges. Very high concentrations exist above Greiz where a large chemical plant discharges.

Lead can be found almost exclusively in the easily-reducible and in the residual fraction (Fig. 8-11). Its main binding forms are obviously the adsorption on iron and manganese hydroxides and mineral binding. The large increase in the lead concentration from Elsterberg to Greiz is not characterized by an increase of an individual binding form.

Zinc can be found in all investigated sediments, 25% in the carbonate fraction and 40% in the easily-reducible fraction (Fig. 8-12). Solely the content of the organic and resi-

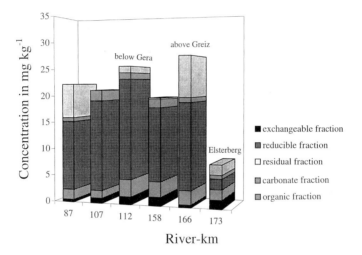

Fig. 8-11. Binding forms of lead in the sediment of the river Weiße Elster

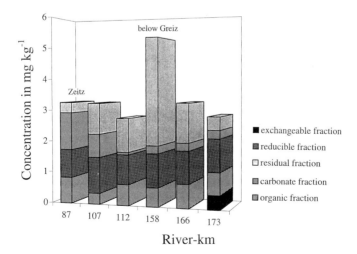

Fig. 8-12. Binding forms of zinc in the sediment of the river Weiße Elster

dual fractions changes. Only in the sediment at Elsterberg zinc could be found in the exchangeable fraction.

These investigations demonstrate that different anthropogenic and geogenic sources of the pollutant loads of river sediments cannot be unambiguously classified by investigation of the binding forms of heavy metals.

8.3.1.3 Factor Analysis of the Total Concentrations of Heavy Metals in River Sediments

In preparing the data matrix of the Elster sediments for FA, the data set from sampling points on the left and the right sides of the river were treated separately. By application of FA to the matrix of 6 features and 14 samples, two factors were extracted which describe 74.3% of the common variance (Tab. 8-9).

Tab. 8-9. Matrix of factor loadings of the sediments from the river Weiße Elster (factor loadings <0.70 are set to 0 for a greater clarity)

Element	Factor 1	Factor 2	Communalities h_i^2
Cd	0.821	0	0.676
Co	0	0.750	0.999
Cr	0	0	0.663
Cu	0	0	0.518
Pb	0.845	0	0.716
Zn	0	0.831	0.885

The first factor, which describes 42.0% of the common variance and which is highly loaded by Pb and Cd, shows highest factor scores at river-km 166 and indicates the influence of waste water discharged by the Greiz/Dölau chemical plant (Fig. 8-13). The scores of the second factor, which describes 32.3% of the common variance of the data set and which is highly loaded by the elements Co and Zn, increase in the lower part of the river (Fig. 8-13). This results from the influence of waste water discharged by the mining industry, particularly in the water catchment basin in the territory of Gera and Zeitz. A similar factor was also found during investigations of water pollution of the river Weiße Elster [GEISS and EINAX, 1991].

8.3.1.4 Comparison of Chemical and Multivariate Statistical Investigations

By applying methods of sequential leaching, direct "chemical information" can be obtained and a rough assessment of the remobilization potential of the sedimented heavy metals is possible. Because of the lack of specificity of the extraction procedures particular species cannot be identified (see also critical hints in [KHEBOIAN and BAUER,

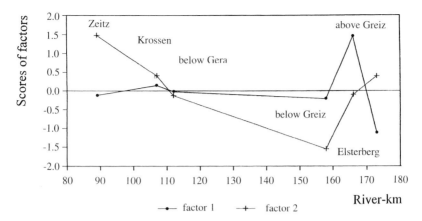

Fig. 8-13. Representation of the scores of the factors for the sediments of the river Weiße Elster as a function of the distance of flow

1987; SAGER et al., 1990]). The merits and limitations of sequential leaching are discussed in detail in the literature [SAGER and VOGEL, 1993]. Otherwise, interpretation of statistical hypotheses found by means of chemometric methods requires definite environmental knowledge. Pollution sources, dischargers, and the degree of pollution [PARDO et al., 1993] can be identified or quantified. Methods of sequential leaching and application of chemometric methods should, therefore, both be applied to obtain more detailed information on binding forms and on pollution sources.

8.3.2 Chemometric Methods in Combination with Electrochemical Analysis Applied to the Investigation of Heavy Metal Binding Forms in Waters

To investigate the influence of pH, and calcium, and fulvic acid (fa) concentration on the binding forms of the heavy metals Cd, Cu, and Zn a combination of statistical planning of experiments, electrochemical determination of the heavy metals, and empirical statistical modeling was applied. The procedure and the results are discussed in detail in the literature [TÜMPLING et al., 1992].

8.3.2.1 Experimental

The electrochemically available concentrations of Cd, Cu, and Zn, as a measure of the free ions, were analyzed for their dependence on pH and on different concentrations of calcium and fulvic acid (fa) by differential pulse anodic stripping voltammetry. A statistical experimental design in the form of a full 2^3-factorial plan was used to minimize the number of experiments required and to investigate the various interactions between the independent variables (pH, Ca, and fa) on the one hand and the electrochemically available part of the heavy metals as the dependent variables on the other hand. The two steps for the independent variables were selected in an environmentally relevant range (Tab. 8-10). Each point in the factorial space was measured in triplicate.

Tab. 8-10. Ion concentrations in simulated water

Ion	Simulated water	Measured concentrations in the river Saale	Concentration range of quality class 2–3 [TGL 22764, 1981]
Sodium*		not analyzed	70–150 mg L^{-1}
Potassium	8 mg L^{-1}	not analyzed	no value
Chloride	49 mg L^{-1}	32–158 mg L^{-1}	100–250 mg L^{-1}
Nitrate*		0– 40 mg L^{-1}	20– 40 mg L^{-1}
Sulfate	39 mg L^{-1}	12–205 mg L^{-1}	150–300 mg L^{-1}
Cadmium	10 µg L^{-1}	0– 6 µg L^{-1}	≤5 µg L^{-1}
Copper	50 µg L^{-1}	1– 63 µg L^{-1}	10–100 µg L^{-1}
Zinc	150 µg L^{-1}	1–158 µg L^{-1}	10–100 µg L^{-1}

* NaOH and HNO$_3$ were used for the adjustment to pH = 8 and pH = 4, respectively.

A full 2^3-factorial plan (Fig. 8-14) [BANDEMER and BELLMANN, 1976] was used for the determination of the influence of pH, calcium, and fulvic acid on the binding of the heavy metals Cd, Cu, and Zn. It includes the minimum number of experiments by

	Plan matrix			Matrix of independent variables							Response matrix		
Variable	1	2	3	0	1	2	3	12	13	23	123	y_1	y_2	y_3
Experiment no.														
1	-	-	-	+	-	-	+	-	+	+	-			
2	-	-	+	+	-	-	+	+	-	-	+			
3	+	-	+	+	+	-	-	+	+	-	-			
4	-	+	-	+	-	+	-	-	+	-	+			
5	-	+	+	+	-	+	-	+	-	+	-			
6	+	-	-	+	+	-	-	-	-	+	+			
7	+	+	-	+	+	+	+	-	-	-	-			
8	+	+	+	+	+	+	+	+	+	+	+			
Effect matrix														

Fig. 8-14. Scheme of the used 2^3-factorial plan

simultaneous variation of the three influencing factors to obtain the required information on the objectives. An effect matrix results from the determination of the relationship between the plan matrix (matrix of the independent influencing quantities) (Tab. 8-11) and the matrix of responses (concentrations of electrochemically available heavy metal ions) (Tab. 8-12). This effect matrix enables the quantitative assessment of the extent to which the independent parameters influence the response. The interactions between the independent parameters, and their effects, are also determinable.

The variable parameters were selected as follows (+/–notation, see Section 3):
- pH values of 4 (–) or 8 (+)
- Ca ion concentrations of 25 (–) or 75 (+) mg L^{-1}
- fa concentrations of 2 (–) or 10 (+) mg L^{-1}

These levels correspond to the concentration ranges of medium polluted river water [TÜMPLING et al., 1992].

Tab. 8-11. Experimental design of the plan matrix for the parameter pH and the concentrations of calcium and fulvic acid

Experiment	pH	Ca concentration in mg L^{-1}	Fulvic acid concentration in mg L^{-1}
1	4	25	2
2	4	25	10
3	8	25	10
4	4	75	2
5	4	75	10
6	8	25	2
7	8	75	2
8	8	75	10

Tab. 8-12. Electrochemically available heavy metal concentrations c (matrix of responses) and standard deviations s in µg L^{-1}

Experiment	c_{Cd}	s_{Cd}	c_{Cu}	s_{Cu}	c_{Zn}	s_{Zn}
1	8.6	0.3	41.8	5.0	146.4	3.4
2	8.5	0.6	15.0	2.8	149.3	4.3
3	6.2	0.5	3.3	0.6	94.7	15.4
4	8.6	0.3	27.8	5.1	148.1	1.3
5	7.9	0.5	12.2	1.4	143.7	3.2
6	7.7	0.2	12.5	1.8	124.2	8.9
7	7.5	0.3	5.3	0.9	121.5	6.8
8	5.0	0.2	1.8	0.2	86.0	10.2

8.3.2.2 Results and Discussion

The matrix of responses as the mean values for the threefold replicate experiments is illustrated in Tab. 8-12.

Firstly, the method of multiple linear regression (MLR) is applied for quantitative description of the investigated system (for the mathematical basis see Section 5.7.1 or [HENRION et al., 1988]):

$$\hat{y} = \sum_{i=0}^{n} (a_i x_i) \tag{8-1}$$

The regression coefficients as an expression of the interactions with the metal ions are determined for the scaled plan matrix according to the following equation (equivalent to Eq. 3-10):

$$a = \frac{\sum xy}{\sum x^2} \tag{8-2}$$

a – regression coefficient
x – scaled independent variable
y – response

The calculated regression coefficients are demonstrated in Tab. 8-13. The regression coefficients are significant if:

$$|a| > t(P, f) \cdot s_a \tag{8-3}$$

$$s_a^2 = \frac{1}{\sum x^2} s^2 \tag{8-4}$$

s_a^2 – variance of the coefficients a
s^2 – variance of the responses y [SCHEFFLER, 1986]

Tab. 8-13. Regression coefficients for the determination of the influence of independent variables on metal concentration (nonsignificant parameters in parentheses)

Influence	Variable	Regression coefficients		
		Cd	Cu	Zn
Constants		7.5	14.9	126.7
Main	pH	−0.9	−9.2	−20.1
influences	Ca	−0.2	−3.2	(−1.9)
	fa	−0.6	−6.9	−8.3
Two-factor	pH-Ca	(−0.1)	(−1.0)	(−0.9)
interactions	pH-fa	−0.4	3.7	−7.9
	Ca-fa	−0.2	2.1	(−1.7)
Three-factor interaction	pH-Ca-fa	(−0.1)	(−0.7)	(0.2)

The following critical values for the regression coefficients indicate a probability of an error of the first kind α = 0.025, and 16 degrees of freedom (3 replicates for each of the 8 experimental points):

Cd: 0.175
Cu: 1.28
Zn: 3.56

For quantitative assessment of the change in the concentration of the electrochemically available metal after increasing one of the main influencing factors a quotient is defined:

$$c_{rel} = \frac{c_{eh}}{c_{el}} \cdot 100 \tag{8-5}$$

c_{rel} – in %
c_{eh} – electrochemically available heavy metal concentration at the high level of the influencing factor
c_{el} – electrochemically available heavy metal concentration at the low level of the influencing factor

Fig. 8-15 illustrates that alteration of the pH strongly influences the binding forms of the heavy metals Cd, Cu, and Zn. This finding for river water corresponds with the ex-

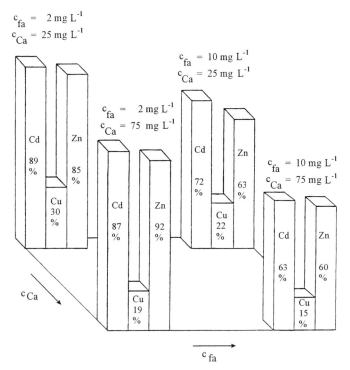

Fig. 8-15. Electrochemically available heavy metal concentrations at pH = 8, relative to values at pH = 4

planation in the literature [FLORENCE, 1986] that in marine water containing only very slight amounts of humic substances 80% of the copper is bound as $CuCO_3$ at a pH of 8.2. Other species, e.g. $Cu(OH)^+$ and $Cu(OH)(CO_3)^-$, also exist [SIGG and STUMM, 1994]. This effect is not so distinct for Cd and Zn. Cadmium forms hardly any hydroxy-complexes and in the investigated pH range hardly forms complexes with the carbonate ion either. At pH = 8 and low concentrations of calcium and fulvic acid more than 80% of the Cd ions exist in the unbound form. Inorganic neutral complexes with chloride ions are possible [TJUTJUNOWA, 1980], therefore the electrochemical availability may be influenced. Similar pH-dependent binding behavior is characteristic of Zn [FÖRSTNER and MÜLLER, 1974], as was verified by the experiments.

As the significance test of the coefficients for the interactions shows, the calcium ion concentration in water does not have a significant influence on the binding of Zn to humic substances. The binding of Cd and Cu to humic substances increases significantly at higher calcium ion concentrations (Fig. 8-16), but to a lesser degree than after increasing the pH. This can be explained by the stabilities of the humic complexes [WILLIAMS, 1967]. Calcium humates are only slightly stable. The binding strength of cadmium and copper compounds to humic substances is somewhat stronger.

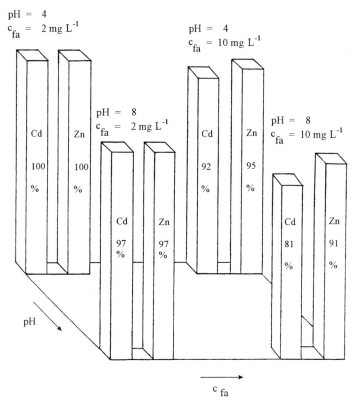

Fig. 8-16. Electrochemically available heavy metal concentrations at a calcium concentration of 75 mg L^{-1}, relative to 25 mg L^{-1}

Fig. 8-17 demonstrates that for Cd and Zn an interaction between the pH and the fulvic acid concentration exists which is mathematically detectable by the 2-factor-interaction. At pH = 4 the binding of the heavy metal ions is relatively slight, whereas 35% of the heavy metals are bound to the fulvic acid at pH = 8.

Modeling of the connections and the binding forms requires the application of multivariate statistical methods. A regression model with latent variables, e.g. the PLS method (see Section 5.7.2 for mathematical details), seems to be useful. The PLS method models not only the influences of the plan matrix on the responses, but also the interactions between the metals in the response matrix. In the matrix of the dependent variables the electrochemically available parts of the heavy metals c_{ea} were filled in (Tab. 8-12):

$$c_{ea} = \frac{c_e}{c_{et}} \quad (8\text{-}6)$$

c_e — measured electrochemically available concentration
c_{et} — total metal concentration

(Cd: 10 µg L^{-1}, Cu: 50 µg L^{-1}, Zn: 150 µg L^{-1})

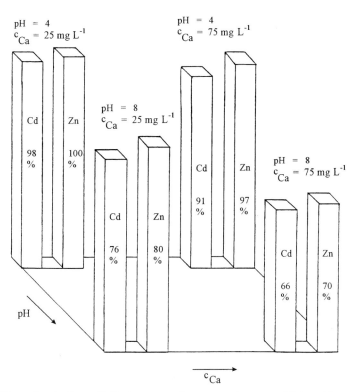

Fig. 8-17. Electrochemically available heavy metal concentrations at a fulvic acid concentration of 10 mg L^{-1}, relative to 2 mg L^{-1}

The matrix of the independent variables corresponds to the plan matrix of the experimental design (Tab. 8-11).

From the modeling (see Section 5.7.2) the latent vectors t and u take values which are difficult to interpret. The matrix of the dependent variables Y_{calc} has been computed for validation of the PLS model (Tab. 8-14).

Tab. 8-14. Matrix of the experimentally determined dependent variables Y and the dependent variables after PLS modeling Y_{calc}

Y			Y_{calc}		
Cd	Cu	Zn	Cd	Cu	Zn
0.86	0.84	0.98	0.86	0.50	0.99
0.85	0.30	0.99	0.77	0.36	0.86
0.62	0.07	0.63	0.60	0.12	0.65
0.86	0.56	0.99	0.91	0.58	1.06
0.79	0.24	0.95	0.82	0.44	0.93
0.77	0.25	0.82	0.70	0.26	0.77
0.75	0.11	0.81	0.74	0.34	0.84
0.50	0.04	0.57	0.65	0.20	0.72

The difference between the matrices Y and Y_{calc} is the absolute modeling error for the intended size on each experimental point: The medium relative modeling error for the electrochemically available part of cadmium in the simulated water sample amounts to 6.6%; for zinc it is 5.3%. In contrast with that, the medium modeling error for copper amounts to 41.0%.

The relative modeling errors demonstrate that modeling for Cd and Zn is feasible. The high modeling error for Cu results from the experiment. The reproducibility of the results from Cu determination is highly dependent on the slightest variations in pH.

For evaluation of the PLS model and for comparison with multiple linear regression the independent parameters were varied in the calibration range and predictions were made. Tab. 8-15 illustrates the comparison of the predicted and the measured values.

At a pH of 6.0 the fulvic acid was adsorbed at the hanging mercury drop electrode. The resulting peak suppressions made the evaluation of the measured results impossible.

The accordance of the modeled values of Cd and Zn with the experimentally determined values is sufficient. So the effects of binding of Cd and Zn to fulvic acid in water are quantitatively interpretable. The error of PLS modeling is smaller than that of MLR modeling because the interactions between the analyzed heavy metals have also been taken into consideration.

Tab. 8-15. Electrochemically available amount of heavy metals compared with the total content

Varied parameters			Predicted part of metals (MLR)			Predicted part of metals (PLS)			Measured part of metals		
pH	Ca in mg L^{-1}	fa in mg L^{-1}	Cd in %	Cu in %	Zn in %	Cd in %	Cu in %	Zn in %	Cd in %	Cu in %	Zn in %
5	40	4	82	51	93	81	43	92	82	20	92
3	15	1	88	101	102	90	56	105	82	96	107
7	60	8	67	13	74	70	27	78	81	39	86
6	15	6	78	39	84	72	30	80	–*	–*	–*
7	100	6	66	35	77	80	43	92	84	72	91
3	50	6	88	69	104	88	53	102	104	94	107
8	50	6	66	11	71	67	23	74	59	97	62
6	50	1	84	50	93	81	44	93	–*	–*	–*
6	50	15	62	0	72	65	19	70	–*	–*	–*
Medium modeling error in %			12	65	8.6	9.7	58	6.0			

* Heavy metal concentrations cannot be determined reproducibly under these conditions

8.3.3 Investigations of the Interaction Between River Water and River Sediment

8.3.3.1 Problem and Experimental

In rivers and streams heavy metals are distributed between the water, colloidal material, suspended matter, and the sedimented phases. The assessment of the mechanisms of deposition and remobilization of heavy metals into and from the sediment is one task for research on the behavior of metals in river systems [IRGOLIC and MARTELL, 1985]. It was hitherto, usual to calculate enrichment factors, for instance the geoaccumulation index for sediments [MÜLLER, 1979; 1981], to compare the properties of elements. Distribution coefficients of the metal in water and in sediment fractions were calculated for some rivers to find general aspects of the enrichment behavior of metals [FÖRSTNER and MÜLLER, 1974]. In-situ analyses or laboratory experiments with natural material in combination with speciation techniques are another means of investigation [LANDNER, 1987; CALMANO et al., 1992]. Such experiments manifest univariate dependencies for the metals and other components, for instance between different metals and nitrilotriacetic acid [FÖRSTNER and SALOMONS, 1991], but the interactions in natural systems are often more complex.

As demonstrated in Section 8.3.2 PLS regression is a useful and powerful tool for description of the relationship between independent and dependent parameters in model aqueous systems. Therefore in the following the applicability of PLS modeling shall be tested for empirical description of the multivariate dependence of the heavy metal partition ratio between river water and sediment on important major and minor components in the water [EINAX and GEISS, 1994].

It shall be remarked that the application of PLS modeling for other case studies in the aquatic environment is also described in the literature. SEIP et al. [1994] applied PLS calibration to determine which physical factors determine the phytoplankton mass in lakes.

The sampling and analytical conditions are described in detail in Section 8.2.1. For estimation of a medium water influence on the sediments the mean values of the water components over the investigation period of two years were used as the first (independent) data matrix; the metal distribution coefficients were taken as the dependent data matrix. For description of deposition-remobilization effects a heavy metal distribution coefficient, DC_i, was defined as:

$$DC_i = \frac{\bar{c}_{water\,i}}{c_{sediment\,i}} \tag{8-7}$$

with $\bar{c}_{water\,i}$ in µg L^{-1} and $c_{sediment\,i}$ in mg kg^{-1} for each sampling point i.

The PLS model was computed according to the algorithm described in detail in Section 5.7.2. It consists of one t-vector verified by means of cross-validation. The prediction errors for the model of the element distribution coefficients described by the 15 major and minor water components investigated are illustrated in Tab. 8-16.

Tab. 8-16. Error in the prediction of metal distribution coefficients from the water components

Element	Prediction error in %
Fe	24.1
Cd	10.8
Cr	23.5
Zn	31.9
Cu	16.6
Ni	13.6

The mean error for all the elements studied is 20.1%, which means that the influence of water components on the variation of the metal distribution coefficients can be predicted with an error of approximately 20%. The errors can be explained by sediment transport processes, inhomogeneities in the water phase during the sampling process, and biochemical processes in the complex river system. Otherwise, it is possible to describe the distribution of heavy metals between the water phase and the sediment in the right magnitude. On the basis of this multivariate calibration selected water conditions were varied and their influence on metal distribution simulated. The variations have to be performed within the range of validity of the PLS model obtained, which means within the range of the calibration space. Such simulations are important for assessment of the implications of changes in the discharger "structure" along a river on the deposition or remobilization of heavy metals.

8.3.3.2 Variation of pH Value

Increasing the pH from 6.85 to 7.05 results in better dissolution of heavy metals out of the sediment (Fig. 8-18) because the binding strength of metals in metal complexes increases with increasing basicity [LIMNIK and NABIVANETS, 1984]. The stability of metal complexes correlates strongly with the order of binding strength to humic and fulvic acids:

$$Cu > Ni > Zn > Cd$$

A pH increase from 6.85 to 7.05 results in a change in copper concentration of 55 µg L^{-1} in river water in contact with sediment containing 55 mg kg^{-1}.

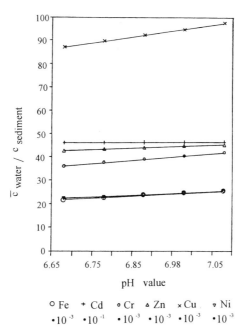

Fig. 8-18. PLS modeling of the effect of pH variation on heavy metal distribution coefficients

8.3.3.3 Variation of Organic Loading

Variation of organic loading in the range $13 \leq$ chemical oxygen demand (in mg L^{-1}) ≤ 17 shows only slight effects. The relatively highest changes result for copper which will be deposited by binding to organic coatings, in accordance with the literature [FÖRSTNER and MÜLLER, 1974].

8.3.3.4 Variation of the Concentration of Condensed, Organic, and Ortho Phosphates

The strong dissolution effects on heavy metals of increasing phosphate concentrations in river water (Fig. 8-19) indicate the strong remobilization effects which result from complex formation. This tendency for complex formation with triphosphates is also described in the literature [GMELIN, 1965]. Increasing the concentration of phosphate in water from 0.38 to 0.68 mg L^{-1} increases the iron concentration by 648 µg L^{-1} in water in contact with sediment in which the iron concentration is 16220 mg kg^{-1}.

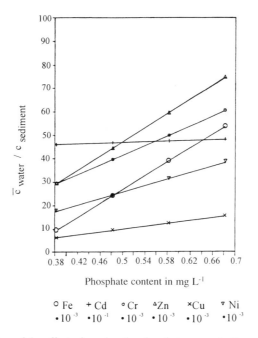

Fig. 8-19. PLS modeling of the effect of varying the phosphate concentration on heavy metal distribution coefficients

8.3.3.5 Variation of the Concentration of Suspended Material

Variation of the concentration of suspended material in the water from 12.5 to 52.5 mg L^{-1} leads to particularly strong deposition of chromium, iron, and zinc in the sedimented material (Fig. 8-20). This is also in accordance with laboratory experiments [FÖRSTNER and WITTMANN, 1983].

Fig. 8-21 presents an overview of the consequences of changes of pH, phosphate concentration, and suspended material content on heavy metal concentrations.

Concluding the section on the PLS modeling it should be pointed out that results from the application of canonical correlation analysis (see also Section 5.5) for the quantitative description of interactions between river water and sediments are comparable to those from PLS analyses [GEISS, 1990].

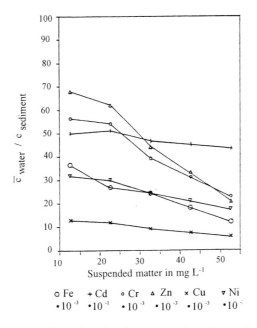

Fig. 8-20. PLS modeling of the effect of varying the concentration of suspended material on heavy metal distribution coefficients

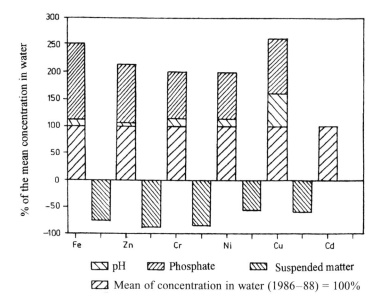

Fig. 8-21. PLS modeling of the consequences of changes in pH, phosphate concentration, and suspended material content on the concentrations of heavy metals in the system river water/river sediment

References

Andrade, J.M., Prada, D., Alonso, E., Lopez, P., Muniategui, S., de la Fuente, P., Quijano, M.A.: Anal. Chim. Acta 292 (**1994**) 253

Bandemer, H., Bellmann, A.: Statistische Planung von Experimenten, B.G. Teubner Verlagsgesellschaft, Leipzig, **1976**

Calmano, W., Hong, J., Förstner, U.: Vom Wasser 78 (**1992**) 245

Einax, J., Geiß, S.: Fresenius' J. Anal. Chem. 350 (**1994**) 14

Fachgruppe Wasserchemie in der Gesellschaft Deutscher Chemiker in Gemeinschaft mit dem Normenausschuß Wasserwesen im DIN Deutsches Institut für Normung e.V. (Ed.): Deutsche Einheitsverfahren zur Wasser-, Abwasser- und Schlammuntersuchung, DIN-Normen, Beuth, Berlin, in Lieferung seit **1960**

Florence, T.M.: Analyst 11 (**1986**) 489

Förstner, U., Müller, G.: Schwermetalle in Flüssen und Seen, Springer, Berlin, Heidelberg, New York, **1974**

Förstner, U., Wittmann, G.T.W.: Metal Pollution in the Aquatic Environment, 2nd Ed., Springer, Berlin, Heidelberg, New York, Tokyo, **1983**

Förstner, U., Salomons, W. in: Merian, E. (Ed.): Metals and their Compounds in the Environment, VCH, Weinheim, New York, Basel, Cambridge, **1991**, pp. 379

Frimmel, F.H., Sattler, D.: Vom Wasser 59 (**1982**) 335

Fuchs, F., Raue, B.: Vom Wasser 57 (**1981**) 95

Geiß, S.: Schwermetalle in Fließgewässern – analytische und chemometrische Untersuchungen, Dissertation, Friedrich-Schiller-Universität, Jena, **1990**

Geiß, S., Einax, J.: Acta hydrochim. hydrobiol. 19 (**1991**) 615

Geiß, S., Einax, J.: Vom Wasser 78 (**1992**) 201
Geiß, S., Einax, J., Danzer, K.: Fresenius' Z. Anal. Chem. 333 (**1989**) 97
Gmelins Handbuch der anorganischen Chemie, Phosphor, Teil C, 8. Aufl., Verlag Chemie, Weinheim, **1965**, pp. 247
Grimault, J.O., Olive, J.: Anal. Chim. Acta 278 (**1993**) 159
Henrion, G., Henrion, A., Henrion, R.: Beispiele zur Datenanalyse mit BASIC-Programmen, Deutscher Verlag der Wissenschaften, Berlin, **1988**, pp. 134
Institut für Wasserwirtschaft Berlin (Ed.): Ausgewählte Methoden der Wasseruntersuchung, Bd. 1 und 2, Gustav Fischer, Jena, **1986**
Irgolic, K.J., Martell, A.E. in: Irgolic, K.J., Martell, A.E. (Eds.): Environmental Inorganic Chemistry, VCH, Deerfield Beach, Florida, **1985**, p. 1
Kheboian, C., Bauer, C.F.: Anal. Chem. 59 (**1987**) 1417
Landner, L. in: Bhattacharji, S., Friedman, G.M., Neugebauer, H.J., Seilacher, H. (Eds.): Lecture Notes in Earth Sciences, Springer, Berlin, Heidelberg, New York, London, Paris, Tokyo, **1987**
Limnik, P.N., Nabivanets, B.J.: Acta Hydrochim. Hydrobiol. 12 (**1984**) 335
Lund, W.: Fresenius' Z. Anal. Chem. 337 (**1990**) 557
Müller. G.: Umschau 79 (**1979**) 465
Müller, G.: Umschau 81 (**1981**) 778
Pardo, R., Barrado, E., Castrillejo, Y., Velasco, M.A., Vega, M.: Anal. Lett. 26 (**1993**) 1719
Sager, M., Vogel, W.: Acta Hydrochim. Hydrobiol. 21 (**1993**) 1
Sager, M., Belocky, R., Pucsko, R.: Acta Hydrochim. Hydrobiol. 18 (**1990**) 157
Salbu, B., Björnstad, H.E., Lindström, N.S.: Anal. Chim. Acta 167 (**1985**) 161
Scheffler, E.: Einführung in die Praxis der statistischen Versuchsplanung, 2. Aufl., Deutscher Verlag für Grundstoffindustrie, Leipzig, **1986**, pp. 118
Schramel, P., Lill, G., Seif, R.: Fresenius' Z. Anal. Chem. 326 (**1987**) 135
Seip, K.L., Sneek, M., Snipen, L.-G.: Chemom. Int. Lab. Syst. 23 (**1994**) 247
Sigg, L., Stumm, W.: Aquatische Chemie, 3. Aufl., Verlag der Fachvereine, Zürich, B.G. Teubner, Stuttgart, **1994**, pp. 211
Stumm, W., Keller, L. in: Merian, E. (Ed.): Metalle in der Umwelt, Verlag Chemie, Weinheim, Deerfield Beach/Florida, Basel, **1984**, pp. 21
Tessier, A., Campbell, P.G.C.: Water Research 24 (**1990**) 1055
Tessier, A., Campbell, P.G.C., Bisson, M.: Anal. Chem. 51 (**1979**) 844
TGL 22764: Nutzung und Schutz der Gewässer – Klassifizierung der Wasserbeschaffenheit von Fließgewässern, Fachbereichsstandard der DDR, **1981**
Tjutjunowa, F.I.: Physiko-chemische Prozesse in Grundwässern, Deutscher Verlag für Grundstoffindustrie, Leipzig, **1980**, pp. 35
Tümpling, W. von, Jr., Geiß, S., Einax, J.: Acta Hydrochim. Hydrobiol. 20 (**1992**) 320
Weber, E.: Grundriß der biologischen Statistik, 9. Aufl., Gustav Fischer, Jena, **1986**, p. 466
Welz, B.: Fresenius' Z. Anal. Chem. 325 (**1986**) 95
Wenning, R., Paustenbach, D., Johnson, G., Ehrlich, R., Harris, M., Badbury, H.: Chemosphere 27 (**1993**) 55
Williams, R.J.P.: Endeavour 26 (**1967**) 96

9 Pedosphere

The pedosphere has a relatively high stress capacity, but has a very low self-purifying capacity; the soil is, therefore, one of the most endangered compartments of our environment and a potential source of remobilization of pollutants. Because of production and utilization in human society, heavy metals in particular have become a group of substances with considerable relevance to the environment, especially the soil. Heavy metals remain in the soil for a long time, they are transported to deeper layers, or removed with the crops to a small extent only. Depending on their concentration, these elements can reduce soil fertility and lead to increased input into the food chain [LEPP, 1981a; 1981b]. Increasing heavy metal contents in foodstuffs can endanger the health of animals and humans [FRIBERG et al., 1986].

The case studies described in Chapter 9 deal with the occurrence and the behaviour of heavy metals in the pedosphere; but the chemometric solutions are, however, in principle transferable to other groups of contaminants, as for example, to organic pollutants, also.

9.1 Representative Soil Sampling

9.1.1 Problem

Effective protection of the soil against excessive exposure to heavy metals is based on the reliable monitoring of their levels. The degree of contamination of large areas should be precisely characterized without requiring too much time and expense. The number of samples necessary depends essentially on the extent of inhomogeneity of the distribution of the heavy metals. The purpose of this section is investigation of the relationship between this inhomogeneity and the distance between the sampling points, by means of chemometric procedures on a test field [EINAX et al., 1992].

9.1.2 Soil Sampling and Heavy Metal Determination

The samples investigated originate from agricultural land (total area = 57000 m^2) for which emission-induced contamination with heavy metals cannot be excluded. The sampling points were equally distributed over the area in a square raster screen. The distance between these sampling points amounted to 25 m and the distance from the outer points to the dead furrow was more than 10 m. At the sampling points, soil from an area of about 10 cm² was collected to a depth of 20 cm. The soil was dried and sieved (grain size < 2 mm) and the heavy metals were extracted with boiling HNO_3 (c = 1.5 mol L^{-1}) [MACHELETT et al., 1986a]. The determination of heavy metals in the extract was performed by flame atomic absorption spectrometry (AAS 3, Carl Zeiss Jena) [GRÜN et al., 1987]. A deuterium lamp was used for background compensation.

9.1.3 Chemometric Investigation of Measurement Results

9.1.3.1 Univariate Analysis of Variance

For the characterization of the selected test area it is necessary to investigate whether there is significant variation of heavy metal levels within this area. Univariate analysis of variance is used analogously to homogeneity characterization of solids [DANZER and MARX, 1979]. Since potential interactions of the effects between rows (horizontal lines) and columns (vertical lines in the raster screen) are unimportant to the problem of local inhomogeneity as a whole, the model with fixed effects is used for the two-way classification with simple filling. The basic equation of the model, the mathematical fundamentals of which are formulated, e.g., in [WEBER, 1986; LOHSE et al., 1986] (see also Sections 2.3 and 3.3.9), is:

$$y_{ij} = \mu + \alpha_i + \beta_j + \alpha_i \beta_j + \varepsilon_{ij} \qquad (9\text{-}1)$$

y_{ij} – measured value
μ – mean value
α_i – effect of stage i of factor A (column)
β_j – effect of stage j of factor B (row)
ε_{ij} – error in the model

According to the position of the sampling points, the results of the analysis are assigned to a value matrix for the practical application of the univariate two-way analysis of variance. Thus, rows and columns can be defined. As a result of the calculation, the $\hat{F}$-values of the corresponding features for rows and columns calculated by

$$\hat{F} = \frac{s_1^2}{s_2^2} \qquad (9\text{-}2)$$

are represented in Tab. 9-1.

The definition of a significance value, SV, analogous to the homogeneity index [SINGER and DANZER, 1984] enables semiquantitative assessment of the significance of inhomogeneity:

$$SV = \frac{\hat{F}}{F(f_1; f_2; \alpha)} \tag{9-3}$$

The significance values are higher than unity for all investigated heavy metals in the rows and columns (Tab. 9-1). In the case of a critical probability of an error of the first kind of $\alpha = 0.05$, all investigated heavy metals show significantly changed concentrations along both rows and columns. The investigated test area must, therfore, be regarded as inhomogeneous.

Tab. 9-1. $\hat{F}$-values and significance values, SV, from univariate two-way analysis of variance (A – column, B – row)

Element	$\hat{F}_A$	$\hat{F}_B$	SV_A	SV_B
Cd	8.36	4.43	3.92	2.31
Ni	9.70	5.28	4.08	2.75
Pb	5.46	3.68	2.56	1.92
Cu	2.83	5.53	1.33	2.88
Zn	7.06	6.82	3.31	3.55
Cr	5.57	7.00	2.62	3.64

9.1.3.2 Multivariate Data Analysis

9.1.3.2.1 Cluster Analysis

The data matrix is subjected to hierarchical agglomerative cluster analysis (CA; for the mathematical fundamentals see Section 5.3; further presentation of the algorithms is given by [HENRION et al., 1987]) in order to find out whether territorial structures with different multivariate patterns of heavy metals exist within the test area.

Applying the average linkage algorithm and taking the EUCLIDean distance as a basis, the number of clusters rises relatively evenly with increasing similarity (Fig. 9-1), i.e. structuring of the data material is not recognizable. The application of the algorithm according to WARD (connected with the application of the squared EUCLIDean distance), however, makes clear structuring of the multivariate data possible, as described in the literature [DERDE and MASSART, 1982; EINAX, 1990].

Fig. 9-1 shows the formation of four clusters which are preserved even in the case of varied similarity over a larger range (30%). The local distribution of cluster points is presented in Fig. 9-2. This hypothesis found with the unsupervised learning technique is checked in the following section by means of multidimensional classification.

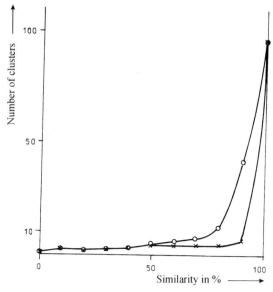

Fig. 9-1. Number of clusters as a function of similarity. (Cluster algorithm: o average linkage, x according to WARD)

	a	b	c	d	e	f	g	h
1	A	A	A	A	B	B	B	C
2	B	B	B	B	B	B	B	B
3	B	B	B	B	B	B	C	B
4	B	B	B	B	B	C	C	A
5	A	B	A	B	C	C	C	A
6	A	B	A	B	C	C	C	D
7	A	B	A	A	C	C	C	D
8	A	B	A	B	C	C	D	D
9	A	B	A	C	C	C	C	D
10	A	C	D	C	C	C	D	C
11	B	B	D	C	D	D	D	C
12	A	C	C	D	C	C	D	D

Fig. 9-2. Results of cluster analysis according to WARD. (A, B, C, D – designation of the four stable clusters within the similarity range 50–80%)

9.1.3.2.2 Multivariate Analysis of Variance and Discriminant Analysis

The result from cluster analysis presented in Fig. 9-2 is subjected to MVDA (for mathematical fundamentals see Section 5.6 or [AHRENS and LÄUTER, 1981]). The principle of MVDA is the separation of predicted classes of objects (sampling points). In simultaneous consideration of all the features observed (heavy metal content), the variance of the discriminant functions is maximized between the classes and minimized within them. The classification of new objects into *a priori* classes or the reclassification of the learning data set is carried out using the values of the discriminant function. These values represent linear combinations of the optimum separation set of the original features. The result of the reclassification is presented as follows:

		Discriminated samples in class				Misclassification rate
		A	B	C	D	
Given	A	16	0	3	0	3/19
class	B	2	29	2	0	4/33
	C	0	0	29	1	1/30
	D	0	0	0	14	0/14

The relatively low rate of false assignments (8.3%) compared with the error of a randomly correct reclassification (72.4%) confirms the cluster analytical hypothesis of four partial areas of the test field in which heavy metal pattern differed detectably.

On the other hand, the presentation of the values of the two discriminant functions with the strongest discriminating power versus each other (Fig. 9-3), including the wide overlapping scattering circles of the F-distribution with a critical probability of an error of the first kind of $\alpha = 0.05$, shows that the multivariate changes at the borderlines of the classes are not very distinct though the class centroids differ significantly.

9.1.3.2.3 Principal Components Analysis

The principle of principal components analysis (PCA) is the transformation of the original features into uncorrelated new variables (principal components) by means of linear combinations. The first principal component explains most of the variance within the data matrix and the second explains most of the remaining variance, and so on (for mathematical fundamentals see Section 5.4.1 or [LEBART et al., 1984]). The graphical representation of the principal components scores of the objects with the highest variance (81.6% of the total variance of the tested data matrix) within the test area shows (Fig. 9-4) that objects with high scores of principal component 1 are mainly found in the right lower part of the area, whereas those with high negative scores are in the left upper part of the area.

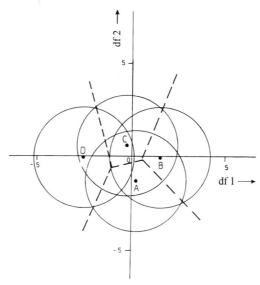

Fig. 9-3. Plot of the scores of discriminant function df 2 (44.9% variance proportion) vs. scores of discriminant function df 1 (55.1% variance proportion) of the four classes obtained by CA. (The circles correspond to the 5% risk of error of the MANOVA; – – separation line of discrimination corresponding to highest probability)

Comparison of Fig. 9-4 with Fig. 9-2 shows that the results of PCA are in qualitative accordance with the results of CA.

The application of different multivariate statistical methods enables clarification of the changes of the heavy metal pattern within the test area examined here. A multivariate change of the heavy metal content of the test field from the upper left to the lower right part of the area can be demonstrated in the example investigated.

Because of the local inhomogeneity of the test area, the sampling and analysis of several soil samples is necessary for representative characterization of the heavy metal exposure of the total area.

In the following section the maximum possible distance between two sampling points necessary for the representative characterization of the total exposure is determined by means of correlation analysis.

9.1.3.3 Autocorrelation Analysis

9.1.3.3.1 Univariate Autocorrelation Analysis

Mathematical methods for calculating correlation are applied to describe the degree of relationship between one or more measuring rows (for mathematical fundamentals see Section 6.6). The theoretical fundamentals of univariate auto- and cross-correlation ana-

	a	b	c	d	e	f	g	h
1	3	1	2	3	1	1	1	2
2	1	1	1	1	2	1	1	1
3	1	1	1	1	2	2	3	2
4	1	1	2	2	1	3	3	3
5	2	1	3	2	2	3	2	3
6	1	1	2	2	3	3	3	4
7	2	1	2	2	3	3	4	4
8	2	1	2	2	2	3	4	4
9	2	1	3	2	4	3	3	4
10	1	3	4	3	4	4	4	2
11	1	1	4	4	4	4	4	3
12	2	1	2	4	3	4	4	4

Fig. 9-4. Plot of the scores of the first principal component in the investigated area

Range of the scores of the first principal component	Class
< −0.50	1
−0.05 to −0.01	2
0 to 0.50	3
> 0.50	4

lysis and their application in analytical chemistry are described by DOERFFEL and WUNDRACK [1986], with improvement of the signal-to-noise ratio (see also [DOERFFEL et al., 1986]) being particularly important for applicability.

By analogy with the procedure by DOERFFEL et al. [1988] for testing homogeneity in analytical investigations of solids, the regular raster screen of the investigated area is divided into a meandering shape, i.e. the two-dimensional local series is transformed into a one-dimensional local series.

The smoothing of the autocorrelation functions is performed by means of regression according to the empirical function:

$$r_{xx} = a_0 + a_1 \, l + a_2 \, l^2 \tag{9-4}$$

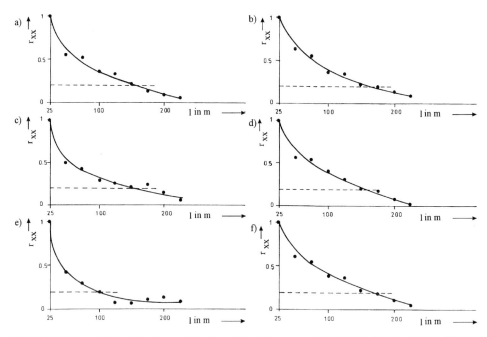

Fig. 9-5. Autocorrelation functions of the investigated heavy metals. (a) Cd, (b) Cr, (c) Cu, (d) Ni, (e) Pb, (f) Zn. (The dashed lines correspond to the highest possible values of a random correlation for the probability of an error of the first kind of $\alpha = 0.05$)

The autocorrelation functions of the investigated features calculated according to Eq. 6-22 and smoothed according to Eq. 9-4 are represented in Fig. 9-5. The increasing distance l of the measuring points results in steadily decaying autocorrelation functions for all features, with that for lead decaying most conspicuously.

The points of intersection between the highest value of a random correlation and the lower limit of the confidence interval of the corresponding empirical model function according to Eq. 9-4 correspond to the lower limits of the confidence range of the critical distances between the sampling points. These values are represented in Tab. 9-2.

Tab. 9-2. Critical distances, l_{crit}, between sampling points and estimated values of the lower limit of the confidence range, $l_{crit\ low}$, for a critical error probability of $\alpha = 0.05$

Element	l_{crit} in m	$l_{crit\ low}$ in m
Cd	140.2	111.2
Ni	146.7	120.3
Pb	96.9	74.3
Cu	132.3	87.4
Zn	152.5	127.1
Cr	150.3	124.6

The autocorrelation function of lead shows the most conspicuous decay in the investigated area. Consequently it follows that the estimated value of the lower limit of the confidence interval for the critical distance between the measuring points for lead is the limiting sampling step.

If the length of basic steps of soil sampling, Δl, in the investigated test area is maintained at 25 m; a distance of $\Delta l \approx 75$ m between the measuring points is sufficient for further sampling from this area to characterize the heavy metal exposure of the soil with sufficient statistical reliability. Thus, according to the mode of sampling described, 33.3% of the investigated soil samples, i.e. 32, are necessary for representative investigation of the test area of 57000 m^2.

It is obvious that the determined critical sampling distance depends on the statistical reliability required. A smaller confidence level (according to a higher probability of an error of the first kind) corresponds to a higher value for the critical sampling distance required. For the discussed case a change in the probability for an error of the first kind from $\alpha = 0.05$ to $\alpha = 0.1$ increases the critical sampling distance for lead from $l_{crit\ low} = 74.3$ m to $l_{crit\ low} = 78.0$ m.

9.1.3.3.2 Multivariate Autocorrelation Analysis

The computation of the multivariate autocorrelation function (MACF) is useful if the simultaneous consideration of all measured variables and their interactions is of interest. The mathematical fundamentals of multivariate correlation analysis are described in detail in Section 6.6.3. The computed multivariate autocorrelation function R_{xx} according to Eqs. 6-30–6-37 is demonstrated in Fig. 9-6. The periodically encountered

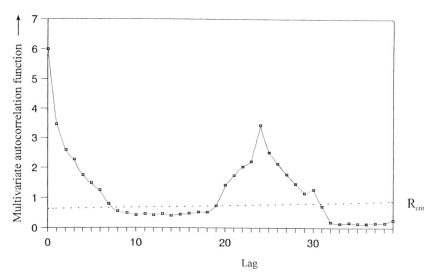

Fig. 9-6. Multivariate autocorrelation function

higher autocorrelation values in the range of *Lag* = 24 may be explained by the meandering course of the local series.

The multivariate autocorrelation function of the measured values compared with the highest randomly possible correlation value shows significant correlation up to *Lag* ≈ 7. So, the range of multivariate correlation is more extended than that of univariate correlation (see Section 9.1.3.3.1). This fact must be understood because the computation of the MACF includes the whole data matrix with all interactions between the measured parameters. For characterization of the multivariate heavy metal load of the test area only 14 samples in the screen are necessary.

9.1.4 Conclusions

The following general conclusions can be drawn from the computations discussed above:

- Univariate analysis of variance enables detection of feature-specific inhomogeneities within an investigated test area.
- The application of methods of multivariate statistics (here demonstrated with examples of cluster analysis, multivariate analysis of variance and discriminant analysis, and principal components analysis) enables clarification of the lateral structure of the types of feature change within a test area.
- Relatively narrow screening sampling is necessary for the first characterization of the heavy metal content of an area of soil.
- The determination of the critical distance between the sampling points can then be performed by autocorrelation analysis on the analytically determined concentrations.
- If the soil is repeatedly investigated, the optimized distance between the sampling points is sufficient for the representative characterization of the area. If a multiple of the original distance between the sampling points is taken as a basis, assessment of temporal changes of exposures of the total area as well as of individual measuring points is still possible.
- If the autocorrelation function decays immediately below the corresponding significance limit, the screen of measuring points may be reduced in size for repeated investigations.
- The application of multivariate autocorrelation analysis is useful for determination of the distance between samples for representative sampling for characterization of multivariate loading by heavy metals.

9.2 Multivariate Statistical Evaluation and Interpretation of Soil Pollution Data

The following two examples [EINAX et al., 1990; KRIEG and EINAX, 1994] demonstrate not only the power, but also the limits of multivariate statistical methods applied to the description of polluted soils loaded with heavy metals from different origins. Case studies with chemometric description of soil pollution by organic compounds are also discussed in the literature. DING et al. [1992], for example, evaluated local sources of chlorobenzene congeners in soil samples by using different methods of multivariate statistical analysis.

9.2.1 Studies on Heavy Metal Pollution of Soils at Different Locations

The purpose of this case study is to investigate the following two questions by means of multivariate statistical methods:

- Are differences between the state of heavy metal pollution of different areas detectable, and wherein are the investigated territories distinct?
- Is it possible to identify the origins of soil pollution?

9.2.1.1 Experimental

Soils were taken as multiple samples from three partial areas within the investigated territory:

- area 1: loaded by particulate emissions from a ferrous metallurgical works
- area 2: loaded by particulate emissions from a chemical works
- area 3: area for comparison (not specially loaded)

The samples were taken from the organic horizon. The locations for soil sampling were selected taking into consideration the influencing emitters, the specific territorial situation on the one hand and the different emission situations and the resulting long-term loading of the pedosphere by heavy metals on the other hand [MÖLLER, 1987]. The samples were dried and sieved as described in Section 9.1.2. The "mobile" toxicologically relevant fraction of the heavy metals was extracted with boiling HNO_3 ($c = 1.5$ mol L^{-1}) [MACHELETT et al., 1986a]. The concentrations of the investigated elements Cd, Co, Cr, Cu, Fe, Mn, Ni, Pb, and Zn were determined with the AAS 3 (Carl Zeiss Jena) atomic absorption spectrometer. The flame technique or electrothermal atomization were used, depending on the concentration range of the metals. After optimizing the analytical conditions for each element the analyses were performed as fourfold determinations of the concentration on the basis of two parallel digestions. The accuracy of the analytical method was tested by analyzing a reference soil material (Soil 7, IAEA Vienna).

To determine soil pH two suspensions were, furthermore, prepared from each sample of soil – one with distilled water and one with KCl solution ($c = 1$ mol L^{-1}). According to the literature [LUX and FABIG, 1987] the mean value of the two quantities is approximately the true pH of the soil.

9.2.1.2 Analytical Results and Chemometric Interpretation

The following results are based on the levels of nine heavy metals and the pH values determined in 54 soil samples for the three subterritories investigated.

9.2.1.2.1 Univariate Aspects

The ranges of variation of the heavy metal content of the territories investigated amounted to two orders of magnitude [EINAX et al., 1990]. The variations in the concentrations of the three selected elements Fe, Cr, and Co as a function of the distance between the emitter and the locations of soil sampling are demonstrated in Fig. 9-7.

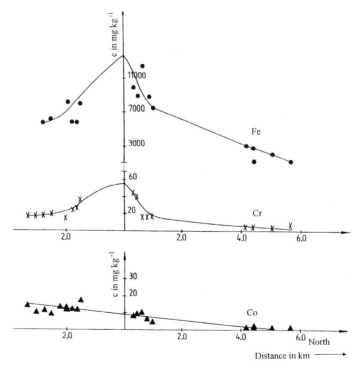

Fig. 9-7. Concentrations of selected heavy metals in the soil vs. the distance of the sampling points from a ferrous metallurgical factory

The emitter (in this case a ferrous metallurgical works) lies at the zero point of the *x*-axis. The concentration of Co shows only a slight, geologically caused, trend. The significant decrease in the Fe and Cr content of the soil in both directions of the *x*-axis depends on the distance from the emitter and on the specific territorial situation. The concentrations of Fe, Cr, and of the other elements in the immediate surroundings of the factory are considerably increased by anthropogenic emissions.

9.2.1.2.2 Cluster Analysis

The principle of cluster analysis (CA) consists in the partitioning of a data set into small groups reflecting unknown groupings [MASSART and KAUFMAN, 1983]. The results of the application of methods of hierarchical agglomerative clustering (for the mathematical fundamentals see also Section 5.3) were obtained from a representative selection of the large number of mathematical algorithms used in CA. The best structuring of the data can be achieved with the method of WARD in accordance with other investigated multivariate problems [DERDE and MASSART, 1982; EINAX, 1990].

The dendrogram obtained by CA from the heavy metal data of the three investigated territories demonstrates the differences between the highly polluted points in the surroundings of the ferrous metallurgical works and the polluted regions near the chemical works (Fig. 9-8). All the heavily polluted points are clearly separated from the widely unloaded sampling points.

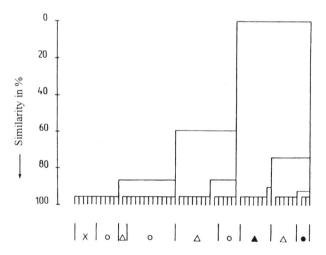

Fig. 9-8. Dendrogram according to the cluster algorithm of WARD for the different investigated areas (▲ heavily polluted soil samples from the surroundings of the ferrous metallurgical factory, △ slightly and unspecifically polluted soil samples from the surroundings of the ferrous metallurgical factory, • heavily polluted soil samples from the surroundings of the chemical factory, ○ slightly and unspecifically polluted soil samples from the surroundings of the chemical factory, × soil samples from unpolluted area for comparison)

The degree of the sharpness of the structuring of the data set using the method of WARD reflects the environmental circumstances because there is only gradual variation in the levels of the heavy metals. CA proves to be a useful tool for soil pollution monitoring and indirectly also for monitoring receptor sites. It is of particular importance to recognize those sampling points which are heavily polluted by special industrial emitters.

9.2.1.2.3 Multivariate Analysis of Variance and Discriminant Analysis

The principle of multivariate analysis of variance and discriminant analysis (MVDA) consists in testing the differences between *a priori* classes (MANOVA) and their maximum separation by modeling (MDA). The variance between the classes will be maximized and the variance within the classes will be minimized by simultaneous consideration of all observed features. The classification of new objects into the *a priori* classes, i.e. the reclassification of the learning data set of the objects, takes place according to the values of discriminant functions. These discriminant functions are linear combinations of the optimum set of the original features for class separation. The mathematical fundamentals of the MVDA are explained in Section 5.6.

Fig. 9-9 demonstrates the results of MVDA for the three investigated territories in the plane of the computed two discriminant functions. The separation line corresponds to the limits of discrimination for the highest probability. The results prove that good separation of the three territories with a similar geological background is possible by means of discriminant analysis. The misclassification rate amounts to 13.0%. The scattering radii of the 5% risk of error of the multivariate analysis of variance overlap considerably. They demonstrate also that the differences in the multivariate data structure of the three territories are only small.

This means that although the multivariate ratio between the anthropogenically and geologically caused differences between the areas is not high, it is statistically reliable and detectable by means of MVDA.

Classification of differences in the state of pollution of the immediate surroundings of the ferrous metallurgical works was the subject of further investigations. A model of three loaded areas has, therefore, been made. The basis of the model was the knowledge of the emission structure, in particular the distance from the works and the specific territorial situation in the region investigated.

The results from MVDA of the heavy metal content of the soil in the surroundings of the ferrous metallurgical works are represented as a 3-class model in Fig. 9-10.

The isolated class of the relatively highly polluted sampling points quite near to the works clearly contrasts with the class of more distant sampling points. The less polluted classes II and III are more similar to one another; they are, nevertheless, clearly distinct, as is proven by the error of discrimination of 0%.

This result shows that application of MVDA gives a good possibility of objective assessment and differentiation of differently polluted areas. The characterization of load-

9.2 Multivariate Statistical Evaluation and Interpretation of Soil Pollution Data   333

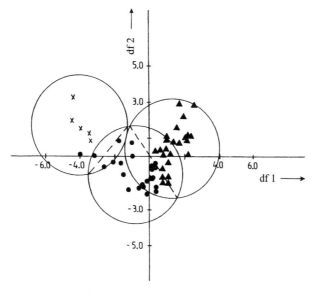

Fig. 9-9. Plot of the scores of discriminant function df 2 vs. scores of discriminant function df 1 of the different areas (▲ surroundings of the ferrous metallurgical factory, ● surroundings of the chemical factory, × unpolluted area for comparison). (The circles correspond to the 5% risk of error of the MANOVA)

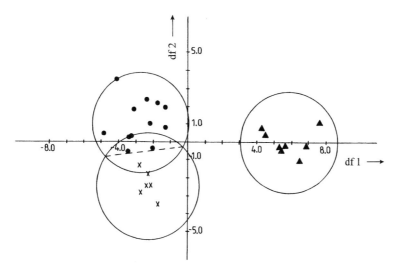

Fig. 9-10. Plot of the scores of discriminant function df 2 vs. scores of discriminant function df 1 of differently polluted areas. Loading state: ▲ heavy (area I), ● moderate (area II), × slight (area III). (The circles correspond to the 5% risk of error of the MANOVA)

ing states thus enables conclusions to be made about the magnitude of the impact of particulate emissions.

The question of whether or not the pattern and the state of pollution of the soils with heavy metals is also reflected in the metal content of the plants is also important. The solution of this question also gives an indication of the biological availability of the investigated metals. Another unpolluted territory with a similar geological background was, therefore, also selected in order to enable valid comparison of the types of plants growing in the surroundings of the ferrous metallurgical and chemical works. Samples of annual plants were taken at the same places as the soil samples. The samples were quickly washed to remove dust from the surface, then dried, homogenized, and digested as described above. The Cd, Cr, Cu, Ni, Pb, and Zn content of the plant samples were determined by flame atomic absorption spectrometry. The pH values were also determined.

The discrimination errors resulting from the application of MVDA to the heavy metal content of the soil and the plant samples from the three investigated territories are presented in Tab. 9-3.

Tab. 9-3. Discrimination results from MVDA of the heavy metal content of soil and plant samples from different regions

Soil samples

	Discriminated samples in class			Misclassification rate
	A	B	C	
Given classes A	13	0	1	1/14
B	2	14	4	6/20
C	0	2	7	2/9

Plant samples

	Discriminated samples in class			Misclassification rate
	A	B	C	
Given class A	12	2	1	3/15
B	2	18	1	3/21
C	0	0	13	0/13

A – surroundings of the ferrous metallurgical factory, B – surroundings of the chemical factory, C – unpolluted area for comparison

The error of the discrimination of the soils amounts to 20.9%; the error of the discrimination of the plants is smaller, only 12.2%.

This means:
- The pattern of soil pollution correlate strongly with contamination of the plants. The metals in the soils of the investigated territories are highly biologically available.
- The heavy metal pollution in a territory can be measured and objectively classified not only in soils but also in annual plants.

- The sampling of comparable species of annual plants requires a detailed biological knowledge. It is, therefore, more promising to characterize the state of loading of polluted territories by investigation of soils, especially in routine environmental monitoring.

9.2.1.2.4 Factor Analysis

The purpose of application of factor analysis (FA) is the characterization of complex changes of all observed features in partial systems of the environment by determination of summarized factors which are more comprehensive and causally explicable. The method extracts the essential information from a data set. The exclusive consideration of common factors in the reduced factor analytical solution seems to be particularly promising for the analytical process. The specific variances of the observed features will be separated from the reduced factor analytical results by means of the estimation of the communalities. They do not falsify the influence of the main pollution sources (see also Tab. 7-2). The mathematical fundamentals of FA are explained in detail in Section 5.4.3 (see also [MALINOWSKI, 1991; WEBER, 1986]).

In the following discussion FA will be used to reduce the dimensionality of the original data space and to extract factors which may make it possible to identify the main sources of pollution.

FA of the heavy metal content of the soil samples of the investigated areas yields a reduced solution with three common factors. These factors, their part of the determinable variance, the loading elements, and their main origins are demonstrated in Tab. 9-4.

Tab. 9-4. Common factors of the heavy metal content of soil samples from the three areas investigated

Factor	Part of determinable variance in %	Components with high factor loadings	Main origins
1	37.8	Pb, Mn, Zn, Cr, Cu, Fe	Emissions from the ferrous metallurgical factory
2	27.0	Ni, Co, Fe, Cu	Emissions from the chemical factory and raised secondary dust from old mine dumps
3	14.7	Cd, pH	Pedogenic factor

The chemometrically obtained interpretation of the connection of the amount of Cd with the pH in the third factor corresponds to the perception of MACHELETT et al. [1986b] of the strong relationship between pH and Cd mobility in soils.

The scores of the original objects in the three-dimensional factor space were computed by the BARTLETT estimate in the second step of FA (see Section 5.4.3).

The interpretation of the matrix of the factor loadings (Tab. 9-4) is verified unambiguously by the plot of the scores of the two factors with the highest eigenvalues (Fig. 9-11).

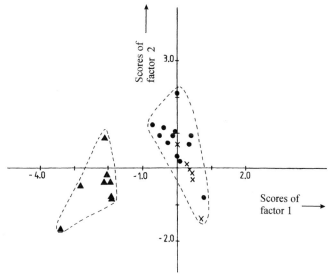

Fig. 9-11. Plot of the scores of factor 2 vs. scores of factor 1 of differently polluted areas in the surroundings of the ferrous metallurgical factory. Loading state: ▲ heavy (area I), ● moderate (area II), × slight (area III)

The heavily polluted sampling points in the immediate surroundings of the ferrous metallurgical factory and the chemical works were clearly separated. On the other hand, the slightly loaded points in the more distant surroundings of the chemical works and the not specifically loaded points of the area for comparison are extremely similar.

The estimate of the factor scores and their graphical representation as a display proves to be a useful tool for the detection of multivariate loads in soils and for giving well-founded hints about their origins.

9.2.2 Soil Depth Profiles

9.2.2.1 Problem and Experimental

In soil science, the empirical description of soil horizons predominates. Only a few applications of statistical methods in this scientific field are described. SCHEFFER and SCHACHTSCHABEL [1992] give an example for the classification of different soils into soil groups using cluster analysis. They claim the objectivity of the results to be one advantage of multivariate statistical methods.

In the following discussion an example is given of the multivariate characterization of soil metal status to distinguish pollution of soil layers from geogenic enrichments from the background. Further case studies and a deeper description are given in detail in the literature [KRIEG and EINAX, 1994].

The sampling point is close to a big nonferrous metallurgical works in Hettstedt, situated in Saxony-Anhalt (Germany). At the sampling point a soil depth profile was taken in 5 cm steps until the C-horizon was reached. The material was dried at 60°C, sieved (grain size < 2 mm) and the heavy metals were extracted with boiling aqua regia according to the official procedure for routine environmental monitoring [FACHGRUPPE WASSERCHEMIE, 1983]. Determination of the metal concentrations was performed by means of atomic emission spectrometry with inductively coupled plasma (Spectroflame, Spectro, Kleve) and, for lead and cadmium, by graphite furnace atomic absorption spectrometry (AAS 3, Carl Zeiss Jena). The soil pH was measured potentiometrically in a suspension of soil in water, by the method according to RUMP [1992].

Total soil carbon was determinated by elemental analysis with an automatic analyzer (CHNS 932, Lego). FOURIER transform infrared spectroscopy (IFS 66, Bruker) was used for analysis of the main soil components clay, feldspar, silicate, carbonate, and sulfate. This method is based on the application of a multi-step iterative spectra exhaustion method in which the soil spectrum is decremented by a small fraction of the spectrum of the most probable component [HOBERT et al., 1993].

9.2.2.2 Results and Discussion

The soil depth profile was sampled 600 m to the Southwest of the Hettstedt metallurgical works from a clayey soil (>35% clay) of the Wipper meadow. Fig. 9-12 shows that the heavy metal emission in the Hettstedt district originated from the copper metallurgical industry [UMWELTBUNDESAMT, 1991]. The heavy metal emission was at a maximum in 1983 and 1985 and is nowadays at a low level because of the closure of most of the emitters. As would be expected from the high values of heavy metal dust emission in recent years, high soil concentrations were found for the elements zinc, copper,

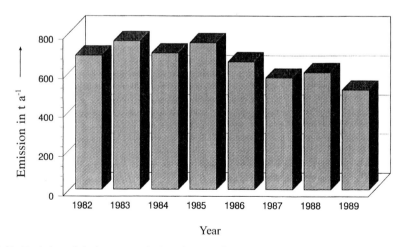

Fig. 9-12. Emission of the heavy metals Cu, Pb, Zn, Cd, and As in the Hettstedt district

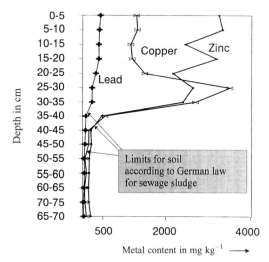

Fig. 9-13. Levels of copper, lead, and zinc versus soil depth at the sampling location in Hettstedt

and lead (Fig. 9-13). Zinc and lead have the same depth behavior, enrichment in the topsoil and normal concentrations in the deeper soil layers. The depth profile of copper shows high values (approximately 1000 mg kg^{-1}) in the 0–25 cm layer and has a large maximum between 25 and 35 cm. The geological background values of the investigated heavy metals are reached in the horizon deeper than 40 cm. The concentrations of the heavy metals copper, lead, and zinc are considerably in excess of the legally fixed thresholds according to the German law for sewage sludge [KLÄRSCHLAMMVER-ORDNUNG, 1992] (Fig. 9-13). According to FILIPINSKI and GRUPE [1990], anthropogenic heavy metal entry can be detected in meadow soils over a long period, and so these heavy metal depth profiles can be used to trace the historical industrial emission of the Hettstedt copper metallurgical industry [UMWELTBUNDESAMT, 1991].

Because the results obtained for the depth distribution of the three tracer metals resulting from anthropogenic pollution, the dendrogram of cluster analysis, computed with the whole data matrix, can clearly divide the soil into two main horizons (high heavy metal pollution: 0–35 cm and lower heavy metal pollution: 35–75 cm) in a multivariate manner (Fig. 9-14). Each of these main horizons can be split into two subgroups. To confirm and to refine the hypothesis arising from cluster analysis, MVDA, as supervised method, was performed assuming a 4-class-model each characterized by a different state of heavy metal pollution. Fig. 9-15 shows the successful separation of the four given classes with correct reclassification of each object.

FA was used to detect the latent information within the data set. The reduced factor analytical model consists of two common factors; they explain 86.6% of the data's variance.

For the interpretation of the common factors it is useful to consider the factor loading matrix after rotation (Tab. 9-5). Factor 1, which describes 69.8% of the common var-

9.2 Multivariate Statistical Evaluation and Interpretation of Soil Pollution Data 339

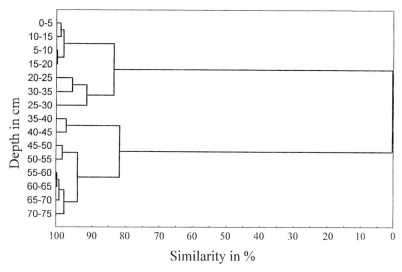

Fig. 9-14. Dendrogram obtained from cluster analysis according to WARD for the soil profile in Hettstedt

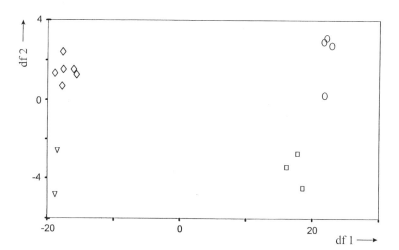

Fig. 9-15. Plot of the scores of discriminant function df 2 vs. scores of discriminant function df 1 for the different soil depth classes at the sampling location in Hettstedt.
  ○ class 1 – topsoil with high anthropogenic heavy metal pollution: 0–20 cm
  □ class 2 – horizon with extremely high copper pollution: 20–35 cm
  ▽ class 3 – transition horizon with rapidly decreasing heavy metal concentrations: 35–45 cm
  ◇ class 4 – subsoil with natural geogenic background concentrations: 45–75 cm

Tab. 9-5. Factor loading matrix of the Hettstedt soil profile after varimax rotation (loadings less than 0.5 in absolute value are set to zero)

Feature	Factor 1	Factor 2
Zn	0.93	0
Pb	0.92	0
As	0.91	0
C content	0.91	0
Ca	0.79	0
pH	0.64	0
Feldspar content	0.60	0
Cu	0.56	0
Mn	–0.84	0
Ba	–0.84	0
Clay content	–0.70	0.61
Fe	–0.69	0.70
Cr	0	0.85
Mg	0	0.82
Ni	0	0.81
Na	0	0.64

iance, has high positive factor loadings in the features Zn, Pb, As, and Cu as well as for carbon content, pH, and the concentrations of Ca and feldspar. Furthermore, the features Mn, Ba, clay, and Fe, determined mainly by geogenic influences, have negative factor loadings. This first factor can be interpreted as the contamination factor. It has high factor scores in the surface soil horizon and decreases in deeper horizons. This means that in the upper layer the anthropogenic influence predominates and more geogenic character is prevalent in the deeper layers. The chemometrically found hypothesis that the heavy metals emitted by anthropogenic activities are bound by soil organic matter has been confirmed by other environmental investigations [BLUME and HELLRIEGEL, 1981; SIMS, 1986].

The second factor, which describes 16.8% of the common variance, has positive loadings for Cr, Mg, Ni, Fe, Na, and clay content. The scores of this factor increase in deeper soil horizons. This factor characterizes the transfer of clay particles to deeper soil layers. Migration of anthropogenic contaminants is not connected with the transport of clay particles. In the upper layer anthropogenic influences predominate and in deeper layers the more geogenic character of the soil is prevalent. These results are in good agreement with investigations in the region of Hettstedt [UMWELTBUNDESAMT, 1991]. Those investigations detected that the anthropogenically emitted heavy metals occur in a form which is only slightly mobile.

9.2.2.3 Conclusions

- Multivariate methods are useful for describing depth profiles of soils and detecting latent information about data sets which could not be easily discovered by univariate analysis.
- The combined use of CA and MVDA yields objective information about the size and the number of soil horizons.
- Applying FA to major, minor, and trace components and elements in anthropogenically polluted soil profiles enables conclusions about main input sources and about the correlations between pollutants and soil parameters such as soil organic matter or clay content.
- Thus, the described analytical procedure in connection with statistical data analysis gives an overview of the metal status and can separate important parameters from those which are less important, which helps in the detection of antropogenic pollution.

9.3 Use of Robust Statistics to Describe Plant Lead Levels Arising from Traffic

9.3.1 Problem

In the field of toxic substances which pollute our environment, special importance must be attached to the heavy metal lead, which, until the late eighties arose primarily from automobile exhaust [EWERS and SCHLIPKÖTER, 1984]. For quantitative measurement of the impact of this lead, it is essential to establish the correlation between the lead content of the plants and soils, the traffic density, and the distance from the road.

Quantitative studies by means of parametric statistical methods are, however, often very unreliable because of high environment-related variations very often amounting to several orders of magnitude [FÖRSTNER and WITTMANN, 1983; EINAX, 1990]. In other words: environmental data sets often contain values which are extremely high or low, i.e. they are outliers in the statistical sense. Also, because environmental data are often not normally distributed, the application of parametric statistical methods results in distorted reflections of reality.

Application of robust statistics, especially methods of median statistics, for quantitative description of widely varying values may give information which can often be interpreted better than the results from normal parametric statistical methods.

The purpose of this study is the use of median statistics for quantitative and interpretable recording of the traffic-related impact of lead on soils [EINAX et al., 1991] .

9.3.2 Mathematical Fundamentals

The methods of robust statistics have recently been used for the quantitative description of series of measurements that comprise few data together with some outliers [DAVIES, 1988; RUTAN and CARR, 1988]. Advantages over classical outlier tests, such as those according to DIXON [SACHS, 1992] or GRUBBS [SCHEFFLER, 1986], occur primarily when outliers towards both the maximum and the minimum are found simultaneously. Such cases almost always occur in environmental analysis without being outliers in the classical sense which should be eliminated from the set of data. The foundations of robust statistics, particularly those of median statistics, are described in detail by TUKEY [1972], HUBER [1981], and HAMPEL et al. [1986] and in an overview also by DANZER [1989]; only a brief presentation of the various computation steps shall be given here.

On the basis of the computing scheme described by ROUSSEEUW and LEROY [1987] the evaluation of data sets to be analyzed was performed using multiple linear regression based on median statistics. For comparison with the following results, normal linear regression on the basis of GAUSSian error squares minimization (LS regression) is computed first. Then least median squares regression (LMS regression) is performed. To minimize the amount of computation for large data sets, "subsamples" are obtained via a random number generator; coefficient vectors a_j are computed from these by applying the minimization criterion:

$$\text{med}\,(y_i - x_i a_{1j} - a_{0j})^2 = \min! \tag{9-5}$$

The elimination of outliers is possible in an additional step. For this purpose the re-weighted least squares regression (RLS regression) is performed, the weighting factor w_i being [ROUSSEEUW and LEROY, 1987]:

$$w_i = \begin{cases} = 1, & \text{if } \left|\dfrac{e_i}{disp}\right| \leq 2.5 \\ = 0, & \text{if } \left|\dfrac{e_i}{disp}\right| > 2.5 \end{cases} \tag{9-6}$$

with the difference between the measured values y_i, and the computed data $\hat{y}_i$:

$$e_i = y_i - \hat{y}_i \tag{9-7}$$

The limit of 2.5 is arbitrary but quite reasonable because in a GAUSSian situation there will be very few residuals larger than 2.5 $disp$ [ROUSSEEUW and LEROY, 1987].
The dispersion:

$$disp = C\,(\text{med}\,e_i^2)^{\frac{1}{2}} \tag{9-8}$$

respresents the robust form of the standard deviation. The constant C is a factor introduced to achieve consistency in GAUSSian error distribution. After reweighting, the traditional LS regression, i.e. freed from outliers, is performed.

9.3.3 Experimental

In routine environmental monitoring of lead loading in the territory of the former GDR, samples were taken from the soil profile at depths between 0 and 20 cm (A_p-horizon) as parallel samples at various distances (0–25 m) from different highways and roads, the mean traffic density of which had also been determined. Plant specimens were also taken at the same locations. After the soil samples had been dried (<40 °C) lead was extracted with HNO_3 (c = 3%) [MACHELETT et al., 1986a]. The plant specimens were incinerated for 3 h at 480 °C. After cooling the residue was moistened with NH_4NO_3 solution (c = 10%), again treated at 480 °C for 3 h, and subsequently dissolved in HCl (c = 2.4%). The lead content of both extracts was determined by flame atomic absorption spectrometry [GRÜN et al., 1987].

9.3.4 Analytical Results and Chemometric Evaluation

The following computations were made on the basis of 226 observations consisting of the plant lead content, soil lead content (extracted with 1.5 mol L^{-1} HNO_3), road traffic density, and distance from the road. The computations were performed according to the procedure described by ROUSSEEUW and LEROY [1987]. Tab. 9-6 shows the univariate correlation coefficients of the linear regression between plant lead content on the one hand, and the factors soil lead content, traffic density, and distance from the road, on the other. Because of the high variability of the data only low correlation coefficients, i.e. statistically very unreliable correlations, are obtained when using LS regression (Eq. 2-53).

Whereas application of LMS regression (Eq. 9-5) results in considerable improvement of the rank correlation coefficients, the correlation coefficients obtained by RLS regression (with outlier elimination; Eq. 9-6) are again markedly lower.

Tab. 9-6. Different kinds of correlation coefficients of plant lead content with the other parameters investigated by LS, LMS, and RLS regression

	LS	LMS	RLS
Plant lead content/soil lead content	0.3013	0.7434	0.5473
Plant lead content/traffic density	0.3654	0.6976	0.4743
Plant lead content/distance from the road	−0.3839	−0.7788	−0.4014

Compared with the slopes of the LMS and RLS regression lines, the slopes of the LS regression lines are significantly higher because all measurement points were taken into account (Fig. 9-16). It is obvious that the variance of the points corresponding to higher values of plant lead content is higher than that of those corresponding to lower values. The results obtained by use of LMS regression and RLS regression are largely similar.

The coefficient of the LMS regression function for the relationship between plant and soil lead content, $a_{o,soil} = 0.90$ mg kg^{-1} Pb, corresponds to the median amount of lead taken up by the plants from atmospheric emissions if the soil lead content were zero,

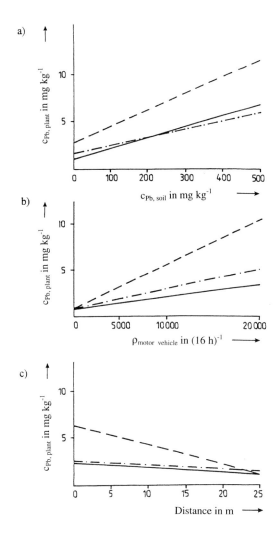

Fig. 9-16. Regressive relationships between plant lead content and (a) soil lead content, (b) traffic density, (c) distance from the road. (--- LS regression, — LMS regression, ⋯⋯ RLS regression). The data points were removed for easier visual comparison of the different regression models.

i.e. for $c_{soil} = 0$ mg kg^{-1} Pb. On the other hand the coefficient of the LMS regression function for the relationship between plant lead content and traffic density, $a_{o,vehicle} = 1.28$ mg Pb kg^{-1}, corresponds to the median lead uptake of the plants from soil if the traffic density were zero, i.e. for $\rho_{vehicle} = 0$ (16 h)$^{-1}$. So it can be demonstrated that description of the origins of the lead uptake of plants is possible by means of LMS regression.

It is obvious that LS regression yields values that are too high and cannot be used for interpretation of plant lead content. The introduction of the reweighted median-related values enables comparison of the results of the various LMS regressions with one another. For this purpose the regression coefficient a_{1j} of the function

$$c_{Pb,plant} = a_{0j} + a_{1j} \cdot x_{ij} \qquad (9\text{-}9)$$

is multiplied by the median $\tilde{x}_j$:

$$a_{1j}^* = a_{1j} \cdot \tilde{x}_j \qquad (9\text{-}10)$$

The resulting normalized equation is:

$$c_{Pb,plant} = a_{0j} + a_{1j}^* \frac{x_{ij}}{\tilde{x}_j} \qquad (9\text{-}11)$$

The plot of the normalized LMS regression results (Fig. 9-17) shows that the effect of the absolute values of the coefficients a_{1j}^* on plant lead content decreases in the order:

traffic density > distance from the road > soil lead content

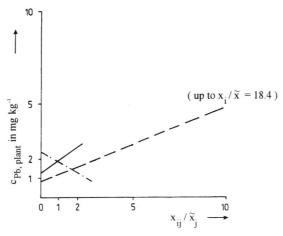

Fig. 9-17. Normalized LMS representation $c_{Pb,plant} = f(x_{ij}/\tilde{x}_j)$
(--- soil lead content, — traffic density, ······ distance from the road)

The very wide variation of the lead content of the investigated soil samples is, moreover, conspicuous. RLS regression yields qualitatively comparable results, but with lower correlation coefficients (see Tab. 9-6). This is primarily because many outliers were eliminated from the data obtained (Tab. 9-7). Since the measuring points weighted with $w_i = 0$ are not outliers in the classical sense, but environment-related variability, preference should be given to LMS regression when environmental data have to be analyzed. The RLS regression should primarily be reserved for cases with real outliers.

Tab. 9-7. Eliminated outliers in application of RLS regression

Regression relation	Number of eliminated observations
Plant lead content/soil lead content	55 ($\triangleq$24.34%)
Plant lead content/traffic density	44 ($\triangleq$19.47%)
Plant lead content/distance from the road	51 ($\triangleq$22.57%)

Because the correlations between the various factors influencing plant lead content are relatively low and of a similar magnitude (Tab. 9-8), the multiple linear model is not distorted by the existing collinearities in respect of the linear functions of the individual parameters.

Tab. 9-8. Matrix of the univariate correlation coefficients

	Plant lead content	Soil lead content	Traffic density	Distance from the road
Plant lead content	1.0000			
Soil lead content	0.3013	1.0000		
Traffic density	0.3654	0.2496	1.0000	
Distance from the road	−0.3839	−0.3965	−0.0444	1.0000

The multiple LMS regression according to the following equation:

$$c_{Pb, plant} = a_0 + a_1 \cdot c_{Pb, soil} + a_2 \cdot \rho_{vehicle} + a_3 \cdot l \qquad (9\text{-}12)$$

yields its normalized form, by analogy with Eq. 9-11:

$$c_{Pb, plant} = a_0 + a_1^* \cdot \frac{c_{Pb, soil}}{\tilde{c}_{Pb, soil}} + a_2^* \cdot \frac{\rho_{vehicle}}{\tilde{\rho}_{vehicle}} + a_3^* \cdot \frac{l}{\tilde{l}} \qquad (9\text{-}13)$$

The normalized coefficients of the variables (in mg kg^{-1} Pb) are $a_o = 1.4804$, $a_1^* = 0.0510$, $a_2^* = 0.4529$, and $a_3^* = -0.3107$. The order of the absolute values of these coefficients is comparable with the above findings from single LMS regression.

The advantage of applying the multiple LMS regression model offers the possibility of making robust estimates of the plant lead content, to be expected from a knowledge of such parameters as soil lead content, traffic density, and distance from the road.

9.4 Geostatistical Methods

9.4.1 Introduction

Anthropogenic activities considerably accelerate biogeochemical heavy metal cycles. The most important sources of heavy metal pollution of soils are emissions from metallurgical and heating factories, and the influences of agricultural fertilization. The heavy metals emitted are fixed predominantly in the upper zones of soil. Subsequent partial remobilization of the heavy metals may cause not only increased entry into groundwater, but also an increased input into useful plants and, therefore, into the human food chain. Heavy metal loading by transport of secondary dust, owing to the action of the wind is, moreover, also possible. These loadings require investigation of soils in the surroundings of actual or former emitters to assess both the concentration and the territorial distribution of the emitted pollutants. Even if the geological background is relatively homogeneous and the emission gradient of the pollutant is only small, the lateral distribution of heavy metals in soils is highly variable. An objective assessment of polluted areas on the basis of visual inspection of their original values is possible but very limited.

Geostatistical methods are suitable for describing and modeling the spatial structure of environmental data. In the following text, the applicability of geostatistical methods will be tested with the example of a big metallurgical works at Thuringia (Germany) [EINAX and SOLDT, 1995]. The purpose of this work is the objective assessment of the extent and the area of soil pollution caused by emissions from the plant. The mathematical basis of geostatistical methods applied in the following case study is described in detail in Section 4.4.

9.4.2 Experimental

The Maxhütte Unterwellenborn in Thuringia was one of the largest metallurgical factories in the former GDR. The neighborhood also contained a cement mill. A map of the region showing the situation is given in Fig. 9-18. The surroundings of the metallurgical factory and the cement mill were heavily polluted by particulate emissions over a period of some decades. Fig. 9-19 illustrates the impact of sedimented airborne particulates arising from the emissions in 1989.

Sixty-two soil samples were taken in a depth of 0–20 cm over an area of ca. 18 km^2 around the metallurgical factory and the cement mill. Samples were dried at 60 °C and

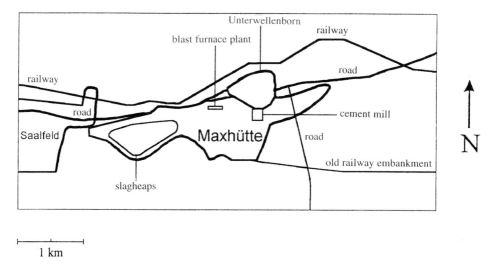

Fig. 9-18. Sketch of the region around the Maxhütte Unterwellenborn (Thuringia)

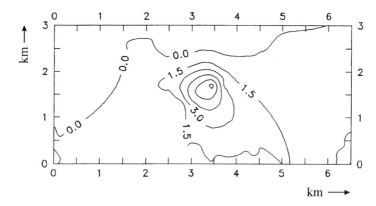

Fig. 9-19. Impact of pollution by sedimented airborne particulate emission in the surroundings of the metallurgical factory and the cement mill in 1989 (deposition rate in g m^{-2} d^{-1})

sieved with a mesh size of ≤ 2 mm. Soil (1 g) was extracted with a mixture of 21 mL HCl ($c = 32\%$, p.A., Merck) and 7 mL HNO$_3$ ($c = 65\%$, p.a., Merck), under reflux for 2 h [FACHGRUPPE WASSERCHEMIE, 1983]. After cooling, the suspension was filtered and the solution was diluted to 100 mL. The concentrations of the elements Ca, Cr, Cu, Fe, K, Mg, Mn, Na, Ni, and Zn were determined by optical emission spectrometry with inductively coupled plasma (Spectroflame; Spectro Kleve) [FACHGRUPPE WASSERCHEMIE, 1988], and Pb [FACHGRUPPE WASSERCHEMIE, 1981] and Cd [FACHGRUPPE WASSERCHEMIE, 1980] by atomic absorption spectrometry (PC 5100 ZL; Perkin Elmer) using the graphite furnace technique with ZEEMAN background correction. The analytical conditions, optimized for each element, are described in detail in the literature [EINAX and SOLDT, 1995].

9.4.3 Results and Discussion

9.4.3.1 Semivariogram Estimation and Analysis

Semivariograms shall be applied for the characterization of the spatial structure of the data. First, the data have to be tested for normal distribution. The results of the test hypothesis are acceptable for potassium only, in all other cases the hypothesis has to be rejected. After logarithmic transformation of the original data, however, a normal distribution can be obtained. Thus, the following calculations must be performed on the logarithms of the data – only for potassium can the untransformed values be used.

The semivariograms have to be computed for all elements, according to Eq. 4-28. In the following text the procedure is described for Ca, Cd, Fe, K, and Pb. Initially, only general all-directional semivariograms over all sampling points were calculated. Fig. 9-20 shows the semivariogram for cadmium. A spherical model can be best fitted to the semivariogram of cadmium.

The results of semivariogram estimation for some selected elements are represented in Tab. 9-9. The model of the semivariogram, nugget effect, sill, and range are given. The applicability of different semivariogram models is tested by cross-validation and the best model is selected. In the case investigated the sampling points for the different elements are correlated to a distance from 1.3 to 3.6 km. For each element the range is larger than the sampling distance, i.e. the sampling distance is close enough to reflect the typical properties of the area with regard to the elements investigated.

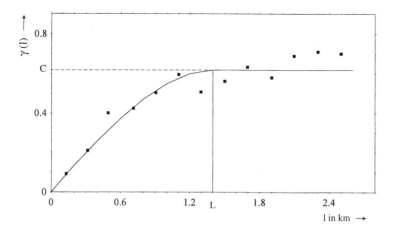

Fig. 9-20. All-directional semivariogram for cadmium

Tab. 9-9. Results of semivariogram estimation for some selected elements

Element	Model	Nugget effect	Sill	Range in km
Ca	Exponential	0.0	1.20	3.6
Cd	Spherical	0.0	0.62	1.4
Fe	Spherical	0.0	0.62	1.5
K	Exponential	1.0	4.30	3.6
Pb	Spherical	0.13	0.30	1.3

In a further step the anisotropy of the spatial distribution is investigated. Semivariograms for different directions are calculated for each element. Only for iron can anisotropy be detected. The semivariograms for the directions 0° (E–W) and 90° (N–S) with a tolerance of 10° are illustrated in Fig. 9-21. The semivariograms in the east–west and in north–south directions differ both in sill and range. The sampling points in the east–west direction correlate over a longer distance than in the north–south direction because the east–west direction is the main wind direction in the area investigated. The sampling variance is larger.

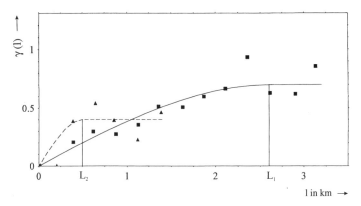

Fig. 9-21. Directional semivariogram for iron (▲ north–south direction, ■ east–west direction)

9.4.3.2 Kriging

On the basis of the metal concentrations at the irregularly distributed sampling points, the pollutant concentrations have been estimated by linear kriging according to Eqs. 4-33 to 4-35 for the points of a regular grid. The spatial distributions of the elements Fe, Cd, Pb, Ca, and K, as given by kriging, are represented as isoline plots in Fig. 9-22 a–e. A typical spatial distribution of each of the elements is visible. Elevated concentrations of iron and cadmium are observed in a narrow limited region only and are quite similar

(Fig. 9-22 a and b). They can be found in the immediate vicinity of the emission source. In a small area (see Tab. 9-10) the concentrations of cadmium exceed the index value of the category B of the "Dutch" list [1988] (Tab. 9-11) which implies a need for more detailed investigation. By means of the kriging estimation the critical polluted area can be well assessed. The locations of the concentration maxima enable conclusions to be drawn about the source of emission. The blast furnace works in the north of the Maxhütte proves to be the main source of iron and cadmium. This also corresponds to the impact of particulate emissions in the territory investigated (see also Fig. 9-19).

Tab. 9-10. Assessment of the polluted areas

Element	Concentration in µg g^{-1}	Category	Polluted area in 10^3 m^2
Cd	5– 20	B	590
Cr	250– 800	B	710
Cu	100– 500	B	5220
Ni	100– 500	B	560
Pb	150– 600	B	9290
	>600	C	460
Zn	500–3000	B	2590

Tab. 9-11. Index values of "Dutch" list [1988]

Element	B in µg g^{-1}	C in µg g^{-1}
Cd	5	20
Cr	250	800
Cu	100	500
Ni	100	500
Pb	150	600
Zn	500	3000

B – category for more detailed investigation; C – category for investigation in the event of redevelopment

The isoline plot of the element lead shows two maxima (Fig. 9-22 c). The first is close to the blast furnace works, similar to Fe and Cd. A small increase is visible in the direction of the slagheaps to the west of the investigated area. Here, the blast furnace slag was stored until it was used in the cement mill attached to the north–east of the Maxhütte. The second maximum is located by an old railway embankment in the south of the investigated area. This railway embankment is where blast furnace slag was deposited. The index value of category B of the "Dutch" list (150 µg g^{-1}) is greatly exceeded. The area extends from the slagheaps over the steelworks as far as the railway embankment. The index value of the category for investigations for sanitation (cate-

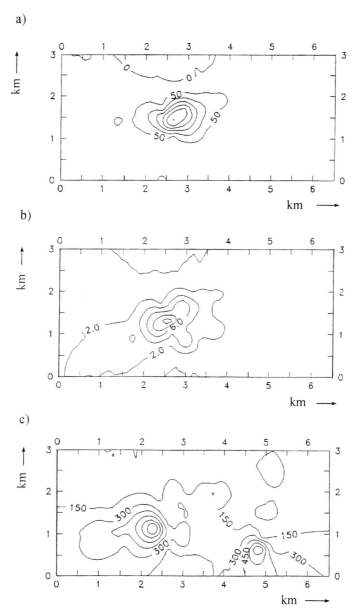

Fig. 9-22. Kriging estimation for some selected elements (isolines in µg g^{-1}): (a) iron, (b) cadmium, (c) lead, (d) calcium, (e) potassium

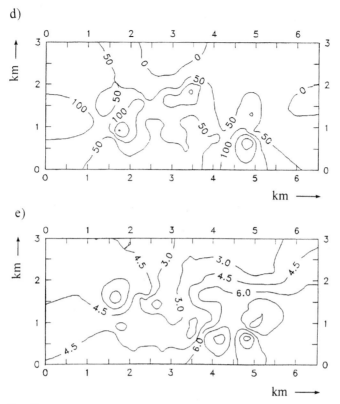

Fig. 9-22 (continued)

gory C) is exceeded in two areas, i.e. near to the blast furnace works and on the railway. The polluted areas are limited in size. The concentrations of lead exceed the index value for tolerable total amounts content in agricultural soils according to KLOKE [HEIN and SCHWEDT, 1993] (100 µg g^{-1}) over almost the entire area investigated. But the soil cannot be described as contaminated (4000 µg g^{-1}).

The element calcium shows a spatial distribution similar to that of lead (Fig. 9-22 d). The locations of the concentration maxima are the same; a small maximum has also arisen from the cement and slag mill in the north of the area investigated. There is no index value for calcium because calcium does not rank among the toxic elements.

The last element to be discussed is potassium. This element shows a completely different spatial distribution (Fig. 9-22 e). The concentration maxima are located in the agricultural areas to the east and the south–west of the region investigated. Increased concentrations cannot be detected in the region of the metallurgical factory. The potassium levels arise as a result of fertilization.

9.4.3.3 Assessment of the Polluted Area

Soil is a very complex mixture of components of geogenic or anthropogenic origin. The geogenic heavy metal content is determined by the material from which the soil is formed, the process of soil formation, and subsequent transport processes. Heavy metals are also contributed by human activities.

The extent of anthropogenic pollution can be assessed by means of index and limit values which are often fixed officially. A comparison between the index values of the KLOKE or "Dutch" lists (Tab. 9-11) and the pollutant values estimated by kriging enables the following conclusions to be drawn:

- The extent of areas with increased concentrations is different for each element and locally restricted.
- The concentrations of the elements Cd, Cr, Ni, and Zn exceed the tolerable contents according to KLOKE only close to the metallurgical factory. The concentrations of Cu and Pb are also high in the area around the slagheaps and an old railway embankment where blast furnace slag was deposited.
- Values higher than the index values indicating contamination in the KLOKE list [HEIN and SCHWEDT, 1993] cannot be found.
- The index values of the "Dutch" list [1988] are lower than those of the KLOKE list [HEIN and SCHWEDT, 1993]. The concentrations of the element Pb exceed the values of the category for investigations for sanitation (category C) at two locations, i.e. the eastern brink of the slagheaps and the above-mentioned railway embankment.

An assessment of the polluted area was performed on the basis of the kriging estimates. The results are presented in Tab. 9-10 and illustrated as isolines with the B value according to the "Dutch" list for the element Cd and both the B and the C values for the element Pb (Fig. 9-22 b and c).

9.4.3.4 Determination of the Minimum Number of Samples for Representative Assessment of the State of Pollutant Loading

The number of samples which must be taken in an area which is to be investigated to record the properties of interest without loss of information is still an interesting question. In other words: how many samples are necessary for representative assessment of the state of pollution? A method for the determination of the minimum number of samples is suggested below. For that purpose the advantages of the kriging method are used.

The kriging estimation is carried out by means of a semivariogram model (see Eqs. 4-29 to 4-31). According to Eq. 4-36 the kriging variance can be calculated for each estimated point. With standardization of the error of estimation as given by Eq. 4-38 it is possible to compare the goodness of fit by different kriging estimations. If fewer points in the investigated area are randomly selected for the kriging estimation, the error of the kriging estimates will also rise. The kriging model does not describe the spatial reality adequately – the kriging variance increases and the theoretical semivariogram model does not exactly describe the points of the semivariogram.

If the number of samples taken is high enough the plot of the mean of the standardized estimated values given by Eq. 4-38 vs. the number of sampling points shows a curve which approaches the required value of zero. As the number of samples is reduced this mean differs increasingly from zero.

The above-described procedure was performed for the element iron. The results are demonstrated in Fig. 9-23. The investigated area can be representatively described by 45 sampling points. This means the number of samples taken was sufficient for representative assessment of iron pollution. Frequent measurements in the future, for instance to investigate temporal changes of the soil pollution, can be performed on the basis of this minimum number of samples.

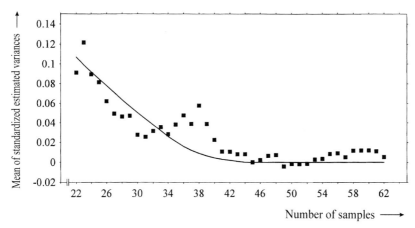

Fig. 9-23. Representation of the mean of standardized estimated variance vs. the number of samples for iron

9.4.4 Conclusions

– The univariate treatment of the results from investigation of the surroundings of a large metallurgical factory enables first conclusions to be drawn on the heavy metal pollution. But the high environmentally derived variability of the spatial distribution of pollutants strongly limits quantitative assessment of the polluted area.
– The state of pollution of the soil can be more objectively described by means of geostatistical methods. The computation of semivariograms and the use of kriging uncover spatial structures which are not discernible by means of simple univariate statistical tests.
– The semivariogram gives information about both the spatial structure of the element distribution and the anisotropy, i.e. the dependence of this distribution on direction.

- The kriging method is a useful tool for objective assessment of the extent and area of lateral distribution of the elements investigated. The advantage is that the estimated variance is minimized and, therefore, the estimate is, in principle, distortion-free. A further advantage, particularly for environmental investigations, is that the sampling points need not necessarily be distributed in regular grids.
- A first assessment of the polluted area is possible on the basis of a relatively small number of samples. The pollution maxima can be resolved in more detail by means of a second sampling process if it is required or necessary.
- The cost of sampling and analysis can be minimized by determining the minimum number of samples required for representative description of the area under investigation.

References

Ahrens, H., Läuter, J.: Mehrdimensionale Varianzanalyse, 2. Aufl., Akademie-Verlag, Berlin, **1981**
Blume, H.P., Hellriegel, T.: Z. Pflanzenernähr. Bodenk. 144 (**1981**) 156, 181
Danzer, K.: Fresenius' Z. Anal. Chem. 335 (**1989**) 869
Danzer, K., Marx, G.: Anal. Chim. Acta 110 (**1979**) 145
Davies, P.L.: Fresenius' Z. Anal. Chem. 331 (**1988**) 513
Derde, M.P., Massart, D.L.: Fresenius' Z. Anal. Chem. 313 (**1982**) 484
Ding, W.-H., Aldons, K.M., Briggs, R.G., Valente, H., Hilger, D.R., Connor, S., Eadon, G.A.: Chemosphere 25 (**1992**) 675
Doerffel, K., Wundrack, A. in: Fresenius, W., Günzler, H., Huber, W., Lüderwald, I., Tölg, G., Wisser, H. (Eds.): Analytiker-Taschenbuch, Bd. 6, Akademie-Verlag, Berlin, **1986**, pp. 37
Doerffel, K., Wundrack, A., Tarigopula, S.: Fresenius' Z. Anal. Chem. 324 (**1986**) 507
Doerffel, K., Meyer, N., Küchler, L.: Kurzreferate der Originalbeiträge zur COMPANA '88, Jena, **1988**, p. 17
"Dutch" list: Leidraad bodemsaniering, 1988 in: Hein, G., Schwedt, G.: Richt- und Grenzwerte, Wasser-Boden-Abfall-Chemikalien-Luft, 3. Aufl., Vogel, Würzburg, **1993**, p. 4-4
Einax, J.: Chemometrische Bewertung analytischer Untersuchungen von Metallen in der Umwelt, Habilitationsschrift, Friedrich-Schiller-Universität, Jena, **1990**
Einax, J., Soldt, U.: Fresenius' J. Anal. Chem. 351 (**1995**) 48
Einax, J., Oswald, K., Danzer, K.: Fresenius' J. Anal. Chem. 336 (**1990**) 394
Einax, J., Machelett, B., Danzer, K.: Fresenius' J. Anal. Chem. 339 (**1991**) 169
Einax, J., Machelett, B., Geiß, S., Danzer, K.: Fresenius' J. Anal. Chem. 342 (**1992**) 267
Ewers, U., Schlipköter, H.-W. in: Merian, E. (Ed.): Metalle in der Umwelt, Verlag Chemie, Weinheim, Deerfield Beach/Florida, Basel, **1984**, pp. 351
Fachgruppe Wasserchemie in der Gesellschaft Deutscher Chemiker in Gemeinschaft mit dem Normenausschuß Wasserwesen im DIN Deutsches Institut für Normung e.V. (Ed.): Deutsche Einheitsverfahren zur Wasser-, Abwasser- und Schlammuntersuchung, DIN 38406 Teil 19; Kationen (Gruppe E), Verlag Chemie, Weinheim, **1980**
Fachgruppe Wasserchemie in der Gesellschaft Deutscher Chemiker in Gemeinschaft mit dem Normenausschuß Wasserwesen im DIN Deutsches Institut für Normung e.V. (Ed.): Deutsche Einheitsverfahren zur Wasser-, Abwasser- und Schlammuntersuchung, DIN 38406 Teil 6; Kationen (Gruppe E), Verlag Chemie, Weinheim, **1981**
Fachgruppe Wasserchemie in der Gesellschaft Deutscher Chemiker in Gemeinschaft mit dem Normenausschuß Wasserwesen im DIN Deutsches Institut für Normung e.V. (Ed.): DIN 38414 Teil 7; Schlamm und Sedimente (Gruppe S), Verlag Chemie, Weinheim, **1983**

Fachgruppe Wasserchemie in der Gesellschaft Deutscher Chemiker in Gemeinschaft mit dem Normenausschuß Wasserwesen im DIN Deutsches Institut für Normung e.V. (Ed.): Deutsche Einheitsverfahren zur Wasser-, Abwasser- und Schlammuntersuchung, DIN 38406 Teil 22; Kationen (Gruppe E), VCH, Weinheim, **1988**

Filipinski, M., Grupe, M.: Z. Pflanzenernähr. Bodenk. 153 (**1990**) 69

Förstner, U., Wittmann, G.T.W.: Metal Pollution in the Aquatic Environment, 2nd Ed., Springer, Berlin, Heidelberg, New York, Tokyo, **1983**

Friberg, L., Nordberg, G.F., Vouk, V.B.: Handbook on the Toxicology of Metals, Elsevier, Amsterdam, New York, Oxford, **1986**

Grün, M., Eschke, H.D., Machelett, B., Kulick, J., Podlesak, W.: Schwermetalle in der Umwelt, Kolloquien des Institutes für Pflanzenernährung, Jena, 2 (**1987**) 13

Hampel, F.R., Ronchetti, E.M., Rousseeuw, P.J., Stahel, W.A.: Robust Statistics: The Approach Based on Influence Functions, Wiley, New York, **1986**

Hein, G., Schwedt, G.: Richt- und Grenzwerte, Wasser-Boden-Abfall-Chemikalien-Luft, 3. Aufl., Vogel, Würzburg, **1993**, p. 4-2

Henrion, A., Henrion, R., Urban, P., Henrion, G.: Z. Chem. 27 (**1987**) 56

Hobert, H., Meyer, K., Weber, I.: Vibrational. Spectrosc. 4 (**1993**) 207

Huber, P.J.: Robust Statistics, Wiley, New York, **1981**

Klärschlammverordnung - AbfKlärV: Bundesgesetzbl. Teil I, **1992**, pp. 912

Krieg, M., Einax, J.: Fresenius' J. Anal. Chem. 348 (**1994**) 490

Lebart, L., Morineau, A., Fenelon, J.-P.: Statistische Datenanalyse, Methoden und Programme, Akademie-Verlag, Berlin, **1984**

Lepp, N.W.: Effect of Heavy Metal Pollution on Plants, Pt. 1, Effects of Trace Metals on Plant Function, Applied Science, London, New Jersey, **1981 a**

Lepp, N.W.: Effect of Heavy Metal Pollution on Plants, Pt. 2, Metals in the Environment, Applied Science, London, New Jersey, **1981 b**

Lohse, H., Ludwig, R., Röhr, M.: Statistische Methoden für Psychologen, Pädagogen und Soziologen, Volk und Wissen, Berlin, **1986**

Lux, J., Fabig, F.: Praxis Naturwiss. Chemie 36 (**1987**) 7

Machelett, B., Grün, M., Podlesak, W.: Mengen- und Spurenelemente, Workshop Leipzig, Tagungsband, **1986 a**, pp. 219

Machelett, B., Grün, H., Podlesak, W. in: Anke, H. et al. (Eds.): 5. Spurenelementsymposium, Karl-Marx-Universität Leipzig, Friedrich-Schiller-Universität Jena, **1986 b**, pp. 293

Malinowski, E.R.: Factor Analysis in Chemistry, 2nd Ed., Wiley, New York, Chichester, Brisbane, Toronto, Singapore, **1991**

Massart, D.L., Kaufman, L.: The Investigation of Analytical Chemical Data by the Use of Cluster Analysis, Wiley, New York, **1983**

Möller, F. in: Podlesak, W., Grün, H. (Eds.): Schwermetalle in der Umwelt, Kolloquien des Institutes für Pflanzenernährung, Jena, 2 (**1987**) 36

Rousseeuw, P.J., Leroy, A.M.: Robust Regression and Outlier Detection, Wiley, New York, Chichester, Brisbane, Toronto, Singapore, **1987**

Rump, H.H., Krist, H.: Laborhandbuch für die Untersuchung von Wasser, Abwasser und Boden, 2. Aufl., VCH, Weinheim, **1992**

Rutan, S.C., Carr, P.W.: Anal. Chim. Acta 215 (**1988**) 131

Sachs, L.: Angewandte Statistik, 7. Aufl., Springer, Berlin, **1992**, pp. 362

Scheffer, F., Schachtschabel, P.: Lehrbuch der Bodenkunde, 13. Aufl., Enke, Stuttgart, **1992**, p. 405

Scheffler, E.: Einführung in die Praxis der statistischen Versuchsplanung, 2. Aufl., Deutscher Verlag für Grundstoffindustrie, Leipzig, **1986**, pp. 118

Sims, J.T.: Soil Sci. Soc. Am. 50 (**1986**) 367

Singer, R., Danzer, K.: Z. Chem. 24 (**1984**) 339

Tukey, J.W.: Robust Estimation of Location: Survey and Advances, Princeton University Press, Princeton, **1972**

Umweltbundesamt: Abschlußbericht zum Forschungs- und Entwicklungsvorhaben: Umweltsanierung des Großraumes Mansfeld, Eisleben, **1991**

Weber, E.: Grundriß der biologischen Statistik, 9. Aufl., Gustav Fischer, Jena, **1986**

10 Related Topics

10.1 Foods

10.1.1 Problem

A general problem in the monitoring and control of human food consists in assessing the real risk to the health of consumers. The following study was, therefore, implemented with the purpose of estimating the dietary intake of 11 mineral nutrients or pollutants in the eastern part of Thuringia (Germany) [HAHN et al., 1992].

The most realistic and reliable way of determining real exposure is investigation of food that is ready for consumption [KRUG, 1988; ELLEN et al., 1990]. For determination of real exposure two possibilities exist:

– analysis of the shopping basket
– analysis of 24-hour duplicate portions (total-day feeding)

Because of the difficulties connected with the representativeness of the shopping basket, 24-hour duplicate portions were analyzed to answer the questions:

– How many single samples are necessary for the representative assessment of the total dietary intake of nutrients and pollutants?
– Are there detectable differences between individual and communal feeding in the territory investigated?
– What is the real dietary load of the population?

10.1.2 Experimental

Seventy total-day portions from the communal kitchen in a hospital and 49 total-day portions of the individual food of 20 coworkers in a hygiene institute were the basis of the following investigations. The food samples were stored under refrigeration at a temperature of $-20\,°C$. All nonedible parts of the food were eliminated and the portions were homogenized by mixing. After homogenization a defined amount of the samples was lyophilized and homogenized for a second time. The so prepared samples were digested under pressure with a mixture of HNO_3 and H_2O_2 for 3 h at $170\,°C$. The concentrations of the elements Na, K, P, Ca, Mg, Fe, Mn, Cu, and Zn were determined by ICP-OES and the concentrations of Pb and Cd by graphite furnace AAS using the platform technique. The accuracies of the applied analytical procedures were proved with suitable reference material (Diät H-9, IAEA Vienna).

10.1.3 Results and Discussion

10.1.3.1 Univariate Aspects

The number of samples required for representative assessment can be determined as follows. The relative confidence interval for a given mean value $\bar{x}$ from the sample size of n measured values is given by:

$$\Delta x_{rel} = \frac{\Delta x}{\bar{x}} = \frac{t}{\sqrt{n}} \cdot \frac{s}{\bar{x}} \tag{10-1}$$

In Tab. 10-1 the calculated number of samples n required is demonstrated for probabilities of an error of the first kind of $\alpha = 0.1$ and $\alpha = 0.25$. With the exception of cadmium and lead the number of samples required is less than or equal to 10 for a probability of an error of the first kind of 25%. When the intake is well below the provisional tolerable weekly maximum, as in the case investigated [HAHN et al., 1992], the sample size for representative assessment can be reduced considerably.

Tab. 10-1. Mean values, standard deviations, and sample size for the daily intake of mineral nutrients in communal feeding

Element	$\bar{x}$ in mg d^{-1}	s in mg d^{-1}	n ($\alpha = 0.1$)	n ($\alpha = 0.25$)
Na	6120	970	10	2
K	2730	475	12	2
Ca	728	241	44	7
Mg	248	29	6	1
P	1140	200	12	2
Fe	16.4	6.3	58	10
Mn	2.86	0.57	16	3
Zn	10.4	2.1	17	3
Cu	0.97	0.29	36	6
Pb	0.023	0.0166	207	33
Cd	0.0135	0.0065	93	15

Another possibility for reducing expenditure in sampling and subsequent analysis is transition from 100 percent inspection to the method of sample control according to FELIX and LEMARIE [1964].

Here, the probability P is derived from:

$$k_i(P) = \frac{(T_{0i} - \bar{x}_i)}{s_i} \tag{10-2}$$

$k_i(P)$ – one-sided quantile of the GAUSSian distribution
T_{0i} – legally fixed threshold for the feature i

The risk of unknown threshold infringements α can be calculated according to:

$$\alpha = 1 - P^{n-1} \tag{10-3}$$

For cadmium and lead values of k_i are > 5, thus $P > 0.99999997$. Consequently the risk for a 10 percent sampling is negligible. It can, therefore, be assumed that a sample size of $n = 7$ is large enough to monitor the population when the intake level is this small [HAHN et al., 1992], assuming that the estimated values of the average $\bar{x}$ and the standard deviation s correspond sufficiently exactly to the characteristics of the population investigated.

10.1.3.2 Multivariate Statistical Data Analysis

The basis of the following data analysis is a row data matrix with 119 samples and 12 features (11 element concentrations and the mass of the dried total-day portions). The feature Cu has to be eliminated because some of the measurements are missing. The feature dried mass must also be deleted, because it is an extensive variable. The original values were then autoscaled.

10.1.3.2.1 Principal Components Analysis

Firstly PCA was performed according to the mathematical principles described in Section 5.4. The loadings of the three most important principal components are presented in Tab. 10-2.

Principal component 1 is highly loaded by elements introduced by salting of the food. The second principal component is highly loaded by Fe and P. The third principal component, which is highly loaded by Cd and Pb, hints at anthropogenically caused entry of these heavy metals.

10.1.3.2.2 Multivariate Analysis of Variance and Discriminant Analysis

MVDA with an *a priori* model of two classes is used to classify multidimensional differences between communal and individual feeding. The significant separation strengths of the individual features are: 1.036 for Na, 0.0928 for K, and 0.0426 for Mg.

Tab. 10-2. Loadings of the three most important principal components
(loadings < 0.500 are set to zero for greater clarity)

Feature	PC 1	PC 2	PC 3
Na	0.569	0	0
K	0.744	0	0
Ca	0.642	0	0
Mg	0.844	0	0
P	0.553	0.776	0
Fe	0	0.781	0
Mn	0	0	0
Zn	0.563	0	0
Pb	0	0	0.732
Cd	0	0	0.794
Eigenvalue λ_i	2.952	1.591	1.402
Variance in %	29.5	15.9	14.0

The other features do not separate the classes significantly. The optimum separation set of MVDA consists of Na only. The separation result of the two classes is:

Class	Centroid	Scatter radius
Communal feeding	−0.01	2.84
Individual feeding	0.54	2.84

Whereas the large overlap of the two classes shows the relatively high similarity between them, the discrimination error according to LACHENBRUCH [HENRION et al., 1988] of 13.4% demonstrates, on the other hand, that a largely definite discrimination is possible on the basis of the sodium content.

10.1.4 Conclusions

- The sample size for representative assessment of an investigated population can be estimated in relation to the exceeding of the individual threshold for a nutrient or pollutant.
- PCA is a useful tool for describing the different entry paths of nutrients and pollutants into human food.
- Differences in the food originating from communal or individual kitchens can be confirmed by means of MVDA. The case study investigated demonstrates the different effects arising from salting of the food.
- No significant differences could be detected between the concentrations of the toxic heavy metals Cd and Pb in communal food and in the food produced for individuals.

10.2 Empirical Modeling of the Effects of Interference on the Flame Photometric Determination of Potassium and Sodium in Water

The optimization of an analytical procedure used to solve a particular problem is an essential task for each analyst. The application of chemometric methods to this task is very useful and the principles and many applications have been described in the literature (see, for example, [SHARAF et al., 1986; MASSART et al., 1988]). In this section optimization of an analytical procedure is demonstrated for a particular case study in the field of routine environmental analysis [EINAX et al., 1989].

10.2.1 Problem

In routine water analysis the concentrations of potassium and sodium in drinking, surface, process, and sewage waters are often analyzed by flame photometry [INSTITUT FÜR WASSERWIRTSCHAFT, 1986]. In contrast with classical chemical methods the flame photometric determination of alkali metals is characterized by high detection power and high selectivity and sensitivity. This analytical method is fast and can be automated to a considerable extent. The influence on emission of other alkali and alkali earth elements present at higher concentrations can be eliminated by addition of a more easily ionizable alkali metal salt solution to the analyte solution as a spectroscopic buffer [SCHUHKNECHT and SCHINKEL, 1963]. It is also possible to obtain a series of calibration curves by addition of defined amounts of interfering ions [INSTITUT FÜR WASSERWIRTSCHAFT, 1986]. A further possibility for eliminating matrix interferences is the method of standard addition [DOERFFEL et al., 1990]. The method of standard addition can, however, compensate quantitatively only for nonspecific multiplicative interferences, for example transport interferences. It compensates only approximately for specific multiplicative interferences, such as differences in matrix composition and different analyte binding forms in the sample and the added solution [WELZ, 1986]. Additive interferences, such as spectral interferences and changes of analyte concentrations as a result of contamination-decontamination processes, cannot be eliminated by means of the method of standard addition.

In the following text it shall be demonstrated that possible interferences on the flame photometric determination of potassium and sodium in water can be described and eliminated by empirical mathematical modeling.

10.2.2 Theoretical Fundamentals

The influence of water components on the flame photometric determination of potassium and sodium can be detected by factorial experiments. By application of multifactorial plans according to PLACKETT and BURMAN the qualitative determination of the influence of various variables is possible with relatively few experiments [SCHEFFLER, 1986]. For mathematical fundamentals see Chapter 3.

The basis of the following modeling is the multifactorial plan, represented in Tab. 10-3.

Tab. 10-3. Applied multifactorial plan

Experiment	Variable i, expressed by the transformed concentration x_i								Response
k	x_0	x_1	x_2	x_3	x_4	x_5	x_6	x_7	y_k
1	1	1	1	1	−1	1	−1	−1	y_1
2	1	1	1	−1	1	−1	−1	1	y_2
3	1	1	−1	1	−1	−1	1	1	y_3
4	1	−1	1	−1	−1	1	1	1	y_4
5	1	1	−1	−1	1	1	1	−1	y_5
6	1	−1	−1	1	1	1	−1	1	y_6
7	1	−1	1	1	1	−1	1	−1	y_7
8	1	−1	−1	−1	−1	−1	−1	−1	y_8

The fundamentals and some case studies of experimental design on the basis of different factorial plans and the following empirical modeling were described by KOSCIELNIAK and PARCZEWSKI [1983; 1985].

A model for the description of the effects of interference can be formulated as follows.

The connection between the response (or a derived function), y, and the concentrations of the i analyzed components is given by the polynomial model:

$$\hat{y} = a_0 + \sum_{i=0}^{n} a_i x_i + \sum_{\substack{i,j=1 \\ (i \leq j)}}^{n} a_{ij} x_i x_j + \cdots \qquad (10\text{-}4)$$

x_i and x_j – transformed concentrations according to $x_i = f(c_i)$

The transformation of the original concentrations, c, can be achieved by use of underlying nonlinear functions, for example parabolic, hyperbolic, or logarithmic dependencies [KOSCIELNIAK and PARCZEWSKI, 1985]. In this case study the concentrations were transformed linearly:

$$x_i = \frac{c_i - \bar{c}_i}{\bar{c}_i} \qquad (10\text{-}5)$$

with

$$\bar{c}_i = \frac{c_i^h - c_i^l}{2} \tag{10-6}$$

c_i^h and c_i^l – the high and the low concentration levels of the multifactorial plan

The application of this polynomial model to responses of factorial plans enables the computation of the calibration hyperplane for the investigated system. Modeling of interference effects on the atomic absorption spectrometric determination of calcium is described by KOSCIELNIAK and PARCZEWSKI [1983; 1985].

It has to be remarked that the model discussed above is valid for the description of linear relationships between concentration and the response. If the concentration range has to be extended further, nonlinear effects cannot be excluded [ZWANZIGER et al., 1988]. Then the whole concentration range should be split into two experimental plans for separate treatment of the high and the low concentration levels. Accordingly, two separate linear models can be calculated for the different experimental plans.

10.2.3 Experimental

The measurements were performed with a FLAPHO 4 flame photometer (Carl Zeiss Jena) using a propane-air flame. The influence of sulfate, chloride, and phosphate ions and the mutual influence of alkali metal ions on the determination of potassium and sodium ions were investigated. Anions were selected on the basis of previous experiments to determine their interference properties. The multifactorial plan was realized in the concentration levels represented in Tab. 10-4.

Tab. 10-4. Concentration levels of the applied multifactorial plan

x_i	Component	c_i^h ($x_i^h = 1$) in mg L^{-1}	c_i^l ($x_i^l = -1$) in mg L^{-1}
1	Sodium	15	0
2	Potassium	5	0
3	Pseudovariable	–	–
4	Sulfate	400	0
5	Pseudovariable	–	–
6	Chloride	350	0
7	Phosphate	7	0

The concentration range for potassium and sodium corresponds to the concentration range of these components in drinking waters in the investigated area over a period of several years. The concentrations of sulfate, chloride, and phosphate correspond to the

366   10 Related Topics

thresholds for drinking water, valid at the time of the investigation [TGL 22433, 1971]. Two pseudovariables were introduced with the purpose of estimating the error in the experiments. The calibration curve was evaluated by linear regression.

10.2.4 Results, Mathematical Modeling, and Discussion

The measured results and the standard deviations of the replicates for the flame photometric determinations of potassium and sodium are presented in Tab. 10-5. The following steps of testing and mathematical modeling correspond to the steps described more detailed in Section 3.3.

Tab. 10-5. Measured results and standard deviations of replicates for the flame photometric determination of potassium and sodium (in µg mL^{-1})

Experiment	y_K	y_{Na}	s_K	s_{Na}
1	5.07	15.66	0.06	0.17
2	5.19	14.42	0.06	0.15
3	0.08	13.94	0.00	0.06
4	4.83	0.16	0.00	0.00
5	0.12	14.35	0.00	0.12
6	0.06	0.15	0.00	0.00
7	4.90	0.41	0.06	0.00
8	0.04	0.05	0.00	0.00

10.2.4.1 Test for Variance Homogeneity

The measured results are tested for variance homogeneity by means of the multiple variance comparison according to COCHRAN [KRAUSE and METZLER, 1983]:

$$G = \frac{s_{max}^2}{\sum_{k=1}^{M} s_k^2} \qquad (10\text{-}7)$$

M – number of calibration levels
s_k – standard deviation at one calibration level k
s_{max}^2 – maximum variance from all s_k^2

For potassium and sodium the variance homogeneity can be assumed with a critical probability of an error of the first kind error of $\alpha = 0.01$.

10.2.4.2 Calculation and Testing of the Regression Coefficients

On the basis of the multifactorial plan (Tab.10-3) the specific model can be formulated as:

$$y = a_0 + \sum_{i=0}^{7} a_i x_i \tag{10-8}$$

The coefficients a_0 to a_7 are calculated by multiplication of the vector x_i by the response vector y, and division by the number of solutions (experiments), for example:

$$a_4 = \frac{-1 \cdot y_1 + 1 \cdot y_2 - 1 \cdot y_3 - 1 \cdot y_4 + 1 \cdot y_5 + 1 \cdot y_6 + 1 \cdot y_7 - 1 \cdot y_8}{8} \tag{10-9}$$

The variables x_3 and x_5 represent pseudovariables. The components of the vector x_3 can be represented by multiplication of the corresponding components of the vectors x_1, x_6, and x_7 or x_2, x_4, and x_6. By analogy, the components of the vector x_5 can be obtained by multiplication of the vector components x_1, x_4, and x_6 or x_2, x_6, and x_7. From this the following equivalent model equations result for the investigated system:

$$y = a_0 + a_1 x_{Na^+} + a_2 x_{K^+} + a_3 x_{Na^+} x_{Cl^-} x_{PO_4^{3-}} + a_4 x_{SO_4^{2-}} \\ + a_5 x_{Na^+} x_{SO_4^{2-}} x_{Cl^-} + a_6 x_{Cl^-} + a_7 x_{PO_4^{3-}} \tag{10-10}$$

and

$$y = a_0 + a_1 x_{Na^+} + a_2 x_{K^+} + a_3 x_{K^+} x_{SO_4^{2-}} x_{Cl^-} + a_4 x_{SO_4^{2-}} \\ + a_5 x_{K^+} x_{Cl^-} x_{PO_4^{3-}} + a_6 x_{Cl^-} + a_7 x_{PO_4^{3-}} \tag{10-11}$$

a_3 and a_5 are the coefficients of the interaction of the investigated water components and the other a_i are the coefficients of the direct influences of the other ions on the response.

(a) Test of the Regression Coefficients

The regression coefficients obtained for potassium and sodium have to be proved for significance. Firstly the average measurement error s_y is calculated according to:

$$s_y = \sqrt{\frac{\sum_{k=1}^{M} s_k^2}{M}} \tag{10-12}$$

Otherwise the error of the regression coefficients s_a is given by:

$$s_a = \frac{s_y}{\sqrt{M \cdot m}} \tag{10-13}$$

m – number of replicates for the k individual solutions ($m = 5$)

The test of the regression coefficients is based on the t-test according to STUDENT:

$$t_i = \frac{|a_i|}{s_a} \tag{10-14}$$

Deleting the nonsignificant coefficients from Eq. 10-10 the following models result:

Potassium:

$$y = 2.536 + 0.079\, x_{Na^+} + 2.461\, x_{K^+} + 0.031\, x_{SO_4^{2-}} \\ - 0.016\, x_{Na^+}\, x_{SO_4^{2-}}\, x_{Cl^-} - 0.054\, x_{Cl^-} \tag{10-15}$$

Sodium:

$$y = 7.392 + 7.200\, x_{Na^+} + 0.270\, x_{K^+} + 0.148\, x_{Na^+}\, x_{Cl^-}\, x_{PO_4^{3-}} - 0.060\, x_{SO_4^{2-}} \\ + 0.188\, x_{Na^+}\, x_{SO_4^{2-}}\, x_{Cl^-} - 0.178\, x_{Cl^-} - 0.225\, x_{PO_4^{3-}} \tag{10-16}$$

By analogy with Eq. 10-11 equivalent expressions to Eqs. 10-15 and 10-16 can be derived for the interactions K^+-SO_4^{2-}-Cl^- and K^+-Cl^--PO_4^{3-}. The calculation of the transformed concentrations x_i takes place according to Eqs. 10-5 and 10-6.

(b) Test for Adequacy of the Models

This test is based on the FISHER test:

$$F = \frac{s_A^2}{s_y^2} \tag{10-17}$$

Two variances have to be considered: the variance of the measurements s_y^2 and the variance of adequacy s_A^2:

$$s_A^2 = \frac{\sum_{k=1}^{M} (\bar{y}_k - \hat{y}_k)^2}{M - o} \tag{10-18}$$

$\bar{y}_k$ – average response of the kth solution
$\hat{y}_k$ – response calculated according to the model equation
o – number of significant coefficients in the model

Potassium:

a) Considering all terms in model 10-15
   $F = 2.9$; critical value $F(2;32;0.01) = 5.4$
b) Neglecting the terms of mixed interactions Na^+-SO_4^{2-}-Cl^- (or K^+-Cl^--PO_4^{3-})
   $F = 3.5$; critical value $F(3;32;0.01) = 4.5$

Sodium:

a) Model 10-16 is adequate, because all terms from the multifactorial plan are considered.
b) Neglecting the terms for mixed interactions Na^+-SO_4^{2-}-Cl^- (or K^+-Cl^--PO_4^{3-})
   and Na^+-Cl^--PO_4^{3-} (or K^+-SO_4^{2-}-Cl^-)
   $F = 131.2$; critical value $F(2;32;0.01) = 5.4$

The coefficient a_5 of model 10-15 is of little importance only, because the model is adequate even without this coefficient of the mixed interactions. Furthermore $t(a_5) = 2.81$ is only slightly higher than the critical value $t(32;0.01) = 2.74$. In contrast, in model 10-16 all interactions are important and significant.

(c) Calculation of the Analytical Error Caused by Interference Effects

Eqs. 10-15 and 10-16 describe the calibration hyperplanes. Filling in the average values of the concentration levels in the applied factorial plan for the interfering components, gives the following equations for the coherence between the response and the analyte concentration:

Potassium:

$$y = 0.075 + 0.984\, c_{K^+} \quad (10\text{-}19)$$

Sodium:

$$y = 0.192 + 0.974\, c_{Na^+} \quad (10\text{-}20)$$

with c_i in µg mL^{-1}

The potassium and sodium concentrations calculated according to these calibration equations and the resulting relative errors are represented in Fig. 10-1.

The effects of interference on the flame photometric determination of potassium can be described by the discussed model with only a slight error (Fig. 10-1 a). The effects of interference on the flame photometric determination of sodium are greater, as the model equation 10-16 shows. Also in this case the deviations of the calculated sodium concentrations from the true values are relatively small (Fig. 10-1 b). The comparison

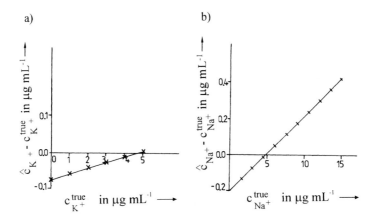

Fig. 10-1. Comparison of empirically modeled and true potassium (a) and sodium (b) concentrations

shows (Tab. 10-6) that the accuracy of the measured results is definitely improved by modeling the effects of interference.

Tab. 10-6. Average absolute errors of analyte concentrations (in μg mL^{-1})

Element	True value	Standard deviation of replicates	Deviation of measured value from true value	Deviation of modeled value from true value
Potassium	0	0.000	0.075	0.076
	5	0.090	0.132	0.002
Sodium	0	0.000	0.192	0.201
	15	0.250	0.738	0.424

10.2.5 Conclusions

- The effects of interference on the flame photometric determination of potassium and sodium can be detected. Quantitative description of these interference effects is possible by empirical modeling on the basis of experimental design, i.e. by applying a multifactorial plan.
- The cost of the experiment (in this case: preparation and analysis of 8 model solutions) is only slightly increased in comparison with the cost of direct calibration, so the described procedure can be used for routine water analysis.
- For the purpose of the analysis the interfering water components have to be determined and the concentration levels for the multifactorial plan have to be defined.

- The experimental design, i.e. the combination of multifactorial plans and empirical modeling, is not limited in application to this specific case study. It can be enlarged to encompass other similar problems in the field of analytical chemistry. Other mathematical methods, for instance LMS regression (see Section 9.3.2) or PLS regression (see Section 5.7.2), can also be used for empirical modeling.

References

Doerffel, K.: Statistik in der analytischen Chemie, 5. Aufl., Deutscher Verlag für Grundstoffindustrie, Leipzig, **1990**

Doerffel, K., Eckschlager, K., Henrion, G.: Chemometrische Strategien in der Analytik, Deutscher Verlag für Grundstoffindustrie, Leipzig, **1990**, pp. 62

Einax, J., Baltes, U., Koscielniak, P.: Z. ges. Hyg. 35 (**1989**) 211

Ellen, G., Egmond, E., van Loon, W., Sahertian, E.T., Tolsma, K.: Food Additives Contam. 7 (**1990**) 207

Felix, M., Lemarie, M.: Chem. Techn. 16 (**1964**) 359

Hahn, K., Günther, G., Einax, J.: Forum Städte-Hygiene 43 (**1992**) 77

Henrion, G., Henrion, A., Henrion, R.: Beispiele zur Datenanalyse mit BASIC-Programmen, Deutscher Verlag der Wissenschaften, Berlin, **1988**, pp. 134

Institut für Wasserwirtschaft Berlin (Ed.): Ausgewählte Methoden der Wasseruntersuchung, Bd. 1, Gustav Fischer, Jena, **1986**, pp. 153 and pp. 169

Koscielniak, P., Parczewski, A.: Anal. Chim. Acta 153 (**1983**) 111

Koscielniak, P., Parczewski, A.: Anal. Chim. Acta 177 (**1985**) 197

Krause, B., Metzler, P.: Angewandte Statistik, Verlag der Wissenschaften, Berlin, **1983**, pp. 277

Krug, W.: Biblthea Nutr. Dieta 41 (**1988**) 84

Massart, D.L., Vandeginste, B.G.M., Deming, S.N., Michotte, Y., Kaufman, L.: Chemometrics: a Textbook, Elsevier, Amsterdam, Oxford, New York, Tokyo, **1988**, pp. 293

Scheffler, E.: Einführung in die Praxis der statistischen Versuchsplanung, 2. Aufl., Deutscher Verlag für Grundstoffindustrie, Leipzig, **1986**, p. 118

Schuhknecht, W., Schinkel, H.: Z. Anal. Chem. 194 (**1963**) 161

Sharaf, M.A., Illman, D.L., Kowalski, B.R.: Chemometrics, Wiley, New York, Chichester, Brisbane, Toronto, Singapore, **1986**, pp. 23

TGL 22 433: Trinkwasser-Gütebedingungen, Fachbereichsstandard der DDR, **1971**

Welz, B.: Fresenius' Z. Anal. Chem. 325 (**1986**) 95

Zwanziger, H., Steinführer, A., Werner, G., Hofmann, R.: Fresenius' Z. Anal. Chem. 330 (**1988**) 478

Appendix

With the few numbers given in this appendix the reader should be enabled to understand the structure of statistical tables and to follow the outlined examples of lower degrees of freedom. For detailed tables the reader is referred to statistical textbooks, as for example [GRAHAM, R.C.: Data Analysis for the Chemical Sciences, VCH, New York, Weinheim, Cambridge, 1993, pp. 381].

Detailed values of distribution functions can be computed using appropriate statistical software.

Useful approximations or even complete sub-programs to be used in homemade programs are given, e.g., by HASTINGS [1955], McCORMICK and ROACH [1987], NOACK and SCHULZE [1980], THIJSSEN and van der WIEL [1984] (for the corresponding references see Chapter 2).

Symbols used in the tables:
f – degree of freedom
k – quantile
P – probability
α – error probability

Appendix A
Selected Quantiles $k(P)$ of the GAUSSian (Normal) Distribution

$\pm k(P)$	P in %	($\alpha = 1 - P$; $q = 1 - \alpha/2$)
1	68.3	
2	95.44	
3	99.73	
1.64	90	(10%; $q = 0.95$)
1.96	95	(5%; $q = 0.975$)
2.58	99	(1%; $q = 0.995$)

Appendix B
Selected Quantiles $k(f; P)$ of the STUDENT t-Distribution

f	$k(f, P = 0.95)$ $q = 1 - \alpha/2 = 0.975$	$k(f, P = 0.95)$ $q = 1 - \alpha = 0.95$	$k(f, P = 0.99)$ $q = 1 - \alpha/2 = 0.995$
1	12.71	6.31	63.66
2	4.30	2.92	9.92
3	3.18	2.35	5.84
4	2.78	2.13	4.60
5	2.57	2.02	4.03
9	2.26	1.83	3.25
10	2.23	1.81	3.17
11	2.20	1.80	3.11
12	2.18	1.78	3.06
15	2.13	1.75	2.95
20	2.09	1.73	2.84
100	1.98	1.66	2.63
∞	1.96	1.65	2.58

Appendix C
Selected Values for Two-Sided Confidence Intervals of Standard Deviations

The table contains lower and upper multipliers for $P = 0.95$ (95%), for calculation of the confidence interval according to:

$$k_{lower} \, s \leq \sigma \leq k_{upper} \, s$$

n	k_{lower}	k_{upper}
3	0.52	6.28
4	0.57	3.73
5	0.60	2.87
6	0.62	2.45
8	0.66	2.04
10	0.69	1.83
12	0.71	1.70
15	0.73	1.58
20	0.76	1.46

Appendix D
Selected Quantiles $k(f_1; f_2; P = 1-\alpha/2)$ of the FISHER F-Distribution

Critical values with $\alpha = 0.05$ ($P = 95\%$), **two-tailed values for F-tests**
(see also next table)

f_1 \ f_2	1	2	3	4	5	6	7	8	9
1	648	800	864	900	922	937	948	957	963
2	38.5	39.0	39.2	39.2	39.3	39.3	39.4	39.4	39.4
3	17.4	16.0	15.4	15.1	14.9	14.7	14.6	14.5	14.5
4	12.2	10.6	9.98	9.60	9.36	9.20	9.07	8.98	8.90
5	10.0	8.43	7.76	7.39	7.15	6.98	6.85	6.76	6.68
6	8.81	7.26	6.60	6.23	5.99	5.82	5.70	5.60	5.52
7	8.07	6.54	5.89	5.52	5.29	5.12	4.99	4.90	4.82
8	7.57	6.06	5.42	5.05	4.82	4.65	4.53	4.43	4.36
9	7.21	5.71	5.08	4.72	4.48	4.32	4.20	4.10	4.03

Selected Quantiles $k(f_1; f_2; P = 1-\alpha)$ of the FISHER F-Distribution

Critical values with $\alpha = 0.05$ ($P = 95\%$), **one-tailed values for analysis of variance**

f_2 \ f_1	1	2	3	4	5	6	7	8	9
1	161	200	216	225	230	234	237	239	241
2	18.5	19.0	19.2	19.2	19.3	19.3	19.4	19.4	19.4
3	10.1	9.55	9.28	9.12	9.01	8.94	8.89	8.85	8.81
4	7.71	6.94	6.59	6.39	6.26	6.16	6.09	6.04	6.00
5	6.61	5.79	5.41	5.19	5.05	4.95	4.88	4.82	4.77
6	5.99	5.14	4.76	4.53	4.39	4.28	4.21	4.15	4.10
7	5.59	4.74	4.35	4.12	3.97	3.87	3.79	3.73	3.68
8	5.32	4.46	4.07	3.84	3.69	3.58	3.50	3.44	3.39
9	5.12	4.26	3.86	3.63	3.48	3.37	3.29	3.23	3.18

Appendix E
Selected Critical Values for DIXON Outlier Tests

Values according to ISO 5725-1986 (E) in [ISO Standards Handbook, Statistical Methods, 3rd Ed., ISO, Geneva, 1989, pp. 459].

In the ISO document mentioned one may find formulae to compute critical values for $n > 40$ or for n values not given below.

n	$\alpha = 0.05$	$\alpha = 0.01$	Index numbers for	
			dix_{left}	dix_{right}
3	0.970	0.994	$A = 2$	$A = N$
4	0.829	0.926	$B = 1$	$B = N - 1$
5	0.710	0.821	$C = N$	$C = N$
6	0.628	0.740	$D = 1$	$D = 1$
7	0.569	0.680		
8	0.608	0.717	$A = 2$	$A = N$
9	0.564	0.672	$B = 1$	$B = N - 1$
10	0.530	0.635	$C = N - 1$	$C = N$
11	0.502	0.605	$D = 1$	$D = 2$
12	0.479	0.579		
13	0.611	0.697	$A = 3$	$A = N$
14	0.586	0.670	$B = 1$	$B = N - 2$
15	0.565	0.647	$C = N - 2$	$C = N$
20	0.489	0.567	$D - 1$	$D = 3$
25	0.443	0.517		
30	0.412	0.483		
35	0.388	0.458		
40	0.371	0.438		

Index

additive component model 212
adequacy, test 62 f.
analysis of variance 46
– design 86 ff.
– fixed effect model 87
– principle 46
– random effect model 87
– sampling, example 111 f.
– type 1 87
– type 2 87
analytical process, steps 4
ANDREWS plot 149 f.
anisotropy 350
ANOVA see analysis of variance
anthropogenic factor 279
apparent error rate 186
ARIMA (autoregressive integrated moving average) 234
– components 244 f.
– error 242 ff.
– example 240 ff.
– forecasting 246
– notation 236
– order 238 ff.
– seasonality 237 f.
– specification 237
– trend 237
ARMA see autoregressive moving average
atmosphere 251 ff.
– interpretation of pollution data 252 ff.
– sampling, impact of emissions 251 f.
autocorrelated variables and errors 225
autocorrelation 222 ff.
– coefficient 222 ff.
– function 223 f.
– multivariate 229 ff.
autoregression 234 ff.
– explanatory variable 225 ff.
autoregressive integrated moving average see ARIMA
autoregressive moving average 234 ff.
autoregressive process 223, 234
– first-order 236
– order 223, 226

autoscaling 141 f., 155
average linkage 158

backshift operator 234 f.
backward strategy 187
BENEDETTI-PICHLER's theory 106
between-class scattering 183
binding forms 300 ff.
biplot 167
blank 50
blank statistics 66 ff.
BOX and WILSON, method 90 ff.
box-whisker plot 150 ff.
– multiple 151 f.

CA see cluster analysis
calibration 64 f.
calibration statistics 66 ff.
canonical correlation analysis 179 ff.
– example 180 ff.
– interpretation 180
CCA see canonical correlation analysis
censoration 43
central value 78 f.
centre of gravity 47
chemometric literature, overview 16 ff.
chemometric methods
– classification 3
– need to apply 8 ff.
– purpose in environmental analysis 12 f.
– typical questions 13 f.
chemometrics
– definition 2 f.
– development 1 f.
– journals 2
– topics 4
CHERNOFF faces 148 f.
city block metrics 154
classification
– by discriminant functions 185 ff.
– error 186 f.
cluster analysis 153 ff.
– example 159 ff.
– impact of emissions 256 f., 271

– soil   321 f.
– soil depth profile   338 f.
– soil pollution   331 f.
cluster solution
– comparison   157
– selection   157
clustering
– agglomerative   156
– divisive   156
– hierarchical   156
– nonhierarchical   156
clustering of features   155
COCHRAN-test   366
coefficient of determination   48
– test   60 f.
communality   172 f.
complete linkage   158
confidence interval
– mean   33
– single measurement   32 f.
confidence level   34
confidence range   64 ff.
confounded estimate   75
cooperative test   43 ff.
correlation   47 ff.
correlation analysis   222 ff.
correlation coefficient   48
covariance   48
critical F-values   376 f.
critical signal value   66
critical t-values   374
cross-correlation
– example   224 f.
– function   224 f.
– multivariate   229 ff.
cross-validation   186, 200 f., 349 f.
– kriging   119 f.
CUSUM   216

DA *see* discriminant analysis
data preprocessing   155
degree of freedom   33
dendrogram   156
density function   26, 32
design *see also* experimental design   73 ff.
– analysis of covariance   88 ff.
– analysis of variance   86 ff.
– balanced   73
– complete   73 f.
– composite   75
– computation scheme   80
– core   79
– cross-classified   75
– extending   80 f.

– fractional factorial   74 f.
– generalizing   77 ff.
– interpretation   85
– levels of factors   76
– main effects   81 ff.
– model adequacy, test   83 f.
– nested   75 f.
– optimum   74
– randomized   75
– regression coefficients   81 ff.
– saturated   74
– symmetrical   73
differencing   214, 237 f.
– seasonal   214
– simple   214
discriminant analysis   184 ff.
– example   189 ff.
– feature reduction   187 f.
– score   185
– standardized coefficient   185
discriminant function   184 ff.
discriminant score   185
dispersion   342 f.
distance   153 f.
distance matrix   155
distribution
– CHEBYCHEV's inequality   34
– environmental data   11
– FISHER   37
– parameters   28 ff.
– STUDENT   37
distribution function, integrated   26
disturbance   236
DIXON outlier tests, critical values   378
DIXON test   41 f.
dummy variable   220 ff.
Dutch list   351

eigenvalue   166
eigenvector   166
– normalized   169 f.
environmental analysis
– characteristics   6 f.
– steps   6
environmental investigation, questions   250
environmetrics   3
error
– calibration offset   55
– predicted values   55
– slope   55
– type I   36
– type II   36
error probability   36
errors, types   25

EUCLIDean distance 154
excess 31
– test 43
experimental design *see also* design 71 ff.
– adequacy, application 368 f.
– application, water 304 ff.
– design matrix 72
– fundamentals 364 f.
– matrix of results 72
– regression coefficient, application 367 f.
exponential smoothing 211 ff.
– application 213 f.
– forecasting 213
– simple 211 f.
– trend and seasonal effects 212
external quality assurance 44
extraction measure 180

FA *see* factor analysis
factor analysis 164 ff.
– common feature 171
– course 175
– environmental analysis, advantages 268
– example 175 ff.
– impact of emissions 264 ff., 278 ff.
– – graphical representation 266 ff.
– – interpretation 266 ff., 273 ff.
– number of factors 173
– river 295 ff.
– river sediments 297 f., 302
– river water 286 ff., 289 ff.
– rotation 174
– scaling 294 f.
– soil depth profile 338 ff.
– soil pollution 335 f.
– specific feature 171
factor loading 164 ff.
factor score 164 ff.
factor solution
– complete 171
– reduced 171 f.
factorial design 73 ff., 93 ff.
– setup 76 ff.
factorial plan, application 304
Fibonacci search 91
FISHER *F*-distribution, quantiles 376 f.
foods 359 ff.
forward strategy 187
fractile (quantile) 32
frequency distribution 25 ff.
F-test 37, 45
functional representation 148 f.

GAUSSian distribution 28
– quantiles 373
geostatistical methods 113 ff.
– development 113
– soil pollution 347 ff.
golden section search 91
goodness of fit, test 61 f.
graphical methods 140 ff.
– limitations 152
grid
– linear or diagonal 126
– polar 126
– rectangular or bottle rack 126
grid plans 124 ff.
GY's theory 103 ff.

half-range 78 f.
heavy metal binding forms, water 303 ff.
heavy metal cycle 7
heavy metal distribution coefficient 312
heteroscedastic 56
homoscedasticity 52
HOTELLING's T^2 187 f.
hydrosphere 285 ff.
– interpretation of data 292 ff.
– sampling 285 ff.
hypothesis 35 ff.
– alternative 36
– null 36
– one-sided 36
– two-sided 36

icons 148 f.
impact of emissions 252 ff.
– MVDA and PLS comparison 263 f.
– one sampling location, example 269 ff., 275 ff.
IND function 173
INGAMELLS' constant 105
interaction between river water and sediment 311 ff.
interaction effects 74
intercept 50
interference effects, analytical error 369 f.
interlaboratory comparison 43 ff.
inter-set loading 180
intra-set loading 180
intrinsic hypothesis 114
isoline plot 350 ff.

kriging 117 ff.
– cross-validation 119 f.
– polluted area, assessment 354

– soil pollution   350 ff.
kriging variance   118
kurtosis   31

lack of fit, test   62 f.
lag   208
latent variable   199 f.
latent vectors   310
least median squares regression   342 ff.
least squares regression   217 ff.
– dummy variable   220 ff.
– explanatory variable   219 ff.
limit of decision   66 ff.
limit of detection   68
limit of quantitation   68
linear regression   51 ff.
– alternative models   52 ff.
– conditions   51

MAHALANOBIS distance   187
main effects   74
Manhattan metrics   154
MANOVA *see* multivariate analysis of variance
matrix plot   145
mean   29
median   29 f., 342 f.
memory effect   211, 222
MINKOWSKI metrics   154
misclassification probability   186
modeling of interference effects   363 ff.
monitoring raster screen   252 ff.
moving average   209 ff., 235 f.
– first-order   236
– smoothing effect   210 f.
multicollinearity   61
multiple regression   196 ff.
– cross-validation   197
– example   198 f.
– feature reduction   197
– multicollinear data   197
– theoretical basis   196 f.
multiplicative component model   212 f.
multivariate analysis of variance   182 ff.
multivariate analysis of variance and discriminant analysis   182 ff.
– course   189
– foods   361 f.
– impact of emissions   258 ff., 271
– – conclusions   262
– river water   286 ff.
– – conclusions   289 ff.
– soil   323 f.
– soil depth profile   338 f.

– soil pollution   332 f.
multivariate autocorrelation
– example   252
– impact of emissions   276 ff.
– soil sampling   327 f.
multivariate correlation
– coefficient   230 f.
– degree of freedom   231
– example   231 ff.
– probability   230 f.
multivariate correlation analysis   229
multivariate cross-correlation, example   229 ff., 232 f.
multivariate data analysis, general remarks   139 ff.
multivariate regression   195 ff.
multivariate statistics, methods   5
multivariate time series   228 ff.
MVDA *see* multivariate analysis of variance and discriminant analysis

nesting techniques   145 ff.
normal distribution   27 f.
nugget effect   115 f., 349

offset   50
– test   55
optimum search   91
ordinary least squares   53
orthogonal regression   53
orthogonal rotation   174
outlier   41 f., 346
outlier test   41 f.

partial autocorrelation
– coefficient   224
– function   224
partial least squares   199 ff.
– application, water   309 ff.
– cross-validation   200 f.
– error   200
– interpretation   201
– modeling, water   312 ff.
path analysis   201
pattern cognition   153 ff.
PCA *see* principal components analysis
pictoral representation   148 f.
PLACKETT and BURMAN   364
PLS *see* partial least squares
PLS modeling, impact of emissions   263
potential methods   157
PRESS   200 f.
principal axes transformation   165 f.
principal components analysis   164 ff.

– example   168 ff.
– foods   361
– interpretation   167
– soil   323 f.
profile plot   147

quality assurance, external   44

range   26, 30, 116, 349
reduced major axis   53
redundancy measure   180
regionalized variables, theory   113
regression   47 ff.
– coefficient   50
– errors   54 f.
– inner   61
– linear   51 ff.
– model   50 ff.
– multiple   61
– multivariate   61
– nonlinear   59
– nonparametric   57 f.
– partial   61
– polynome   59
– quasi linear   59
– robust   56 f.
– tests of the coefficients   54 f.
– weighted   56 f.
relative standard deviation   28
repeatability   44 f.
reproducibility   44 f.
residual variance   54
reweighted least squares regression   342 ff.
risk   34
robust statistics   341 ff.
– fundamentals   342 f.
– plant lead   341 ff.

sample   97 f.
– composite   101
– main types   100 f.
– random   100
– representative   100
– stratified   100
– systematic   100
sample mass   104 ff.
sample size   40 f.
sampling   95 ff.
– basic considerations   97 ff.
– critical distance   326 f.
– detailed description and control   121
– gross description   121
– important norms   97
– judgmental   124

– mass of samples required   103 ff.
– minimization of variance   109 ff.
– monitoring   121
– number of samples
– – foods   360 f.
– – kriging   354 f.
– number of samples required   101 ff.
– origin of variance   111 f.
– planning   96
– primary   127 f.
– purpose   95 f., 121, 130 f.
– required mass, example   107 ff.
– samples required   103
– secondary   128 f.
– simple random   122 f.
– stratified random   123
– systematic   123
– terms   99
– types   122 ff.
– variance   109
sampling distance, river   291 f.
sampling error   95, 98
– example   109 ff.
sampling frequency   112 f.
sampling location   112 f.
sampling operations   99
sampling plan   98, 122
sampling plan and program   120 ff.
sampling process, problem-adapted planning   120
sampling program   129 ff.
– acceptable   96 f.
– instruction and training of personnel   131
– optimized   96
– quality assurance   131
– steps   131
sampling strategy   132
scree plot   265
scree test   173 f.
sea spray factor   279
seasonal decomposition
– additive model   216
– multiplicative model   217
seasonal sub-series plot   208 f.
seasonality   220 ff.
– harmonic term   222
sediment loadings
– chemical differentiation   299 ff.
– multivariate data analysis   299 ff.
sedimented airborne particulates   252
– origins of variances   265
semivariance   115
semivariogram
– indicator   116

– soil pollution 349 f.
semivariogram analysis 114 ff.
sensitivity 50
sequential design 90 ff.
sequential leaching 300 ff.
significance level 34
significance of the regression 61
sill 116, 349
similarity measure 153 f.
simple structure 174
simplex method 92 f.
simultaneous equations 201
single linkage 158
skewness 30 f.
– test 43
slope 50
– test 55
smoothing and filtering, purpose 209
software 15
soil depth profile 336 ff.
– multivariate methods, conclusions 341
soil pollution 329 ff.
– geostatistical methods, conclusions 355 f.
soil sampling 319 ff.
– conclusions 328
speciation of heavy metals 298 f.
specimen 100
standard deviation 29, 48
– confidence interval 375
standardization 141 f.
– pitfalls 142
star plot 147 f.
statistical test 36
strip sampling 126 f.
STUDENT t-distribution, quantiles 374
subsample 101
sums of the squares of deviations 47 f.

Taggart's nomogram 103 f.
target factor rotation 174 f.
taxi driver metrics 154
test
– coefficient of determination 60 f.

– critical value 37 f.
– excess 43
– goodness of fit 61 f.
– offset 55
– one-sided 37
– skewness 43
– slope 55
– two-sided 37
time series analysis 205 ff.
– example: storage reservoir 206 f.
– purpose 205
time series plot 207 f.
transformation 79, 140 ff.
– centering 141
– standardization 141 f.
– variance scaling (autoscaling) 141 f.
trend 220 ff.
– evaluation 217 ff.
t-test 38 ff.

univariate analysis of variance, soil sampling 320 f.
univariate autocorrelation, soil sampling 324 ff.
unsupervised pattern cognition 153 ff.

variability
– anthropogenic 8
– caused by experimental error 9
– natural 8
– spatial and/or temporal 9
variance 29, 48
variance component 46
variance homogeneity, application 366
VISMAN's theory 105
visualization
– correlations 144 f.
– groups of objects 145 ff.
– similar objects 145 ff.

WARD linkage 158
WILKS lambda 188
winsorization 43
within-class scattering 183